P9-DUB-131

Hydroponic Food Production

Fifth Edition

Hydroponic Food Production

A Definitive Guidebook Of Soilless Food-Growing Methods

By Howard M. Resh, Ph.D.

Formerly, Department of Plant Science
University of British Columbia, Vancouver
International Horticultural Consultant
President, International Aquaponics, Inc.*

For the Professional and Commercial Grower and the Advanced Home Hydroponics Gardener

Published by
Woodbridge Press Publishing Company
Santa Barbara, California 93102

* See next page

A Spanish edition is available.

Fifth Edition 1998 Printing

Published by
Woodbridge Press Publishing Company
Post Office Box 209
Santa Barbara, California 93102

Copyright © 1995, 1989, 1987, 1985, 1981, 1978 by Howard M. Resh

All rights reserved.

World rights reserved. This book or any part thereof may not be reproduced in any form whatsoever without the prior written permission of Woodbridge Press Publishing Company, except for brief passages embodied in critical reviews or articles.

Distributed simultaneously in the United States and Canada
Printed in the United States of America

Library of Congress Cataloging-in-Publication Data

Resh, Howard M.
Hydroponic food production : a definitive guidebook of soilless food-growing methods / by Howard M. Resh. — 5th ed.
"For the professional and commercial grower and the advanced home hydroponics gardener."
Includes bibliographical references (p.) and index.
ISBN 0-88007-212-1
1. Hydroponics. 2. Food crops. I. Title.
SB126.5.R47 1995
635' . 0485—dc20 95-22823
CIP

Illustrations on pages 156, 158, 159, 162 and 431 are by Janice Blair.
Page composition by Sigma Graphics.

Product, company, or trade names shown capitalized or in quotation marks should be presumed to be trademarks or service marks of their respective companies.
Comments and suggestions for future printings or editions are invited.

Notice to the Reader: Please check your local regulations concerning the approval of and application recommendations of existing and new pesticides and biological agents. Neither the author nor the publisher can be held responsible for exact pesticide and biological agent application rates, as these come from third parties. The grower must always check the label of the product he is using in determining the correct application rates and safe procedures. Rates and methods for similar products from different manufacturers may vary. All such information as that given herein should be used as a general guideline. The reader is expected to obtain specific information from the sources of products being used, and to determine specific cultural practices, through growing experience, under his or her particular conditions, as well as to be familiar with legal requirements of his particular state, province, or other regulatory jurisdiction. —*The Author*

Disclaimer of warranty: The author and the publisher have used their best efforts in the preparation of this book. Woodbridge Press, Woodbridge Press Publishing Company, and the author make no representations or warranties as to the accuracy or completeness of the book's contents. It is sold "as-is," without warranty of any kind, either express or implied, including but not limited to implied warranties for the book's quality, merchantability, or fitness for any particular purpose. Neither the publisher nor distributors or dealers of the book shall be liable to any person, purchaser, or user concerning any loss of profit or any other damage, including but not limited to special, incidental, consequential, or other damages directly or indirectly related to the book or its contents.

*International Aquaponics, Inc., are greenhouse and hydroponics specialists, 819 W. 20th Ave., Vancouver, B.C., Canada V5Z 1Y3: (604) 874-2605 or (604) 764-7756.

Acknowledgments

This book is based on personal experience and information acquired from numerous sources. Books, scientific journals and government publications have all contributed. Recognition of such sources is given in the references following each chapter and in the general bibliography.

Personal working experience, visits with many growers, discussions with scientists and growers at conferences such as the Hydroponic Society of America (HSA) and the International Society for Soilless Culture (ISOSC) over the past twenty years have added to the information presented.

I wish to thank a few of the people who in the past provided me with photographs and additional information that was included in this writing: Bob Adamson, Michael Anselm, Carlos Arano, Andreas Bruppacher, Sheldon Pomer, Allen Cooper, Mickey Fontes, Merle Jensen, Herbert Corte, Franco Bernardi, Ted Maas, P. A. Schippers, Michele Tropea, Alessandro Vincenzoni, and Bent Vestergaard.

Special thanks to Dr. Silvio Velandia of Hidroponias Venezolanas C.A. of Caracas, Venezuela, for the hospitality and inspiration given me during our association over the past years in developing his sand-culture operation. He gave me the opportunity to gain experience in tropical hydroponics and encouraged me to write a section about it.

My sincere thanks to Arne McRadu for working patiently with me in doing the drawings which add greatly to the interest and understanding of the text.

Thanks also to the business people who have provided me with the opportunity of developing projects for them. To mention a few: Peter Hoppmann of Hoppmann Corporation, Chantilly, VA, Tom Thayer of Environmental Farms, Dundee, FL and Alfred Beserra of California Watercress, Inc., Fillmore, CA.

Also, a very special thanks to the many commercial greenhouse growers who have been very generous in providing me with information on their operations and allowing me to take photographs, many of which appear in this book. To mention a few: Casey Houweling, Houweling Nurseries Ltd., Delta, B.C., David Ryall, Gipaanda Greenhouses Ltd., Surrey, B.C., Harry Otsuki, Otsuki Greenhouses Ltd., Surrey, B.C., Steen Nielsen, Gourmet Hydroponics, Inc., Lake Wales, FL, Frank Armstrong, F.W. Armstrong, Inc., Oak View, CA and Ken Gerhart, Gerhart Greenhouses, Daggett, CA.

My sincere gratitude to all of these people and to my family who have had to remain behind while I have been working on distant projects for extended periods of time.

In no way is the use of trade names intended to imply approval of any particular source or brand name over other similar ones not mentioned in this book.

—The Author

Contents

List of Tables

List of Figures

Chapter 1

Introduction

1.1 The Past

Hydroponics, the growing of plants without soil, has developed from the findings of experiments carried out to determine what substances make plants grow and the composition of plants. Such work on plant constituents dates back as early as the 1600s. However, plants were being grown in a soilless culture far earlier than this. The hanging gardens of Babylon, the floating gardens of the Aztecs of Mexico and those of the Chinese are examples of "hydroponic" culture. Egyptian hieroglyphic records dating back to several hundred years B.C. describe the growing of plants in water.

Before the time of Aristotle, Theophrastus (372–287B.C.) undertook various experiments in crop nutrition. Botanical studies by Dioscorides date back to the first century A.D.

The earliest recorded scientific approach to discover plant constituents was in 1600 when Belgian Jan van Helmont showed in his classical experiment that plants obtain substances from water. He planted a 5-pound willow shoot in a tube containing 200 pounds of dried soil that was covered to keep out dust. After 5 years of regular watering with rainwater he found the willow shoot increased in weight by 160 pounds, while the soil lost less than 2 ounces. His conclusion that plants obtain substances for growth from water was correct. However, he failed to realize that they also require carbon dioxide and oxygen from the air. In 1699, an Englishman, John Woodward, grew plants in water containing various types of soil, and found that the greatest growth occurred in water which contained the most soil. He thereby concluded that plant growth was a result of certain substances in the water, derived from soil, rather than simply from water itself.

Further progress in identifying these substances was slow until more sophisticated research techniques were developed and advances were made in the field of chemistry. In 1804, De Saussure proposed that plants are composed of chemical elements obtained from water, soil and air. This proposition was verified later by Boussingault (1851), a French chemist, in his experiments with plants grown in sand, quartz and charcoal to which were added solutions of known chemical composition. He concluded that water was essential for plant growth in providing hydrogen and that plant dry matter consisted of hydrogen plus carbon and oxygen which came from the air. He also stated that plants contain nitrogen and other mineral elements.

Various research workers had demonstrated by that time that plants could be grown in an inert medium moistened with a water solution containing minerals required by the plants. The next step was to eliminate the medium entirely and grow the plants in a water solution containing these minerals. This was accomplished by two German scientists, Sachs (1860) and Knop (1861). This was the origin of "nutriculture" and similar techniques are still used today in laboratory studies of plant physiology and plant nutrition. These early investigations in plant nutrition demonstrated that normal plant growth can be achieved by immersing the roots of a plant in a water solution containing salts of nitrogen (N), phosphorus (P), sulfur (S), potassium (K), calcium (Ca) and magnesium (Mg), which are now defined as the macroelements or macronutrients (elements required in relatively large amounts).

With further refinements in laboratory techniques and chemistry, scientists discovered seven elements required by plants in relatively small quantities—the microelements or trace elements. These include iron (Fe), chlorine (Cl), manganese (Mn), boron (B), zinc (Zn), copper (Cu) and molybdenum (Mo).

In following years, researchers developed many diverse basic formulae for the study of plant nutrition. Some of these workers were Tollens (1882), Tottingham (1914), Shive (1915), Hoagland (1919), Trelease (1933), Arnon (1938), and Robbins (1946). Many of their formulae are still used in laboratory research on plant nutrition and physiology.

Interest in practical application of this "nutriculture" did not develop until about 1925 when the greenhouse industry expressed interest in its use. Greenhouse soils had to be replaced frequently to overcome problems of soil structure, fertility and pests. As a result, research workers became aware of the potential use of nutriculture to replace conventional soil cultural methods. Between 1925 and 1935, extensive development took place in modifying the laboratory techniques of nutriculture to large-scale crop production.

In the early 1930s, W. F. Gericke of the University of California put laboratory experiments in plant nutrition on a commercial scale. In doing so he termed these nutriculture systems *hydroponics.* The word was derived from two Greek words *hydro* ("water") and *ponos* ("labor")—literally "water working."

Hydroponics can be defined as the science of growing plants without the use of soil, but by use of an inert medium, such as gravel, sand, peat, vermiculite, pumice or sawdust, to which is added a nutrient solution containing all the essential elements needed by the plant for its normal growth and development. Since many hydroponic methods employ some type of medium it is often termed "soilless culture," while water culture alone would be true hydroponics.

Gericke grew vegetables hydroponically, including root crops, such as beets, radishes, carrots and potatoes, and cereal crops, fruits, ornamentals and flowers. Using water culture in large tanks, he grew tomatoes to such heights that he had to harvest them with a ladder. The American press made many irrational claims, calling it the discovery of the century. After an unsettled period in which unscrupulous people tried to cash in on the idea by selling useless equipment, more practical research was done and hydroponics became established on a sound scientific basis in horticulture, with recognition of its two principal advantages, high crop yields and its special utility in nonarable regions of the world.

Gericke's application of hydroponics soon proved itself by providing food for troops stationed on nonarable islands in the Pacific in the early 1940s. In 1945 the U.S. Air Force solved its problem of providing its personnel with fresh vegetables by practicing hydroponics on a large scale on the rocky islands normally incapable of producing such crops.

After World War II the military command continued to use hydroponics. For example, the U.S. Army established a 22-hectare project at Chofu, Japan. The commercial use of hydroponics expanded throughout the world in the 1950s to such countries as Italy, Spain, France, England, Germany, Sweden, the USSR and Israel.

1.2 The Present

With the development of plastics, hydroponics took another large step forward. Plastics freed growers from the costly construction associated with the concrete beds and tanks previously used. Beds are scraped out of the underlying medium and simply lined with a heavy vinyl (20 mil), then filled with the growing medium. With the development of suitable pumps, time clocks, plastic plumbing, solenoid valves and other equipment, the entire hydroponic system can now be automated, reducing both capital and operational costs.

Hydroponics has become a reality for greenhouse growers in virtually all climate areas. Large hydroponic installations exist throughout the world for the growing of both flowers and vegetables. Many hydroponic vegetable production greenhouses exist in North America that are 10 acres or larger in area. For example, Bonita Nurseries, Bonita, Arizona (20 acres—

tomatoes); Ringgold Nurseries, Ringgold, Pennsylvania (10 acres—tomatoes); SunGro Greenhouses, Las Vegas, Nevada (12 acres—tomatoes); Bernac, Fort Pierce, Florida (30 acres—European cucumbers); Houweling Nurseries Ltd., Delta, British Columbia, Canada (30 acres—European cucumbers, tomatoes, peppers); Rifle, Colorado (10 acres—tomatoes); Wolf Creek Produce, Brush, Colorado (36 acres—tomatoes). In the Canary Islands, hundreds of acres of land are covered with polyethylene supported by posts to form a single continuous structure housing tomatoes. The structure has open walls so that the prevailing wind blows through to cool the plants. The structure helps to reduce transpirational loss of water from the plants and to protect them from sudden rainstorms. Such structures can also be used in areas as the Caribbean and Hawaii. Almost every state in the United States has a substantial hydroponic greenhouse industry. Canada also uses hydroponics extensively in the growing of greenhouse vegetable crops.

Recent estmates of hydroponics in the following countries are: Holland, 10,000 acres (4,050 H.); England, 4,200 acres (1,700 H.); Canada, 1,000 acres (400 H.); United States, 600 acres (240 H.).

In arid regions of the world, such as Mexico and the Middle East, hydroponic complexes combined with desalination units are being developed to use sea water as a source of fresh water. The complexes are located near the ocean and the plants are grown in the existing beach sand.

In the former USSR, large greenhouse, soilless farms exist at Moscow and Kiev, while in Armenia an Institute of Hydroponics has been established at Erevan in the Caucasus region. Other countries where hydroponics is used include Australia, New Zealand, South Africa, the Bahama Islands, Central and East Africa, Kuwait, Brazil, Poland, the Seychelles, Singapore, Malaysia and Iran.

1.3 The Future

Hydroponics is a very young science. It has been used on a commercial basis for only 50 years. However, even in this relatively short period of time it has been adapted to many situations, from outdoor field culture and indoor greenhouse culture to highly specialized culture in atomic submarines to grow fresh vegetables for crews. It is a space-age science, but at the same time can be used in developing countries of the Third World to provide intensive food production in a limited area. Its only restraints are sources of fresh water and nutrients. In areas where fresh water is not available, hydroponics can use seawater through desalination. Therefore, it has potential application in providing food in areas having vast regions of nonarable land, such as deserts. Hydroponic complexes can be located along coastal regions in combination with petroleum-fueled or atomic desalination units, using the beach sand as the medium for growing the plants.

Hydroponics is a valuable means of growing fresh vegetables in countries having little arable land and in those which are very small in area yet have a large population. It could also be particularly useful in some smaller countries whose chief industry is tourism. In such countries tourist facilities, such as hotels, have often taken over most arable areas of the country, forcing local agriculture out of existence. Hydroponics could be used on the remaining nonarable land to provide sufficient fresh vegetables for the indigenous population as well as the tourists. Typical examples of such regions are the West Indies and Hawaii, which have a large tourist industry and very little farm land in vegetable production. To illustrate the potential use of hydroponics, tomatoes grown in this way could yield 150 tons per acre annually. A 10-acre site could produce 3 million pounds annually. In Canada the average per capita consumption of tomatoes is 20 pounds. Thus, with a population of 20 million, the total annual consumption of tomatoes is 400 million pounds (200,000 tons). These tomatoes could be produced hydroponically on 1,300 acres of land!

More hydroponic greenhouse operations will be linked with industries having waste heat. Such cogeneration projects already exist in California, Colorado, Nevada and Pennsylvania. Electrical power generating stations use water in their cooling towers. This heated water can be used both for heating the greenhouse and providing distilled water free of minerals for the growing of plants in recirculating systems. This clean water is of particular advantage to growers in areas having normally hard raw water. In most of the sun-belt locations where sunlight is favorable to high production of vegetables, waters are very hard with high levels of minerals which are often in excess of normal plant requirements. The hard water also creates problems with corrosion of equipment, plugging of cooling pads, fogging systems and structural breakdown of growing media.

With the introduction of new technology in artificial lighting, the growing of plants using artificial lighting will become economically feasible, especially in the more northern latitudes where sunlight is limited during the year from late fall to early spring. During this period, of course, the prices for produce are much higher than in summer months. Insulated warehouses could be used to grow vegetable crops. Heat generated from the lights could be used to heat the growing operation.

There are many locations in western North America having geothermal sources of heat. Such sites exist in Alaska, California, Colorado, Idaho, Montana, Oregon, Utah, Washington, Wyoming and British Columbia. One site near Alturas, California, is capable of supporting 20 acres of greenhouses. In the future, large greenhouses should be located close to geothermal sites to utilize the heat, as is presently done in Hokkaido, Japan.

At present a lot of research is being carried out to develop hydroponic systems for the growing of vegetables on the space station to be constructed in the future. Closed-loop recirculating systems are being designed and tested to operate under micro-gravity (very low gravity) environments. Such hydroponic systems will grow food to nourish astronauts on long space missions.

1.4 Suitable Site Characteristics

When considering a site location, a grower should try to meet as many of the following requirements as possible to reduce any risks of failure:

1. Full east, south and west exposure to sunlight with windbreak on north.
2. Level area or one that can be easily leveled.
3. Good internal drainage with minimum percolation rate of 1 inch/hour.
4. Have natural gas, three-phase electricity, telephone and good quality water capable of supplying at least one-half gallon of water per plant per day.
5. On a good road close to a population center for wholesale market and retail market at greenhouses if you choose to sell retail.
6. Close to residence for ease of checking the greenhouse during extremes of weather.
7. North-south oriented greenhouses with rows also north-south.
8. A region which has a maximum amount of sunlight.
9. Avoid areas having excessively strong winds.

1.5 Soil versus Soilless Culture

The large increases in yields under hydroponic culture over that of soil may be due to several factors. In some cases the soil may lack nutrients or have poor structure; therefore soilless culture would be very beneficial. The presence of pests or diseases in the soils greatly reduces overall production. Under greenhouse conditions when environmental conditions other than the medium are similar for both soil and soilless culture, the increased production of tomatoes grown hydroponically is usually 20–25 percent. Such greenhouses practice soil sterilization and use adequate fertilizers; as a result, many of the problems encountered under field conditions in soil would be overcome. This would account for the smaller increases in yields using soilless culture under greenhouse conditions over the very striking 4 to 10 times increase in yields obtained by soilless culture outdoors over conventional soil-grown conditions.

Specific tomato varieties have been bred to produce higher yields under greenhouse culture than field-grown varieties could under the same conditions. These greenhouse varieties of tomatoes cannot tolerate the daily temperature fluctuations of outdoor culture; therefore their use is restricted to greenhouse growing. Nonetheless, given optimum growing

Table 1.1 Advantages of Soilless Culture versus Soil Culture

Cultural Practice	*Soil*	*Soilless*
1. Sterilization of growing medium	Steam, chemical fumigants; labor-intensive; time required is lengthy, minimum of 2–3 weeks.	Steam, chemical fumigants with some systems; others can use simply bleach or HCl; short time needed to sterilize.
2. Plant nutrition	Highly variable, localized deficiencies, often unavailable to plants due to poor soil structure or *p*H, unstable conditions, difficult to sample, test and adjust.	Completely controlled, relatively stable, homogeneous to all plants, readily available in sufficient quantities, good control of *p*H, easily tested, sampled and adjusted.
3. Plant spacing	Limited by soil nutrition and available light.	Limited only by available light; therefore closer spacing is possible; increased number of plants per unit area, therefore more efficient use of space which results in greater yields per unit area.
4. Weed control, cultivation	Weeds present, cultivate regularly.	No weeds, no cultivation.
5. Diseases and soil inhabitants	Many soil-borne diseases, nematodes, insects and animals which can attack crop, often use crop rotation to overcome buildup of infestation.	No diseases, insects, animals in medium; no need for crop rotation.
6. Water	Plants often subjected to water stress due to poor soil-water relations, soil structure and low water-holding capacity. Saline waters cannot be used. Inefficient use of water; much is lost as deep percolation past the plant root zone and also by evaporation from the soil surface.	No water stress. Complete automation by use of moisture-sensing devices and a feed-back control mechanism; reduces labor costs, can use relatively high saline waters, efficient water use, no loss of water to percolation beyond root zone or surface evaporation; if managed properly, water loss should equal transpirational loss.

Cultural practice	***Soil***	***Soilless***
7. Fruit Quality	Often fruit is soft or puffy due to potassium and calcium deficiencies. This results in poor shelf life.	Fruit is firm with long shelf life. This enables growers to pick vine-ripened fruit and still be able to ship it relatively long distances. Also little, if any, spoilage occurs at the supermarket. Some tests have shown higher Vitamin A content in hydroponically grown tomatoes than those grown in soil.
8. Fertilizers	Broadcast large quantities over the soil, non-uniform distribution to plants, large amount leached past plant root zone (50–80%), inefficient use.	Use small quantities, uniformly distributed to all plants, no leaching beyond root zone, efficient use.
9. Sanitation	Organic wastes used as fertilizers on to edible portions of plants cause many human diseases.	No biological agents added to nutrients; no human disease organisms present on plants.
10. Transplanting	Need to prepare soil, uproot plants which leads to transplanting shock. Difficult to control soil temperatures, disease organisms which may retard or kill transplants.	No preparation of medium required prior to transplanting; transplanting shock minimized, faster "take" and subsequent growth. Medium temperature can be maintained optimum by flooding with the nutrient solution. No diseases present.
11. Plant maturity		With adequate light conditions, plant can mature faster under a soilless system than under soil.
12. Permanence of medium	Soil in a greenhouse must be changed regularly every several years since fertility and structure break down. Under field conditions, must fallow.	No need to change medium in gravel, sand or water cultures; no need to fallow. Sawdust, peat, vermiculite may last for several years between changes.
13. Yields	Greenhouse tomatoes 15–20 lb/year/plant.	25–35 lb/year/plant.

conditions of hydroponic greenhouse culture, they will far out-yield field varieties. Similarly, European cucumbers have been developed as fast-growing, high-yielding varieties suitable only to greenhouse culture. These crops are trained to grow vertically in greenhouses, not to spread along the ground as are field varieties. Tomatoes are indeterminate (staking), that is they grow continually not determinate or bush-type tomatoes as are commonly grown commercially in the field for single harvesting. These greenhouse tomatoes and cucumbers are harvested throughout the entire season, which is generally a one-year crop period, whereas, under field culture three to four single harvest crops annually would be the normal cropping cycle in the southern areas of Florida, Arizona, California and Mexico.

For the above reasons, hydroponic culture of greenhouse tomatoes, cucumbers, peppers and lettuce will significantly increase production over similar field-grown crops.

The main disadvantages of hydroponics are the high initial capital cost, some diseases such as *Fusarium* and *Verticillium* which can spread rapidly through the system, and the encountering of complex nutritional problems. Most of these disadvantages can be overcome. Capital cost and complexity of operating the system can be reduced by use of new simplified hydroponic methods, such as the nutrient film technique. Many varieties resistant to the above diseases have been bred.

Overall, the main advantages of hydroponics over soil culture are more efficient nutrition regulation, availability in regions of the world having nonarable land, efficient use of water and fertilizers, ease and low cost of sterilization of the medium, and higher-density planting, leading to increased yields per acre.

Table 1.2 Comparative Yields Per Acre in Soil and Soilless Culture

Crop	*Soil*	*Soilless*
Soya	600 lb	1,550 lb
Beans	5 tons	21 tons
Peas	1 ton	9 tons
Wheat	600 lb	4,100 lb
Rice	1,000 lb	5,000 lb
Oats	1,000 lb	2,500 lb
Beets	4 tons	12 tons
Potatoes	8 tons	70 tons
Cabbage	13,000 lb	18,000 lb
Lettuce	9,000 lb	21,000 lb
Tomatoes	5–10 tons	60–300 tons
Cucumbers	7,000 lb	28,000 lb

Chapter 2

Plant Nutrition

2.1 Plant Constituents

The composition of fresh plant matter includes about 80 to 95 percent water. The exact percentage of water will depend on the plant species and on the turgidity of the plant at the time of sampling, which will be a result of the time of day, the amount of available moisture in the soil, the temperature, wind velocity and other factors. Because of the variability in fresh weights, chemical analyses of plant matter are usually based on the more stable dry matter. Fresh plant material is dried at 70° C for 24 to 48 hours. The dry matter remaining will be roughly 10 to 20 percent of the initial fresh weight. Over 90 percent of the dry weight of most plant matter consists of the three elements: carbon (C), oxygen (O), and hydrogen (H). Water supplies hydrogen and oxygen, and oxygen also comes from carbon dioxide from the atmosphere, as does carbon.

If only 15 percent of the fresh weight of a plant is dry matter, and 90 percent of this is represented by carbon, oxygen, and hydrogen, then all the remaining elements in the plant account for roughly 1.5 percent of the fresh weight of the plant ($0.15 \times 0.10 = 0.015$).

2.2 Mineral and Essential Elements

While a total of 92 natural mineral elements are known, only 60 elements have been found in various plants. Although many of these elements are not considered essential for plant growth, plant roots probably absorb to some extent from the surrounding soil solution any element existing in a soluble form. However, plants do have some ability to select the rate at which they absorb various ions, so absorption is usually not in direct proportion to nutrient availability. Furthermore, different species vary in their ability to select particular ions.

An element must meet each of three criteria to be considered essential to plant growth (Arnon and Stout 1939; Arnon 1950; 1951). (1) The plant cannot complete its life cycle in the absence of the element. (2) Action of the element must be specific; no other element can wholly substitute for it. (3) The element must be directly involved in the nutrition of the plant, that is, be a constituent of an essential metabolite or, at least, required for the action of an essential enzyme and not simply causing some other element to be more readily available or antagonize a toxic effect of another element.

Only 16 elements are generally considered to be essential for growth of higher plants. They are arbitrarily divided into the macronutrients (macroelements), those required in relatively large quantities, and the micronutrients (trace or minor elements), those needed in considerably smaller quantities.

The macroelements include carbon (C), hydrogen (H), oxygen (O), nitrogen (N), phosphorus (P), potassium (K), calcium (Ca), sulfur (S) and magnesium (Mg). The microelements include iron (Fe), chlorine (Cl), manganese (Mn), boron (B), zinc (Zn), copper (Cu), and molybdenum (Mo). The relative concentrations of these elements found in most higher plants are given in Table 2.1.

TABLE 2.1 Elements Essential for Most Higher Plants

Element	*Symbol*	*Available Form*	*Atomic Weight*	*ppm*	*Concentration in dry tissue %*
Hydrogen	H	H_2O	1.01	60,000	6
Carbon	C	CO_2	12.01	450,000	45
Oxygen	O	O_2, H_2O	16.00	450,000	45
			Macronutrients		
Nitrogen	N	NO_3^-, NH_4^+	14.01	15,000	1.5
Potassium	K	K^+	39.10	10,000	1.0
Calcium	Ca	Ca^{++}	40.08	5,000	0.5
Magnesium	Mg	Mg^{++}	24.32	2,000	0.2
Phosphorus	P	$H_2PO_4^-$, $HPO_4^=$	30.98	2,000	0.2
Sulfur	S	$SO_4^=$	32.07	1,000	0.1
			Micronutrients		
Chlorine	Cl	Cl^-	35.46	100	0.01
Boron	B	$BO_3^=$, $B_4O_7^=$	10.82	20	0.002
Iron	Fe	Fe^{+++}, Fe^{++}	55.85	100	0.01
Manganese	Mn	Mn^{++}	54.94	50	0.005
Zinc	Zn	Zn^{++}	65.38	20	0.002
Copper	Cu	Cu^{++}, Cu^+	63.54	6	0.0006
Molybdenum	Mo	$MoO_4^=$	95.96	0.1	0.00001

Source: Modified after P. R. Stout, Proc. of the 9th Ann. Calif. Fert. Conf. 1961. pp. 21–23.

Although most higher plants require only those 16 essential elements, certain species may need others. They may, at least, accumulate these other elements even if they are not essential to their normal growth.

Silicon (Si), nickel (Ni), aluminum (Al), cobalt (Co), vanadium (V), selenium (Se) and platinum (Pt) are a few of these elements absorbed by plants and used in their growth. Cobalt is used by legumes for N-fixation. Nickel is now believed to be an essential element. It is essential for the urease enzyme. Silicone is needed for support. It adds strength to tissues, giving resistance to fungal infection, especially in cucumbers, where it is now common practice to include 100 ppm in the nutrient solution through the use of potassium silicate. Vanadium works with molybdenum (Mo) and can substitute for it. Platinum can increase growth of plants by 20 percent if pure chemicals (laboratory reagents) which have no impurities as do the fertilizer grades, are used. But, it is toxic at very low levels, so care must be taken when using it. Normally, commercial growers use fertilizer salts containing many of the above elements in trace amounts.

The roles of the essential elements are summarized in Table 2.2. All of them play some role in the manufacture and breakdown of various metabolites required for plant growth. Many are found in enzymes and coenzymes which regulate the rate of biochemical reactions. Others are important in energy-carrying compounds and food storage.

TABLE 2.2 Functions of the Essential Elements within the Plant

1. Nitrogen

Part of a large number of necessary organic compounds, including amino acids, proteins, coenzymes, nucleic acids and chlorophyll.

2. Phosphorus

Part of many important organic compounds including sugar phosphates, ATP, nucleic acids, phospholipids and certain co-enzymes.

3. Potassium

Acts as a coenzyme or activator for many enzymes (e.g., pyruvate kinase). Protein synthesis requires high potassium levels. Potassium does not form a stable structural part of any molecules inside plant cells.

4. Sulfur

Incorporated into several organic compounds including amino acids and proteins. Coenzyme A and the vitamins thiamine and biotin contain sulfur.

5. Magnesium

An essential part of the chlorophyll molecule and required for activation of many enzymes including steps involving ATP bond breakage. Essential to maintain ribosome structure.

6. Calcium

Often precipitates as crystals of calcium oxalate in vacuoles. Found in cell walls as calcium pectate which cements together primary walls of adjacent cells. Required to maintain membrane integrity and is part of the enzyme α-amylase. Sometimes interferes with the ability of magnesium to activate enzymes.

7. Iron

Required for chlorophyll synthesis and is an essential part of the cytochromes which act as electron carriers in photosynthesis and respiration. Is an essential part of ferredoxin and possibly nitrate reductase. Activates certain other enzymes.

8. Chlorine

Required for photosynthesis where it acts as an enzyme activator during the production of oxygen from water. Additional functions are suggested by effects of deficiency on roots.

9. Manganese

Activates one or more enzymes in fatty-acid synthesis, the enzymes responsible for DNA and RNA formation, and the enzyme isocitrate dehydrogenase in the Krebs cycle. Participates directly in the photosynthetic production of O_2 from H_2O and may be involved in chlorophyll formation.

10. Boron

Role in plants not well understood. May be required for carbohydrate transport in the phloem.

11. Zinc

Required for the formation of the hormone indoleacetic acid. Activates the enzymes alcohol dehydrogenase, lactic acid dehydrogenase, glutamic acid dehydrogenase and carboxypeptidase.

12. Copper

Acts as an electron carrier and as part of certain enzymes. Part of plastocyanin which is involved in photosynthesis, and also of polyphenol oxidase and possible nitrate reductase. May be involved in N_2 fixation.

13. Molybdenum

Acts as an electron carrier in conversion of nitrate to ammonium and is also essential for N_2 fixation.

14. Carbon

Constituent of all organic compounds found in plants.

15. Hydrogen

Constituent of all organic compounds of which carbon is a constituent. Important in cation exchange in plant-soil relations.

16. Oxygen

Constituent of many organic compounds in plants. Only a few organic compounds, such as carotene, do not contain oxygen. Also involved in anion exchange between roots and the external medium. It is a terminal acceptor of H^+ in aerobic respiration.

2.3 Plant Mineral and Water Uptake

Plants normally obtain their water and mineral needs from the soil. In a soilless medium, the plants must still be provided with water and minerals. Therefore, in order to understand the plant relations in a hydroponic system we must understand the soil-plant relations under which they normally grow.

2.3.1 The Soil

Soil provides four needs to the plant: (1) a supply of water, (2) a supply of essential nutrients, (3) a supply of oxygen and (4) support for the plant root system. Mineral soils consist of four major components: mineral elements, organic matter, water and air. For example, a volume composition of a representative silt loam soil in optimum condition for plant growth may consist of: 25 percent water space, 25 percent air space, 45 percent mineral matter and 5 percent organic matter. The mineral (inorganic) matter is made up of small rock fragments and of minerals of various kinds. The organic matter represents an accumulation of partially decayed plant and animal residues. The soil organic matter consists of two general groups: (1) original tissue and its partially decomposed equivalents; and (2) humus. The original tissue includes undecomposed plant and animal matter which is subject to attack by soil organisms, both plant and animal, which use it as a source of energy and as tissue-building material. Humus is the more resistant product of decomposition, both that synthesized by the microorganism and that modified from the original plant tissue.

The soil water is held within the soil pores and together with its dissolved salts makes up the soil solution which is so important as a medium for supplying nutrients to growing plants. The soil air located in the soil pores has a higher carbon dioxide and lower oxygen content than that found in the atmosphere. The soil air is important in providing oxygen and carbon dioxide to all the soil organisms and to plant roots.

The ability of the soil to provide adequate nutrition to the plant depends on four factors: (1) the amounts of the various essential elements present in the soil; (2) their forms of combination; (3) the processes by which these elements become available to plants; and (4) the soil solution and its *p*H. The amounts of the various elements present in the soil will depend on the nature of the soil and on its organic- matter content since it is a source of several nutrient elements. Soil nutrients exist both as complex, insoluble compounds and as simple forms usually soluble in soil water and readily available to plants. The complex forms must be broken down through decomposition to the simpler and more available forms in order to benefit the plant. These available forms are summarized in Table 2.1. The reaction of the soil solution (*p*H) will determine the availability of the various elements to the plant. The *p*H value is a measure of acidity or alkalinity. A soil is acid if the *p*H is less than 7, neutral if at 7 and alkaline if the *p*H is above 7. Since *p*H is a logarithmic function, a one-unit change in *p*H is a 10-fold change in H^+ concentration. Therefore, any unit change in *p*H can have a large effect on ion availability to plants. Most plants prefer a *p*H level between 6.0 and 7.0 for optimum nutrient uptake. The effect of *p*H on availability of essential elements is shown in Figure 2.1.

Iron, manganese, and zinc become less available as the *p*H is raised from 6.5 to 7.5 or 8.0. Molybdenum and phosphorus availability, on the other hand, is affected in the opposite way, being greater at the higher *p*H levels. At very high *p*H values the bicarbonate ion (HCO_3^-) may be present in sufficient quantities to interfere with the normal uptake of other ions and thus is detrimental to optimum growth.

When inorganic salts are placed in a dilute solution they dissociate into electrically charged units called ions. These ions are available to the plant from the surface of the soil colloids and from salts in the soil solution. The positively-charged ions (*cations*) such as potassium (K^+) and calcium (Ca^{++}) are mostly absorbed by the soil colloids, whereas the negatively-charged ions (*anions*), such as chloride (Cl^-) and sulfate ($SO_4^=$) are found in the soil solution.

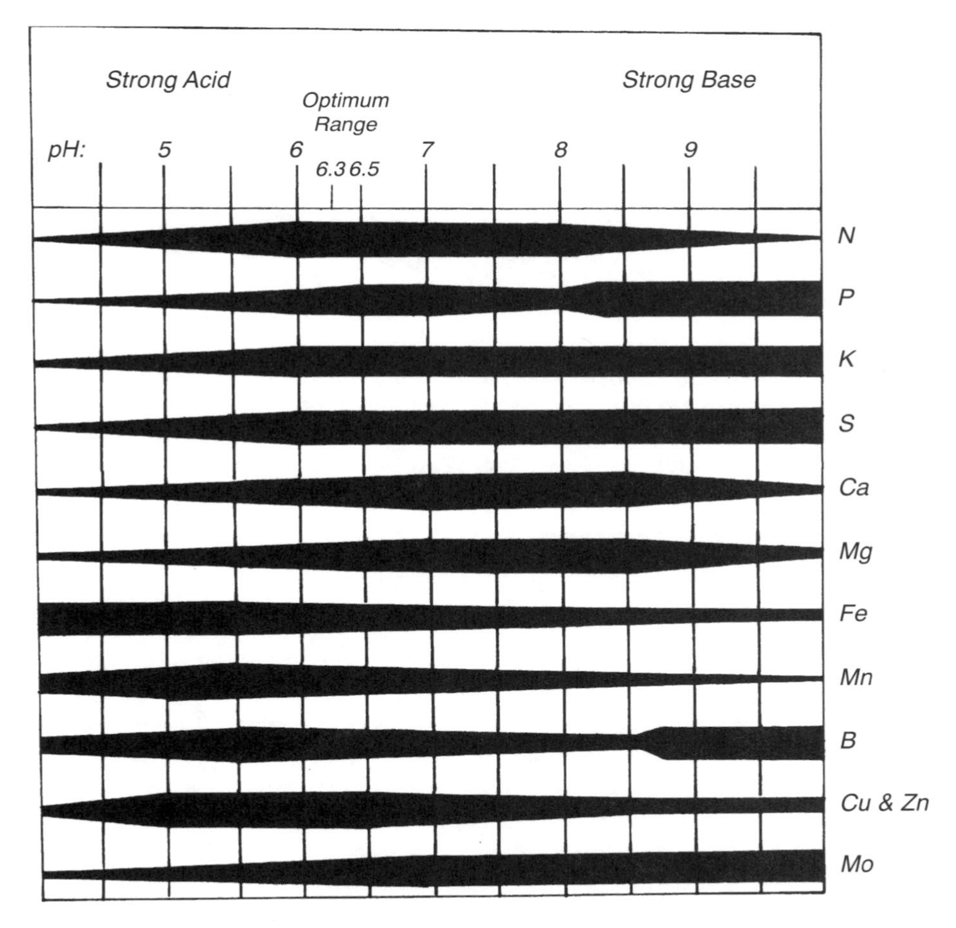

Figure 2.1. The effect of soil pH on the availability of plant nutrient uptake. (Modified from Hunger Signs in Crops, *ed. by H. B. Sprague, 1964, page 18).*

2.3.2 Soil and Plant Interrelations

Plant rootlets and root hairs are in very intimate contact with the soil colloidal surfaces. Nutrient uptake by the plants' roots takes place at the surface of the soil colloids and through the soil solution proper, as shown in Figure 2.2. Ions are interchanged between the soil colloids and the soil solution. Movement of ions takes place between the plant root surfaces and soil colloids and between the plant root surfaces and soil solution in both directions.

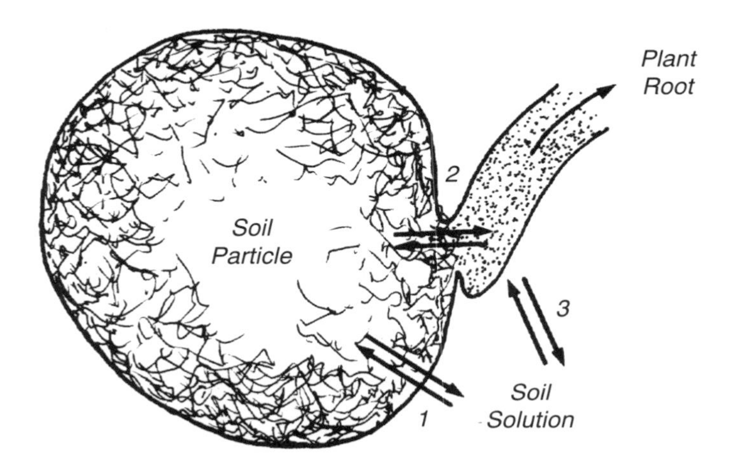

Figure 2.2. The movement of nutrients between plant roots and soil particles. (1) Interchange between soil particles and the soil solution. (2) Movement of the ions from the soil colloids (particles) to the plant root surface and vice versa. (3) Interchange between the soil solution and the absorbing surface of plant root systems.

2.3.3 Cation Exchange

The soil solution is the most important source of nutrients for absorption by plant roots. Since it is very dilute, as plants deplete the nutrients from the soil solution they must be replenished from the soil particles. The solid phase of the soil releases mineral elements into the soil solution partly by solubilization of soil minerals and organic matter, partly by solution of soluble salts, and partly by cation exchange. The negatively-charged clay particles and solid organic matter of the soil hold cations, such as calcium (Ca^{++}), magnesium (Mg^{++}), potassium (K^+), sodium (Na^+), aluminum (Al^{+++}), and hydrogen ions (H^+). Anions, such as nitrate (NO_3^-), phosphate ($HPO_4^=$), sulfate ($SO_4^=$), chloride (Cl^-) and others are found almost exclusively in the soil solution. Cations are also found in the soil solution and their ability to exchange freely with cations adsorbed on the soil colloids enable cation exchange to take place.

2.3.4 Soil versus Hydroponics

There is no physiological difference between plants grown hydroponically and those grown in soil. In soil both the organic and inorganic components must be decomposed into inorganic elements, such as calcium, magnesium, nitrogen, potassium, phosphorus, iron and others before they are available to the plant (Fig. 2.3). These elements adhere to the soil particles and are exchanged into the soil solution where they are absorbed by the plants. In hydroponics the plant roots are moistened with a nutrient solution containing the elements. The subsequent processes of mineral uptake by the plant are the same, as detailed in 2.3.5 and 2.3.6.

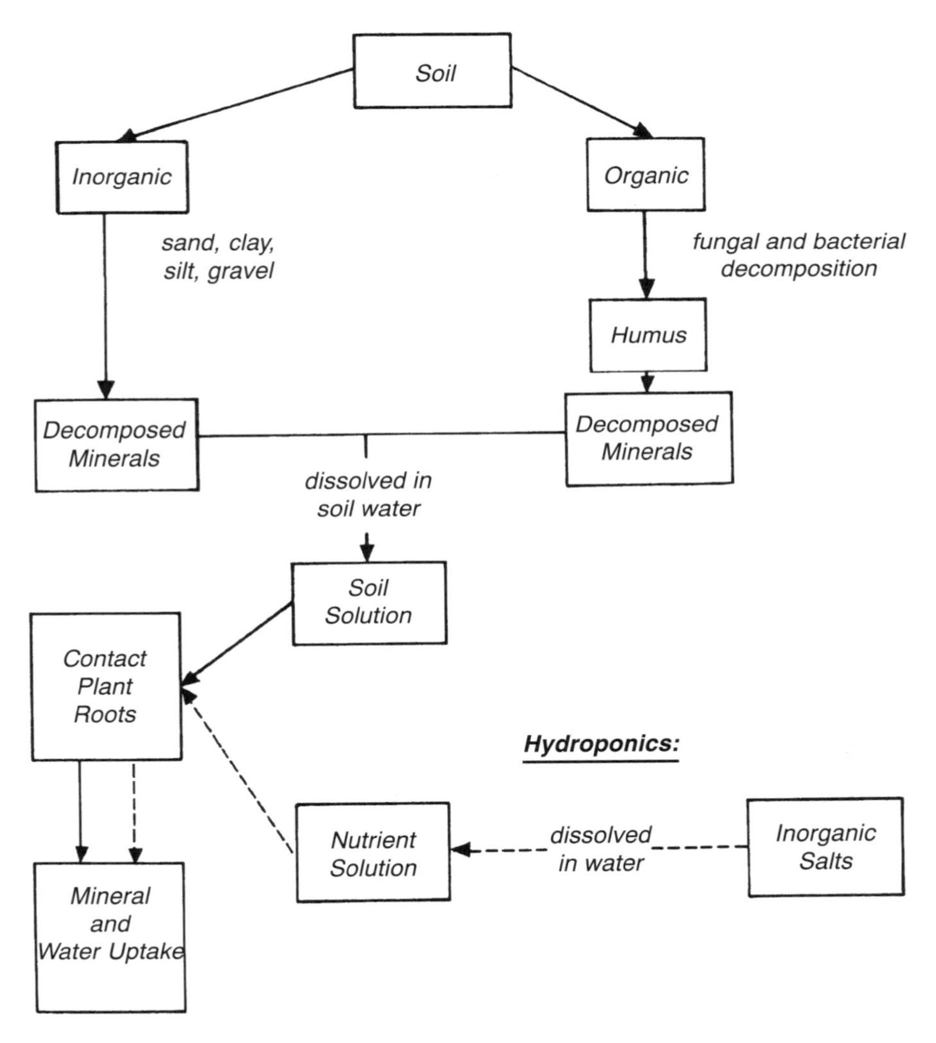

Figure 2.3. Origin of essential elements in soil and hydroponics.

2.3.5 Transfer of Water and Solutes from Soil (or Nutrient Solution) to Root

The question of organic versus inorganic gardening can be clarified by a discussion of mineral uptake by the plant.

In 1932, E. Münch of Germany introduced the apoplast-symplast concept to describe water and mineral uptake by plants. He suggested that water and mineral ions move into the plant root via the interconnecting cell walls and intercellular spaces, including the xylem elements, which he called the *apoplast*, or via the system of interconnected protoplasm (excluding the vacuoles) which he termed the *symplast*. However, whatever the movement may be, its uptake is regulated by the endodermal layer of cells around the stele which constitutes a barrier to free movement of water and solutes through the cell walls. There is a waxy strip, the Casparian strip, around each endodermal cell which isolates the inner portion of the root (stele) from the outer epidermal and cortex regions in which water and mineral movement is relatively free.

If the root is in contact with a soil (nutrient) solution, ions will diffuse into the root via the apoplast across the epidermis, through the cortex and up to the endodermal layer. Some ions will pass from the apoplast into the symplast by an active respiration-requiring process. Since the symplast is continuous across the endodermal layer, ions can move freely into the pericycle and other living cells within the stele (Fig. 2.4).

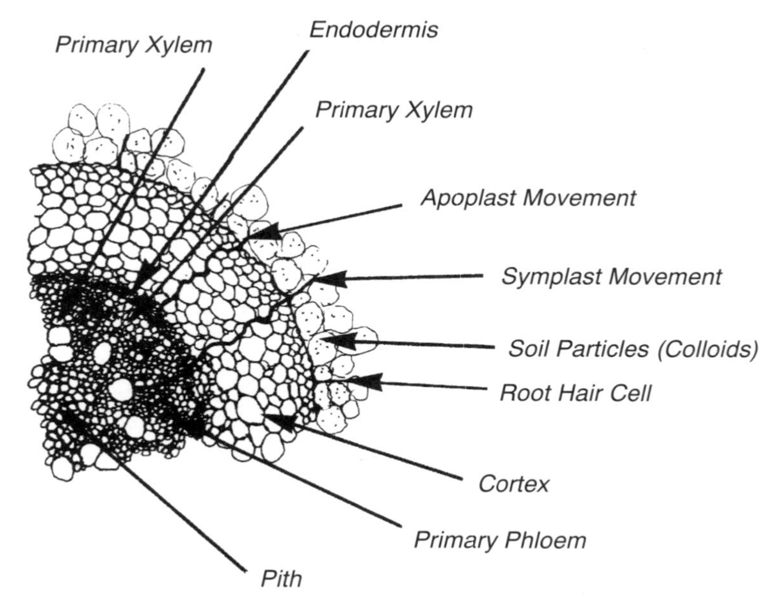

Figure 2.4. A cross section of a root with movement of water and minerals from the soil (nutrient) solution into the plant vascular system.

2.3.6 Movement of Water and Minerals across Membranes

If a substance is moving across a cell membrane, the number of particles moving per unit of time through a given area of the membrane is termed the *flux*. The flux is equal to the permeability of the membrane multiplied by the driving force causing diffusion. The driving force is due to the difference in concentration (the chemical potential) of that ion on the two sides of the membrane. If the chemical potential of the solute is higher outside the membrane than inside, the transport inward is passive. That is, energy is not expended by the plant to take up the ion. If, however, a cell accumulates ions against a chemical potential gradient, it must provide energy sufficient to overcome the difference in chemical potential. Transport against a gradient is active since the cells must actively metabolize in order to carry out the solute uptake.

When ions are transported across membranes, the driving force is composed of both a chemical and electrical potential difference. That is, an electrochemical potential gradient exists across the membrane. The electrical potential difference arises from cations diffusing across more rapidly than their corresponding anions of a salt. Thus, the inside will become positive with respect to the outside. Whether the transport of an ion is active or passive depends upon the contribution of both the electrical potential difference and the chemical potential difference. Sometimes these two factors will act in the same direction, while in other cases they act oppositely. For example, a cation might have a higher concentration inside the cell and yet be transported inward passively with no energy expenditure on the part of the cell if the electrical potential is sufficiently negative. On the other hand, anion absorption against both a chemical potential gradient and a negative electrical potential would always be an active process.

There are a number of theories proposed to explain how respiration and active absorption are coupled, but most of them employ the mechanism of a carrier. For example, when an ion contacts the outside of a membrane of a cell, neutralization may occur as the ion is attached to some molecular entity that is a part of the membrane. The ion attached to this carrier might then diffuse readily across the membrane, being released on the opposite side. The attachment may require the expenditure of metabolic energy and can occur on only one side of the membrane, while release can occur only on the other side of the membrane. The ions separate and move into the cell and the carrier becomes available to move more ions (Fig. 2.5). Selectivity in ion accumulation could be controlled by differences in ability of carriers to form specific combinations with various ions. For example, potassium absorption is inhibited competitively by rubidium, indicating that the two ions use the same carrier or the same site on the carrier.

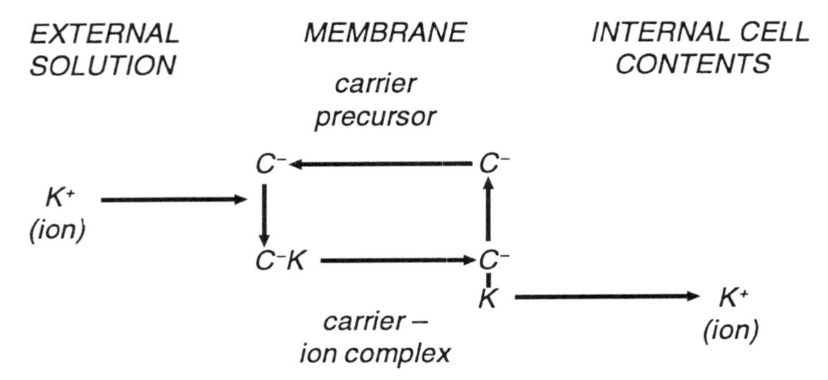

Figure 2.5. Movement of ions across cell membranes by a carrier.

As previously indicated, the foregoing explanation of mineral uptake by plant roots has been presented in an effort to clarify the question of organic versus inorganic gardening. The existence of specific relationships between ions and their carriers which enables their transport across cell membranes to enter the cell demonstrates that mineral uptake functions in the same manner whether the source of such minerals is from organic matter or fertilizers. Large organic compounds making up soil humus are not absorbed by the plant, but must first undergo decomposition into the basic inorganic elements. They can be accumulated by their contact with plant cell membranes only in their ionic form. Thus, organic gardening cannot provide any compounds to the plant which could not exist in a hydroponic system. The function of organic matter in soil is to supply inorganic elements for the plant and at the same time maintain the structure of the soil in optimum condition so that these minerals will be available to the plant. Thus, the indiscriminate application of large amounts of fertilizers to soil without addition of organic matter results in a breakdown of the soil structure and subsequently makes the abundant supply of minerals unavailable to the plant. This is not the fault of the fertilizer but the misuse of it in soil management.

2.4 The Upward Movement of Water and Nutrients

Water with its dissolved minerals moves primarily upward in the plant through xylem tissue. The xylem, composed of several cell types, forms a conduit system within the plant. This vascular tissue is commonly termed *veins*. The veins are actually composed of xylem and phloem tissue. The phloem tissue is the main conduit of manufactured food. The actual phloem translocation of photosynthates is still not fully understood. In general, water and minerals move upward in the xylem to the sites of photosynthesis, and photosynthates move from this source of manufacture to other parts of the plant.

The ascent of sap within a plant was suggested by Dixon (1914) and Renner (1911) in their cohesion hypothesis. They claimed that the force sucking up water and nutrients from the soil into the plant root comes from water evaporation from cell walls in the leaves. The binding force is in the inherent tensile strength of water, a property arising from cohesion of the water molecules (the forces of attraction between water molecules). This cohesion of water in the xylem is due to the capillary dimensions of the xylem elements. The water uptake from the soil comes from the negative water potential which is transferred down the plant to the root cells and soil by the upward driving force of evaporation.

The leaves of plants have a waxy cuticle cover on the outside surface to prevent excessive water loss by evaporation (Fig. 2.6). Small pores (*stomates*) in the epidermis, particularly numerous in the lower epidermis, regulate the passage of carbon dioxide and oxygen in and out of the leaf. Water vapor also moves through these openings. Therefore, water loss is regulated largely by the stomates. Water moves from the xylem vessels in the veins to the leaf mesophyll cells, evaporates and diffuses through the stomates into the atmosphere. This water lost in evapotranspiration must be replaced by water entering the plant roots or water stress will result, which if continued, will lead to death of the plant. In the process of water uptake, minerals are transported to chlorophyll-containing cells (palisade parenchyma, spongy mesophyll and bundle-sheath cells, if present), where they are used in the manufacture of foods through the process of photosynthesis.

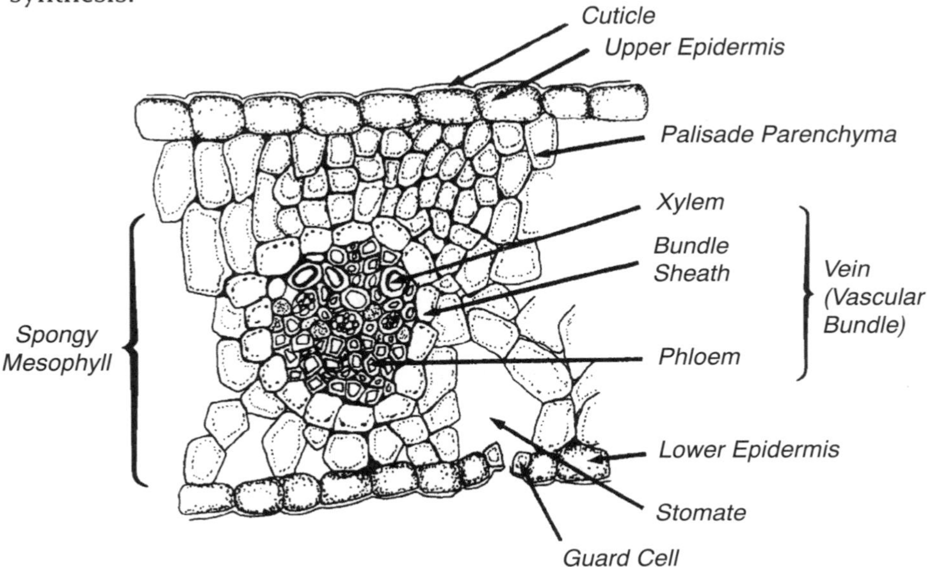

Figure 2.6. A cross section of a typical broad-leaved plant leaf showing the pathway of water movement.

2.5 Plant Nutrition

As mentioned in Chapter 1, hydroponics developed through studies of plant constituents, which led to the discovery of plant essential elements. Plant nutrition is therefore the basis of hydroponics. Anyone intending to employ hydroponic techniques must have a thorough knowledge of plant nutrition. The management of plant nutrition through management of the nutrient solution is the key to successful hydroponic growing.

The absorption and transport of plant nutrients within the plant has already been discussed. The next question is how to maintain the plant in optimum nutrient status. Hydroponics enables us to do this, but it also presents the risk of error which could result in rapid starvation or other adverse effects on the plant. A diagnostic program to determine the nutritional level of the plant at any time is extremely important to avoid nutritional stresses which limit plant growth.

The ideal method of diagnosing the nutritional status of plants is to take tissue analyses of plant leaves periodically (once or twice a week) and, in conjunction with these tests, nutrient solution analyses. The level of each essential element in the plant tissue and nutrient solution must be determined and correlated so that, if needed, adjustments can be made in the nutrient solution to avoid potential nutritional problems. Of course, such a program is costly in time and labor and is not always economically feasible. To do such tests requires a proper laboratory fully equipped with a muffle furnace, atomic absorption analyzer, glassware and other materials. The cost of building and adequately furnishing such a laboratory would be quite substantial and could only be justified by a very large greenhouse complex having a minimum area of 4–5 acres. These analyses can be done by commercial laboratories (see Appendix 2), but sometimes results are slow and crop damage may occur before recommendations are received.

The alternative to such laboratory analyses is visual diagnosis of nutrient stress symptoms expressed by the plant. However, it must be emphasized that once the plant shows symptoms it has already undergone severe nutritional stress and will take some time to regain its health after remedial steps have been taken. Therefore, it is important to correctly identify the nutrient problem immediately in order to prevent the plant from losing vigor.

2.5.1 Nutritional Disorders

A nutritional disorder is a malfunction in the physiology of a plant resulting in abnormal growth caused by either a deficiency or excess of a mineral element(s). The disorder is expressed by the plant externally and/or internally in the form of symptoms. Diagnosis of a nutritional disorder involves accurate description and identification of the disorder.

A deficiency or excess of each essential element causes distinct plant symptoms which can be used to identify the disorder.

Elements are grouped basically into those which are mobile and those which are immobile with some having gradations of mobility. Mobile elements are those which can be retranslocated. They will move from their original site of deposition (older leaves) to the actively-growing region of the plant (younger leaves) when a deficiency occurs. As a result, the first symptoms will appear on the older leaves on the lower portion of the plant. The mobile elements are magnesium, phosphorus, potassium, zinc and nitrogen. When a shortage of the immobile elements occurs, they are not retranslocated to the growing region of the plant, but remain in the older leaves where they were originally deposited. Deficiency symptoms, therefore, first appear on the upper young leaves of the plant. The immobile elements include calcium, iron, sulfur, boron, copper and manganese.

It is important to detect nutritional disorders early, since as they increase in severity the symptoms spread rapidly over the entire plant, resulting in the death of much of the plant tissue. Then symptom characteristics become very general, such as chlorosis (yellowing) and necrosis (browning) of the plant tissue. In addition, disorders of one element often upset the plant's ability to accumulate other elements, and shortly two or more essential elements are simultaneously deficient or in excess. This is particularly true of nutrient deficiencies. When two or more elements are deficient simultaneously, the composite picture or syndrome expressed by the symptoms may resemble no given deficiency. Under such conditions it is generally impossible to determine visually which elements are responsible for the symptoms.

Often a deficiency of one element leads to antagonism toward the uptake of another element. For instance, boron deficiency can also cause calcium deficiency. Calcium deficiency may lead to potassium deficiency and vice versa. The need for accurate and fast identification of a symptom expression cannot be overemphasized. It is often beneficial to grow an indicator plant along with the regular crop. The susceptibility of different plant species to various nutritional disorders varies greatly. For example, if a crop of tomatoes is being grown, plant a few cucumbers, lettuce or even a weed or two if it is known to be very sensitive to nutritional disorders. Cucumbers are very sensitive to boron and calcium deficiency. If such a deficiency occurs, the cucumbers will express symptoms from several days to a week before they appear on the tomatoes. Such early warning enables the grower to adjust his nutrient solution to prevent a deficiency in the tomato crop. In addition, weaker plants of the same species will show symptoms prior to the more vigorous ones. Every possible tactic must be employed to avoid a nutritional disorder in the main crop, since once symptoms are expressed in the crop, some reduction in yield is inevitable.

Once a nutritional disorder has been identified, steps to remedy it can be taken. In a hydroponic system, the first step is to change the nutrient solution. This should be done as soon as a nutritional disorder is suspected, even before identifying it. If the disorder has been diagnosed as a deficiency, a foliar spray can be applied for a rapid response. However, care must be taken not to use a concentration high enough to burn the plants. It is best to try the recommended foliar spray on a few plants and then observe the results for several days before treating the whole crop. The nutrient formulation probably will have to be adjusted (see Chapter 3) to overcome the disorder. If a nutrient deficiency is present, the level of the deficient nutrient should be increased to an above-normal level (up to 25 to 30 percent). As the plants come out of the deficiency, the increase in the identified nutrient could be lowered to about 10 to 15 percent above the level at which the deficiency occurred. Depending upon the severity of disorder, weather conditions and the element itself, the plant may take 7 to 10 days before response toward the control measure is evident.

If a toxicity has occurred, the soilless medium will have to be flushed with water alone to reduce residual levels in the medium. Flushing may have to be done over a period of a week or so depending again on the severity of the disorder. However, nutrient deficiencies are far more common than toxicities in hydroponics. For this reason, nutrient deficiencies will be emphasized in the following discussion on symptomatology.

2.5.2 Symptomatology

One of the first steps in identifying a nutritional disorder is to describe the symptoms in distinct, accurate terms. Table 2.3 summarizes terms commonly used in symptomatology. When observing a disorder, determine what plant part or organ is affected. Does it occur on the lower older leaves or the younger upper leaves? Are the symptoms on the stem, fruit, flowers, and/or growing point of the plant? What is the appearance of the whole plant? Is it dwarfed, deformed, or branched excessively? What is the nature of the ailment? Is the tissue chlorotic (yellow), necrotic (brown) or deformed? Then describe the color pattern and location of this chlorosis, necrosis or deformation using the terms given in Table 2.3.

TABLE 2.3 Terminology used in the Description of Symptoms on Plants

Term	*Description*
Localized:	Symptoms limited to one area of plant or leaf.
Generalized:	Symptoms not limited to one area but spread generally over entire plant or leaf.
Drying (firing):	Necrosis—scorched, dry, papery appearance.
Marginal:	Chlorosis or necrosis—on margins of leaves initially; usually spreads inward as symptom progresses.
Interveinal chlorosis:	Chlorosis (yellowing) between veins of leaves only.
Mottling:	Irregular spotted surface—blotchy pattern of indistinct light and dark areas; often associated with virus diseases.
Spots:	Discolored area with distinct boundaries adjacent to normal tissue.
Color of leaf undersides:	Often a particular coloration occurs mostly or entirely on the lower surface of the leaves, e.g., phosphorus deficiency—purple coloration of leaf undersides.
Cupping:	Leaf margins or tips may cup or bend upward or downward, e.g.,copper deficiency—margins of leaves curl into a tube; potassium deficiency—margins of leaves curl inward.
Checkered (reticulate):	Pattern of small veins of leaves remaining green while interveinal tissue yellows—manganese deficiency.
Brittle tissue:	Leaves, petioles, stems may lack flexibility, break off easily when touched—calcium or boron deficiency.
Soft tissue:	Leaves very soft, easily damaged—nitrogen excess.
Dieback:	May be leaves or growing point that dies rapidly and dries out—boron or calcium deficiencies.
Stunting:	Plant shorter than normal.
Spindly:	Growth of stem and leaf petioles very thin and succulent.

After the symptoms have been observed closely and described, it should be determined whether the disorder may be caused by something other than a nutritional imbalance. The following list of other possible disorders should be checked: insect damage; parasitic diseases; pesticide damage; pollution damage; water stress; light and temperature injury. Pesticide damage may cause burning if greater than recommended dosages are used on the plants. Also, the use of herbicides such as 2,4-D near a greenhouse may cause deformation of plant leaves closely resembling the symptoms of tobacco mosaic virus (TMV). Pollution damage may cause burning or bleaching of leaf tissue or a stippled effect (pinpoint-sized chlorotic spots) on leaves. Water stress, either lack of or excess of, will cause wilting (loss of turgidity) of leaves. Excessive sunlight or temperature may burn and dry leaf tissue, particularly on the margins.

TABLE 2.4 A Key to Mineral Deficiency Symptoms

Symptoms	*Deficient Element*
A. Older or lower leaves affected.	
B. General chlorosis and/or drying of lower leaves, retarded growth.	
C. Chlorosis progresses from light green to yellow, from older leaves up to new growth. Growth is restricted, spindly, loss of older leaves	Nitrogen
CC. Leaves remain dark green, growth restricted, distinctive purple coloration of undersides of leaves. Lower leaves dry. Root growth is restricted. Fruit-set delayed	Phosphorous
BB. Localized mottling or chlorosis with or without dead spots, no drying of lower leaves.	
C. Interveinal chlorosis, mottled effect with green veins. Margins curl upward, necrotic spots, stalks slender	Magnesium
CC. Mottled or chlorotic leaves with spots of dead tissue.	
D. Small dead spots at tips and between veins. Margins cup downward with brown spots. Growth is restricted, slender stalks. Roots poorly developed	Potassium
DD. Spots generalized, rapidly enlarging to include veins, leaves thick, stalks with shortened internodes. Young leaves small, interveinal chlorosis, mottled, curl downward	Zinc
E. Mottling of older leaves with veins remaining light green. Leaf margins become necrotic, may curl upward, necrotic spots of leaf tips and margins. Symptoms spread to younger leaves as deficiency progresses	Molybdenum

Symptoms	***Deficient Element***
AA. Symptoms appear first in young leaves—terminal growth.	
B. Growing tip distorted, young leaves at tips chlorotic with necrotic spots expanding to browning of leaf margins and dieback.	
C. Brittle tissues not present in growing tips, young leaves chlorotic, old leaves remain green, stems thick and woody, growing tip necrotic followed by dieback, blossom-end-rot of fruit (especially tomatoes)	Calcium
CC. Growing tip—leaves and petioles—light green to yellow, brittle tissue, often deformed or curled. Rosetting of terminal growth due to shortening of internodes. Terminal bud dies, new growth may form at lower leaf axils, but these suckers (especially tomatoes) show similar symptoms of chlorosis, necrosis, browning and brittleness. Internal browning, open locules, blotchy ripening of tomato fruit	Boron
BB. Growing tip alive, not distorted, wilting or chlorosis with or without dead spots, veins light or dark green.	
C. Young leaves wilt, chlorosis, necrosis, retarded growth, lodging of growing tip	Copper
CC. Young leaves not wilted, chlorosis with or without necrosis and dead spots.	
D. Interveinal chlorosis, veins remain green to give checkered pattern. Chlorotic areas become brown to later form necrotic spots of dead tissue	Manganese
DD. Interveinal dead spots not present, chlorosis of tissue may or may not involve veins.	
E. Leaves uniformly light green becoming yellow, veins not green, poor, spindly growth, hard and woody stems	Sulfur
EE. Tissue between veins yellows, veins green, eventually veins become chlorotic. Yellow interveinal tissue becomes white, but no necrosis. Stems slender, short. Flowers abort and fall off, tomato flower clusters are small, thin-stemmed	Iron

Once the above factors have been checked and eliminated as potential causes, a nutritional disorder would be suspected. In general, in hydroponics, nutritional disorders will show up on all plants at the same time, whereas many of the nonnutritional disorders would begin on a few plants and progress to neighboring ones. The next step is to identify the nutrient disorder(s) by the use of a "key" (Table 2.4). Deficiency and toxicity symptoms for all the essential elements are given in Table 2.5. These descriptions can be used in conjunction with the key to assist in identification of the nutrient disorder(s). These two tables describe symptoms which could be expected generally on most plants. As discussed earlier, different plant species will express various symptoms to a lesser or greater degree than others. Therefore, in addition to the general table (Table 2.5) information is included on specific symptoms and remedies for tomatoes and cucumbers (Table 2.6).

2.5.3 Use of a Key

The key presented here is a dichotomous table. A decision must be made at each alternative route and finally a single explanation is given at the end. The key to nutritional disorders is based on symptoms observed on the plant, hence the importance of accurate symptom description. The key (Table 2.4) is for use in determining mineral deficiencies only, not toxicities.

The first decision to make is effects on older leaves (A) versus effects on younger upper leaves (AA). Once this choice is made a series of choices follows. The next step is (B) versus (BB) under the previous choice of (A) or (AA). Then, (C) versus (CC), (D) versus (DD) and so on. For example, use the key to find the following symptoms: The young (upper) leaves of a plant are chlorotic, the veins green and no dead spots are visible. The terminal bud is alive and not wilted. The older (lower) leaves show no symptoms. The terminal growing area is somewhat spindly and there is some abortion of flowers. Since the upper leaves are affected the first choice is (AA). The terminal bud is alive, therefore the next choice is (BB). The next decision is (C) versus (CC). Since chlorosis is present, but no wilting, the correct choice would be (CC). The alternative of (D) or (DD) can be made on the basis of the lack of dead spots. The proper choice is (DD). The choice between sulfur (E) versus iron (EE) is a little difficult. In the case of sulfur, the interveinal tissue is a light green, not the bright yellow that appears with iron deficiency. The spindly stalks and flower abortion also indicate iron over sulfur deficiency. As a result, the final choice is iron deficiency.

Table 2.5 Deficiency and Toxicity Symptoms for the Essential Elements

1. Nitrogen

Deficiency Symptoms: Growth is restricted and plants are generally yellow (chlorotic) from lack of chlorophyll, especially older leaves. Younger leaves remain green longer. Stems, petioles and lower leaf surfaces of corn and tomato can turn purple.

Toxicity Symptoms: Plants usually dark green in color with abundant foliage but usually with a restricted root system. Potatoes form only small tubers, and flowering and seed production can be retarded.

2. Phosphorus

Deficiency Symptoms: Plants are stunted and often a dark green color. Anthocyanin pigments may accumulate. Deficiency symptoms occur first in more mature leaves. Plant maturity often delayed.

Toxicity Symptoms: No primary symptoms yet noted. Sometimes copper and zinc deficiency occurs in the presence of excess phosphorus.

3. Potassium

Deficiency Symptoms: Symptoms first visible on older leaves. In dicots, these leaves are initially chlorotic but soon scattered dark necrotic lesions (dead areas) develop. In many monocots, the tips and margins of the leaves die first. Potassium-deficient corn develops weak stalks and is easily lodged.

Toxicity Symptoms: Usually not excessively absorbed by plants. Oranges develop coarse fruit at high potassium levels. Excess potassium may lead to magnesium deficiency and possible manganese, zinc or iron deficiency.

4. Sulfur

Deficiency Symptoms: Not often encountered. Generally yellowing of leaves, usually first visible in younger leaves.

Toxicity Symptoms: Reduction in growth and leaf size. Leaf symptoms often absent or poorly defined. Sometimes interveinal yellowing or leaf burning.

5. Magnesium

Deficiency Symptoms: Interveinal chlorosis which first develops on the older leaves. The chlorosis may start at leaf margins or tip and progress inward interveinally.

Toxicity Symptoms: Very little information available on visual symptoms.

6. Calcium

Deficiency Symptoms: Bud development is inhibited and root tips often die. Young leaves are affected before old leaves and become distorted and small with irregular margins and spotted or necrotic areas.

Toxicity Symptoms: No consistent visible symptoms. Usually associated with excess carbonate.

7. Iron

Deficiency Symptoms: Pronounced interveinal chlorosis similar to that caused by magnesium deficiency but on the younger leaves.

Toxicity Symptoms: Not often evident in natural conditions. Has been observed after the application of sprays where it appears as necrotic spots.

8. Chlorine

Deficiency Symptoms: Wilted leaves which then become chlorotic and necrotic, eventually attaining a bronze color. Roots become stunted and thickened near tips.

Toxicity Symptoms: Burning or firing of leaf tip or margins. Bronzing, yellowing and leaf abscission and sometimes chlorosis. Reduced leaf size and lower growth rate.

9. Manganese

Deficiency Symptoms: Initial symptoms are often interveinal chlorosis on younger or older leaves depending on species. Necrotic lesions and leaf shedding can develop later. Disorganization of chloroplast lamellae.

Toxicity Symptoms: Sometimes chlorosis, uneven chlorophyll distribution and iron deficiency (pineapple). Reduction in growth.

10. Boron

Deficiency Symptoms: Symptoms vary with species. Stem and root apical meristems often die. Root tips often become swollen and discolored. Internal tissues sometimes disintegrate (or discolor) (e.g. "heart rot" of beets). Leaves show various symptoms including thickening, brittleness, curling, wilting and chlorotic spotting.

Toxicity Symptoms: Yellowing of leaf tip followed by progressive necrosis of the leaf beginning at tip or margins and proceeding toward midrib.

11. Zinc

Deficiency Symptoms: Reduction in internode length and leaf size. Leaf margins are often distorted or puckered. Sometimes interveinal chlorosis.

Toxicity Symptoms: Excess zinc commonly produces iron chlorosis in plants.

12. Copper

Deficiency Symptoms: Natural deficiency is rare. Young leaves often become dark green and twisted or misshapen, often with necrotic spots.

Toxicity Symptoms: Reduced growth followed by symptoms of iron chlorosis, stunting, reduced branching, thickening and abnormal darkening of rootlets.

13. Molybdenum

Deficiency Symptoms: Often interveinal chlorosis developing first on older or midstem leaves, then progressing to the youngest (similar to nitrogen deficiency). Sometimes marginal scorching or cupping of leaves.

Toxicity Symptoms: Rarely observed. Tomato leaves turn golden yellow, cauliflower seedlings turn bright purple.

TABLE 2.6 Summary of Mineral Deficiencies in Tomatoes and Cucumbers and their Controls

Mobile elements (first symptoms on older leaves)

1. Nitrogen

Tomatoes

spindly plant
lower leaves—yellowish green
severe cases—entire plant pale green
major veins—purple color
small fruit

Cucumbers

stunted growth
lower leaves—yellowish green
severe cases—entire plant pale green
younger leaves stop growing
fruit—short, thick, light green, spiny

Remedies:

1. Use a foliar spray of 0.25 to 0.5 percent solution of urea.
2. Add calcium nitrate or potassium nitrate to nutrient solution.

2. Phosphorus

Tomatoes

shoot growth restricted
thin stem
severe cases—leaves small, stiff, curved downward
leaf upper sides—bluish green
leaf undersides—including veins—purple
older leaves—yellow with scattered purple dry spots —premature leaf drop

Cucumbers

stunted
severe cases—young leaves small, stiff, dark green
older leaves and cotyledons—large water-soaked spots including both veins and interveinal areas
affected leaves fade, spots turn brown and desiccate, shrivel—except for petiole

Remedies:

1. Add monopotassium phosphate to nutrient solution.

3. Potassium

Tomatoes

older leaves—leaflets scorched, curled margins, interveinal chlorosis, small dry spots
middle leaves—interveinal chlorosis with small dead spots
plant growth—restricted, leaves remain small
later stages—chlorosis and necrosis spreads over large area of leaves, also up plant, leaflets die back
fruit—blotchy, uneven ripening, greenish areas

Cucumbers

older leaves—discolored yellowish green at margins, later turn brown and dry
plant growth—stunted, short internodes, small leaves
later stages—interveinal and marginal chlorosis extends to center of leaf, also progresses up plant, leaf margins desiccate, extensive necrosis, larger veins remain green

Remedies:

1. Foliar spray of 2 percent potassium sulfate.
2. Add potassium sulfate or if no sodium chloride present in water, can add potassium chloride to nutrient solution.

4. Magnesium

Tomatoes

older leaves—marginal chlorosis progressing inwards as interveinal chlorosis, necrotic spots in chlorotic areas

small veins—not green

severe starvation—older leaves die, whole plant turns yellow, fruit production reduced

Cucumbers

older leaves—interveinal chlorosis from leaf margins inward—necrotic spots develop

small veins—not green

severe starvation—symptoms progress from older to younger leaves, entire plant yellows, older leaves shrivel and die

Remedies:

1. Foliar spray—high-volume spray with 2 percent magnesium sulfate or low-volume spray with 10 percent magnesium sulfate.
2. Add magnesium sulfate to nutrient solution.

5. Zinc

Tomatoes

older leaves and terminal leaves—smaller than normal

little chlorosis, but irregular shriveled brown spots develop, especially on petiolules (small petioles of leaflets) and on and between veins of leaflets

petioles—curl downward, complete leaves coil up

severe starvation—rapid necrosis, entire foliage withers

Cucumbers

older leaves—interveinal mottling, symptoms progress from older to younger leaves, no necrosis

internodes—at top of plant stop growing, leading to upper leaves closely spaced, giving bushy appearance

Remedies:

1. Foliar spray with 0.1 to 0.5 percent solution of zinc sulfate.
2. Add zinc sulfate to nutrient solution.

Immobile elements (first symptoms on younger leaves)

1. Calcium

Tomatoes

upper leaves—marginal yellowing, undersides turn purple-brown color especially on margins, leaflets remain tiny, deformed, margins curl up

progression to later stages—leaf tips and margins wither, curled petioles die back

growing point dies

older leaves—finally, chlorosis and necrotic spots form

fruit—blossom-end-rot (leather-like decay at blossom ends of fruit)

Cucumbers

upper leaves—white spots near edges and between veins, marginal interveinal chlorosis progresses inward

youngest leaves (growing point region)—remain small, edges deeply incised, curl upwards, later shrivel from edges inwards and growing point dies

plant growth—stunted, short internodes, especially near apex

buds—abort, finally plant dies back from apex

older leaves—curve downwards

Remedies:

1. Foliar spray in acute cases with 0.75 to 1.0 percent calcium nitrate solution. Can also use 0.4 percent calcium chloride.
2. Add calcium nitrate to nutrient solution, or calcium chloride if do not want to increase nitrogen level, but be sure there is little, if any, sodium chloride in the nutrient solution if calcium chloride is used.

2. Sulfur

Tomatoes

upper leaves—stiff, curl downward, eventually large irregular necrotic spots appear, leaves become yellow

stem, veins, petioles—purple

older leaves, leaflets—necrosis at tips and margins, small purple spots between veins

Cucumbers

upper leaves—remain small, bend downwards, pale green to yellow, margins markedly serrate

plant growth—restricted

older leaves—very little yellowing

Remedies:

1. Add any sulfates to the nutrient solution. Potassium sulfate would be safest since the plants require high levels of potassium.
   Note: Sulfate deficiency rarely occurs since sufficient amounts are added by use of potassium, magnesium and other sulfate salts in the normal nutrient formulation.

3. Iron

Tomatoes

terminal leaves—chlorosis starts at margins and spreads through entire

Cucumbers

young leaves—fine pattern of green veins with yellow interveinal

leaf, initially smallest veins remain green, giving reticulate pattern of green veins on yellow leaf tissues; leaf eventually turns completely pale yellow, no necrosis
progression—symptoms start from terminal leaves and work down to older leaves
growth—stunted, spindly, leaves smaller than normal
flowers—abortion

tissue, later chlorosis spreads to veins and entire leaves turn lemon-yellow; some necrosis may develop on margins of leaves
progression—from top downward
growth—stunted, spindly
axillary shoots and fruits—also turn lemon-yellow

Remedies:

1. Foliar spray with 0.02 to 0.05 percent solution of iron chelate (FeEDTA) every 3 to 4 days.
2. Add iron chelate to nutrient solution.

4. Boron

Tomatoes

growing point—shoot growth restricted, leads to withering and dying of growing point
upper leaves—interveinal chlorosis, mottling of leaflets, remain small, curl inward, deformed, smallest leaflets turn brown and die
middle leaves—yellow-orange tints, veins yellow or purple
older leaves—yellowish-green
lateral shoots—growing points die
petioles—very brittle, break off easily, clogged vascular tissue

Cucumbers

apex—growing point plus youngest unexpanded leaves curl up and die
axillary shoots—wither and die
older leaves—cupped upward beginning along margins, stiff, interveinal mottling
shoot tip—stops growing, leads to stunting

Remedies:

1. Apply a foliar spray as soon as detected of 0.1 to 0.25 percent solution of borax.
2. Add borax to nutrient solution.

5. Copper

Tomatoes

middle and younger leaves—margins curl into a tube toward the midribs, no chlorosis or necrosis, bluish green color,terminal leaves small, stiff and folded up
petioles—bend downward, directing opposite tubular leaflets toward each other

Cucumbers

young leaves—remain small
plant growth—restricted, short internodes, bushy plant
older leaves—interveinal chlorosis in blotches

stem—growth stunted
progression—later stages get necrotic spots adjacent to and on midribs and larger veins

progression—leaves turn dull green to bronze, necrosis, entire leaf withers, chlorosis spreads from older to younger leaves

Remedies:

1. Foliar spray with 0.1 to 0.2 percent solution of copper sulfate to which 0.5 percent hydrated lime has been added.
2. Add copper sulfate to nutrient solution.

6. Manganese

Tomatoes

middle and older leaves—turn pale, later younger leaves also, characteristic checkered pattern of green veins and yellowish interveinal areas, later stages get small necrotic spots in pale areas, chlorosis less severe than in iron deficiency, also chlorosis is not confined to younger leaves as is the case with iron

Cucumbers

terminal or young leaves— yellowish interveinal mottling, at first even the small veins remain green, giving a reticular green pattern on a yellow background
progression—later, all except main veins turn yellow with sunken necrotic spots between the veins
shoots—stunted, new leaves remain small
older leaves—turn palest and die first

Remedies:

1. Foliar spray using high-volume spray of 0.1 percent or low-volume spray of 1 percent solution of manganese sulfate.
2. Add manganese sulfate to nutrient solution.

7. Molybdenum

Tomatoes

all leaves—leaflets show a pale green to yellowish interveinal mottling, margins curl upward to form a spout, smallest veins do not remain green, necrosis starts in the yellow areas, at the margins of the top leaflets and finally includes entire composite leaves which shrivel
progression—from the older to the younger leaves, but the cotyledons stay green for a long time

Cucumbers

older leaves—fade, particularly between veins, later leaves turn pale green, finally yellow and die

progression—from older leaves up to young leaves, youngest leaves remain green
plant growth—normal, but flowers are small

Remedies:

1. Foliar spray with 0.07 to 0.1 percent solution of ammonium or sodium molybdate.
2. Add ammonium or sodium molybdate to nutrient solution.

References

Arnon, D.I. 1950. Inorganic micronutrient requirements of higher plants. *Proc. 7th Int. Bot. Cong.,* Stockholm.

———. 1951. Growth and function as criteria in determining the essential nature of inorganic nutrients. *Mineral nutrition of plants,* ed. E. Truog, pp. 313–41. Madison, WI: Univ. Wisconsin Press.

Arnon, D.I., and P. R. Stout. 1939. The essentiality of certain elements in minute quantity for plants with special reference to copper. *Plant physiol.* 14:371–75.

Buckman, H.O., and N.C. Brady. 1984. *The nature and properties of soils.* 9th ed. New York, NY: Macmillan.

Epstein, E. 1972. *Mineral nutrition of plants: principles and perspectives.* New York, NY: Wiley.

Gauch, H.G. 1972. *Inorganic plant nutrition.* Stroudsburg, PA: Dowden, Hutchinson and Ross.

Kramer, P.J. 1969. *Plant and soil water relationships: a modern synthesis.* New York, NY: McGraw-Hill.

Roorda van Eysinga, J.P.N.L., and K.W. Smilde. 1969. *Nutritional disorders in cucumbers and gherkins under glass.* Wageningen: Center for Agric. Publ. and Documentation.

——. 1971. *Nutritional disorders in glasshouse lettuce.* Wageningen: Center for Agric. Publ. and Documentation.

Salisbury, F.B., and C. Ross. 1969. *Plant physiology.* Belmont, CA: Wadsworth.

Smilde, K.W., and J.P.N.L. Roorda van Eysinga. 1968. *Nutritional diseases in glasshouse tomatoes.* Wageningen: Center for Agric. Publ. and Documentation.

Sprague, H.B., ed. 1964. *Hunger signs in crops: a symposium.* 3rd ed. New York, NY: David McKay.

Chapter 3

The Nutrient Solution

3.1 Inorganic Salts (Fertilizers)

In hydroponics all the essential elements are supplied to the plants by dissolving fertilizer salts in water to make up the nutrient solution. The choice of salts to be used depends on a number of factors. The relative proportion of ions a compound will supply must be compared with that required in the nutrient formulation. For example, one molecule of potassium nitrate (KNO_3) will yield one ion of potassium (K^+) and one of nitrate (NO_3^-), whereas one molecule of calcium nitrate ($Ca(NO_3)_2$) will yield one ion of calcium (Ca^{++}) and two ions of nitrate $2(NO_3^-)$. Therefore, if a minimum number of cations is wanted while supplying sufficient nitrate (anions) then calcium nitrate should be used. That is, half as much calcium nitrate as potassium nitrate would be required to satisfy the needs of the nitrate anion.

The various fertilizer salts which can be used for the nutrient solution have different solubilities (see Appendix 4). Solubility is a measure of the concentration of the salt which will remain in solution when dissolved in water. If a salt has a low solubility, only small amounts of it will dissolve in water. In hydroponics, fertilizer salts having high solubility must be used since they must remain in solution to be available to the plants. For example, calcium may be supplied by either calcium nitrate or calcium sulfate. Calcium sulfate is cheaper but its solubility is very low. Therefore, calcium nitrate should be used to supply the entire calcium requirements.

The cost of a particular fertilizer must be considered in deciding on its use. In general, a greenhouse grade should be used. The cost is somewhat greater than a standard grade, but the purity and solubility will be greater. A poor grade will contain a large amount of inert carriers (clay, silt particles) which can tie up nutrients and plug feeder lines. Regular grade calcium nitrate is bulk shipped to North America and packaged on this continent. For bulk shipping it is coated with a greasy plasticizer to prevent it from accumulating water as it is hydroscopic (attracts water).

Unfortunately, for use in nutrient solutions this greasy coating creates a thick scum that floats on the solution surface and plugs irrigation lines and makes cleaning of tanks and equipment difficult. To avoid this problem use the special grade for greenhouse nutrient solutions called "Solution Grade." It is packaged in a green bag, not the blue & red or red bags of the regular grades ("Viking Ship" brand).

The availability of nitrate versus ammonium compounds to plants is important in promoting either vegetative or reproductive growth. Plants can absorb both the cationic ammonium ion (NH_4^+) and the anion nitrate (NO_3^-). Ammonium, once absorbed, can immediately serve in the synthesis of amino acids and other compounds containing reduced nitrogen. Absorption of ammonium can therefore cause excessive vegetative growth, particularly under poor light conditions. Nitrate nitrogen, on the other hand, must be reduced before it is assimilated; therefore, vegetative growth will be held back. Ammonium salts could be used under bright summer conditions when photosynthetic rates are high or if a nitrogen deficiency occurs and a rapid source of nitrogen is needed. In all other cases nitrate salts should be used.

A summary of some salts which could be used for a hydroponic nutrient solution is given in Table 3.1. The particular choice of salt will depend on the above factors and market availability. If a dry pre-mix is to be used, such as in sawdust, peat, or vermiculite media, some of the more insoluble salts could be used, while if a nutrient solution is to be made up in advance, the more soluble compounds should be used (those marked with an asterisk). Potassium chloride and calcium chloride should be used only to correct potassium and calcium deficiencies respectively. These, however, can be used only if insignificant amounts of sodium chloride (less than 50 ppm) are present in the nutrient solution. If chlorides are added in the presence of sodium, poisoning of the plants will result.

The use of chelates (of iron, manganese, and zinc) is highly recommended since they remain in solution and are readily available to plants even under *p*H shifts. A chelated salt is one having a soluble organic component to which the mineral element can adhere until uptake by the plant roots occurs. The organic component is EDTA (ethylene-diaminetetra acetic acid). EDTA has a high affinity for calcium ions and is thus a poor chelating agent for calcareous media (limestone, coral sand). In this case it should be replaced with EDDHA (ethylene-diamine dihydroxyphenyl acetic acid).

TABLE 3.1 Summary of Fertilizer Salts for Use in Hydroponics

Chemical Formula	Chemical Name	Mol. Wt.	Elements Supplied	Solubility Ratio of Solute to Water	Cost	Other Remarks
			A. Macroelements			
*KNO_3	Potassium Nitrate (Saltpeter)	101.1	K^+, NO_3^-	1:4	Low	Highly soluble, high purity.
*$Ca(NO_3)_2$	Calcium nitrate	164.1	Ca^{++}, $2(NO_3^-)$	1:1	Low-medium	Highly soluble, but is shipped with a greasy plasticizer coat—must be skimmed off nutrient solution—use "Solution Grade."
$(NH_4)_2SO_4$	Ammonium sulfate	132.2	$2(NH_4^+)$, $SO_4^=$	1:2	Medium	These compounds should be used only under very good light conditions or to correct N-deficiencies.
$NH_4H_2PO_4$	Ammonium dihydrogen phosphate	115.0	NH_4^+, $H_2PO_4^-$	1:4	Medium	
NH_4NO_3	Ammonium nitrate	80.05	NH_4^+, NO_3^-	1:1	Medium	
$(NH_4)_2HPO_4$	Ammonium mono-hydrogen phosphate	132.1	$2(NH_4^+)$, $HPO_4^=$	1:2	Medium	
*KH_2PO_4	Mono-potassium phosphate	136.1	K^+, $H_2PO_4^-$	1:3	Very expensive	An excellent salt, highly soluble and pure, but costly.
KCl	Potassium chloride (Muriate of potash)	74.55	K^+, Cl^-	1:3	Expensive	Should only be used for K-deficiencies and when no sodium chloride is present in nutrient solution.
*K_2SO_4	Potassium sulfate	174.3	$2K^+$, $SO_4^=$	1:15	Inex-pensive	It has low solubility, must dissolve in hot water.
$Ca(H_2PO_4)_2H_2O$	Monocalcium phosphate	252.1	Ca^{++}, $2(H_2PO_4^-)$	1:60	Inex-pensive	Very difficult to obtain a soluble grade.
$CaH_4(PO_4)_2$	Triple super phosphate	Variable	Ca^{++}, $2(PO_4^{\equiv})$	1:300	Inex-pensive	Very low solubility, good only for dry pre-mixes, not for nutrient solutions.
*$MgSO_4 \cdot 7H_2O$	Magnesium sulfate (Epsom salts)	246.5	Mg^{++}, $SO_4^=$	1:2	Inex-pensive	Excellent cheap, highly soluble, pure salt.
$CaCl_2 \cdot 6H_2O$	Calcium chloride	219.1	Ca^{++}, $2Cl^-$	1:1	Expensive	Highly soluble, good to overcome Ca-deficiencies, but use only if no NaCl present in nutrient solution.
$CaSO_4 \cdot 2H_2O$	Calcium sulfate (Gypsum)	172.2	Ca^{++}, $SO_4^=$	1:500	Inex-pensive	Very insoluble, cannot be used for nutrient solutions.
H_3PO_4	Phosphoric acid (Orthophosphoric acid)	98.0	$PO_4^{\equiv}$	Concentrated acid solution	Expensive	Good use in correction of P-deficiencies.

Chemical Formula	*Chemical Name*	*Mol. Wt.*	*Elements Supplied*	*Solubility Ratio of Solute to Water*	*Cost*	*Other Remarks*
			B. Microelements			
$FeSO_4 \cdot 7H_2O$	Ferrous sulfate (Green vitriol)	278.0	Fe^{++}, $SO_4^{=}$	1:4		
$FeCl_3 \cdot 6H_2O$	Ferric chloride	270.3	Fe^{+++}, $3Cl^{-}$	1:2		
*FeEDTA	Iron chelate (Sequestrene) (10.5% iron)	382.1	Fe^{++}	Highly soluble	Expensive	Best source of iron; dissolve in hot water.
*H_3BO_3	Boric acid	61.8	B^{+++}	1:20	Expensive	Best source of boron; dissolve in hot water.
$Na_2B_4O_7 \cdot 10H_2O$	Sodium tetra-borate (Borax)	381.4	B^{+++}	1:25		
*$CuSO_4 \cdot 5H_2O$	Copper sulfate (Bluestone)	249.7	Cu^{++}, $SO_4^{=}$	1:5	Inexpensive	
*$MnSO_4 \cdot 4H_2O$	Manganese sulfate	223.1	Mn^{++}, $SO_4^{=}$	1:2	Inexpensive	
$MnCl_2 \cdot 4H_2O$	Manganese chloride	197.9	Mn^{++}, $2Cl^{-}$	1:2	Inexpensive	
*$ZnSO_4 \cdot 7H_2O$	Zinc sulfate	287.6	Zn^{++}, $SO_4^{=}$	1:3	Inexpensive	
$ZnCl_2$	Zinc chloride	136.3	Zn^{++}, $2Cl^{-}$	1:1.5	Inexpensive	
$(NH_4)_6Mo_7O_{24}$	Ammonium molybdate	1163.9	NH_4^{+}, Mo^{+6}	1:2.3 Highly soluble	Moderately expensive	
*ZnEDTA	Zinc chelate	431.6	Zn^{++}	Highly soluble	Expensive	
*MnEDTA	Manganese chelate	381.2	Mn^{++}	Highly soluble	Expensive	

* These more soluble compounds should be used for preparing nutrient solutions.

3.2 Recommended Compounds for Complete Nutrient Solutions

Calcium should be provided by calcium nitrate. Calcium nitrate will also provide nitrate nitrogen. Any additional nitrogen required should be provided by potassium nitrate, which also provides some potassium. All phosphorus may be obtained from monopotassium phosphate, which also provides some potassium. The remaining potassium requirement can be obtained from potassium sulfate, which also supplies some sulfur. Additional sulfur comes from other sulfates such as magnesium sulfate, which is used to supply the magnesium needs.

Micronutrients may be obtained from commercial pre-mixes. While these are relatively expensive, they save the substantial labor of weighing the individual compounds contained in the mix.

Hobby growers may wish to use pre-mixes for the macronutrients, but commercial growers should use the basic compounds listed in Table 3.1. The reason for this is that it is very difficult to get a homogeneous mixture of fertilizers when hundreds of pounds of individual compounds are blended with a mechanical mixer. Many of the compounds are in powder or fine grain form, and often lumpy, which do not mix evenly mechanically. Experience with such pre-mixes has revealed shortages in magnesium, almost always a shortage of iron and often excesses of manganese. In addition, pre-mixes do not offer flexibility in manipulating a nutrient formulation which is necessary during different stages of plant growth and under changing sunlight hours and daylength. This ability to make changes in the nutrient formulation is imperative in the optimization of crop yields.

3.3 Fertilizer Chemical Analyses

The amounts of available nitrogen, phosphorus, and potassium are given on fertilizer bags as the percentage of nitrogen (N), phosphoric anhydride (P_2O_5) and potassium oxide (K_2O). It has traditionally been expressed in these terms, not in percentage of N, P, or K alone. For example, potassium nitrate is given as 13-0-44, indicating 13 percent N, 0 percent P_2O_5, and 44 percent K_2O.

Nutrient formulations for hydroponics express: nitrogen as N, NH_4^+, or NO_3^-; phosphorus as P or $PO_4^{\equiv}$, not P_2O_5; and potassium as K^+, not K_2O. Therefore, it is necessary to convert N to NO_3^-, P_2O_5 as P or $PO_4^{\equiv}$, and K_2O as K^+ or vice versa in each case. Conversions of this nature can be made by calculating the fraction of each element within its compound source. Table 3.2 lists conversion factors to determine the fraction of an element within a compound and vice versa. It was derived by the use of atomic and molecular weights as follows: The fraction of N in NO_3^- is the atomic weight of nitrogen (14) divided by the molecular weight of nitrate (62). The calculation is: 14/62 = 0.226. In Table 3.2 this factor appears under the conversion factor B to A on the second line. That is because we have nitrate and want to find the amount of nitrogen in it. To determine the amount of nitrate needed for a unit of nitrogen, the reciprocal (inverse) of the fraction is used as the following ratio is applied and solved for "x."

$$\frac{1}{\underset{14}{N}} = \frac{x}{\underset{62}{NO_3}} \qquad \text{cross multiply: } 14x = 62$$

$$\text{solve for x: } x = 62/14 = 4.429$$

This is the conversion factor A to B on the second line of Table 3.2. Understanding this concept for the derivation of these factors will enable you to calculate other factors should other compounds than those appearing in Table 3.2 be used.

TABLE 3.2 Conversion Factors for Fertilizer Salts

Column A *	*Column B* *	*Conversion factor*	
		A to B	*B to A*
Nitrogen (N)	Ammonia (NH_3)	1.216	0.822
	Nitrate (NO_3)	4.429	0.226
	Potassium nitrate (KNO_3)	7.221	0.1385
	Calcium nitrate ($Ca(NO_3)_2$)	5.861	0.171
	Ammonium sulfate ($(NH_4)_2SO_4$)	4.721	0.212
	Ammonium nitrate (NH_4NO_3)	2.857	0.350
	Diammonium phosphate ($(NH_4)_2HPO_4$)	4.717	0.212
Phosphorus (P)	Phosphoric anhydride (P_2O_5)	2.292	0.436
	Phosphate (PO_4)	3.066	0.326
	Monopotassium phosphate (KH_2PO_4)	4.394	0.228
	Diammonium phosphate ($(NH_4)_2HPO_4$)	4.255	0.235
	Phosphoric acid (H_3PO_4)	3.164	0.316
Potassium (K)	Potash (K_2O)	1.205	0.830
	Potassium nitrate (KNO_3)	2.586	0.387
	Monopotassium phosphate (KH_2PO_4)	3.481	0.287
	Potassium chloride (KCl)	1.907	0.524
	Potassium sulfate (K_2SO_4)	2.229	0.449
Calcium (Ca)	Calcium oxide (CaO)	1.399	0.715
	Calcium nitrate ($Ca(NO_3)_2$)	4.094	0.244
	Calcium chloride ($CaCl_2 \cdot 6H_2O$)	5.467	0.183
	Calcium sulfate ($CaSO_4 \cdot 2H_2O$)	4.296	0.233
Magnesium (Mg)	Magnesium oxide (MgO)	1.658	0.603
	Magnesium sulfate ($MgSO_4 \cdot 7H_2O$)	10.14	0.0986
Sulfur (S)	Sulfuric acid (H_2SO_4)	3.059	0.327
	Ammonium sulfate ($(NH_4)_2SO_4$)	4.124	0.2425
	Potassium sulfate (K_2SO_4)	5.437	0.184
	Magnesium sulfate ($MgSO_4 \cdot 7H_2O$)	7.689	0.130
	Calcium sulfate ($CaSO_4 \cdot 2H_2O$)	5.371	0.186
Iron (Fe)	Ferrous sulfate ($FeSO_4 \cdot 7H_2O$)	4.978	0.201
	Iron chelate (FeEDTA) (Sequestrene - 10% iron)	10.00	0.100
Boron (B)	Boric acid (H_3BO_3)	5.717	0.175
	Sodium tetraborate ($Na_2B_4O_7 \cdot 10H_2O$) (Borax)	8.820	0.113
Copper (Cu)	Copper sulfate ($CuSO_4 \cdot 5H_2O$)	3.930	0.254
Manganese (Mn)	Manganese sulfate ($MnSO_4 \cdot 4H_2O$)	4.061	0.246
	Manganese chloride ($MnCl_2 \cdot 4H_2O$)	3.602	0.278
	Manganese chelate (MnEDTA) (5% liquid)	20.00	0.050
Zinc (Zn)	Zinc sulfate ($ZnSO_4 \cdot 7H_2O$)	4.400	0.227
	Zinc chloride ($ZnCl_2$)	2.085	0.480
	Zinc chelate (ZnEDTA) (14% powder)	7.143	0.140
	Zinc chelate (9% liquid)	11.11	0.090
Molybdenum (Mo)	Ammonium Molybdate ($(NH_4)_6Mo_7O_{24}$)	1.733	0.577
	Sodium Molybdate ($Na_6Mo_7O_{24}$)	1.777	0.563

Note: These factors are derived from the fraction of an element present in a compound based upon the atomic weight of the element and molecular weight of the compound.

* To convert from an element (column A) to the ccompound supplying it (column B), use conversion factor A to B. To determine amounts of an element (column A) present in a compound (column B), multiply by factor B to A.

3.4 Fertilizer Impurities

Most fertilizer salts are not 100 percent pure. They often contain inert "carriers" such as clay, silt, and sand particles which do not supply ions. Therefore, a percentage purity or a guaranteed analysis is often given on the fertilizer bag. Some of the percentage purities of common fertilizers are given in Table 3.3. These impurities must be taken into consideration when calculating the fertilizer requirements for a particular nutrient formulation.

TABLE 3.3 Percentage Purities of Commercial Fertilizers

Salt	*% Purity*
Ammonium phosphate ($NH_4H_2PO_4$) (food grade)	98
Ammonium sulfate ($(NH_4)_2SO_4$)	94
Ammonium nitrate, pure (NH_4NO_3)	98
Potassium nitrate (KNO_3)	95
Calcium nitrate ($Ca (NO_3)_2$)	90
Monocalcium phosphate ($Ca (H_2PO_4)_2$) (food grade)	92
Potassium sulfate (K_2SO_4)	90
Potassium chloride (KCl)	95
Magnesium sulfate ($MgSO_4 \cdot 7H_2O$)	98
Calcium chloride ($CaCl_2$)	75
Calcium sulfate ($CaSO_4$) (Gypsum)	70
Monopotassium phosphate (KH_2PO_4)	98

The purity is calculated on the basis of the designated formula. Water of crystallization is not considered as impurity.

Note: In Table 3.2, the water of crystallization has been accounted for, in the calculations of the conversion factors.

Many fertilizers have synonyms or common names. A list of these common names is given in Table 3.4.

TABLE 3.4 Chemical Names and Synonyms of Compounds Generally Used in Nutrient Solutions

Chemical Name	*Synonyms or Common Name*
Potassium nitrate KNO_3	Saltpeter.
Sodium nitrate $NaNO_3$	Chile saltpeter; niter; Chile niter.
Ammonium acid phosphate $NH_4H_2PO_4$	Ammonium biphosphate; ammonium bihydrogenphosphate; monobasic ammonium phosphate.

Chemical Name	***Synonyms or Common Name***
Urea $CO(NH_2)_2$	Carbamide; carbonyldiamide.
Potassium sulfate K_2SO_4	Sulfate of potash.
Potassium acid phosphate KH_2PO_4	Potassium biphosphate; potassium dihydrogen phosphate; monobasic potassium phosphate.
Potassium chloride KCl	Muriate of potash; chloride of potash.
Monocalcium phosphate $Ca(H_2PO_4)_2H_2O$	Calcium "superphosphate" (usually 20 percent pure); calcium "treble superphosphate" (usually 75 percent pure); calcium biphosphate; calcium acid phosphate.
Phosphoric acid H_3PO_4	Orthophosphoric acid (U.S.P. grade is 85–88 percent H_3PO_4; commercial technical grade is 70–75 percent H_3PO_4).
Calcium sulfate $CaSO_4{\cdot}2H_2O$	Gypsum.
Calcium chloride $CaCl_2{\cdot}2H_2O$	Calcium chloride dihydrate. Also available as hexahydrate ($CaCl_2{\cdot}6H_2O$).
Magnesium sulfate $MgSO_4{\cdot}7H_2O$	Epsom salts.
Ferrous sulfate $FeSO_4{\cdot}7H_2O$	Green vitriol; iron vitriol.
Zinc sulfate $ZnSO_4{\cdot}7H_2O$	White or zinc vitriol.
Boric acid H_3BO_3	Boracic acid; orthoboric acid.
Copper sulfate $CuSO_4{\cdot}5H_2O$	Cupric sulfate; bluestone.

Source: Adapted from Withrow, R. B. and A. P. Withrow, 1948. *Nutriculture.* Lafayette, Ind.: Purdue Univ. Agr. Expt. Stn. Publ. S.C. 328.

3.5 Nutrient Formulations

Nutrient formulations are usually given in parts per million (ppm) concentrations of each essential element. One part per million is one part of one item in 1 million parts of another. It may be a weight measure, for example, 1 μg/g (one microgram per gram), a weight-volume measure, for example, 1 mg/l (one milligram per liter) or a volume-volume measure, for example 1 μl/l (one microliter per liter). Proofs for these are as follows:

$$(1)\quad 1\mu g/g = \frac{\frac{1}{1{,}000{,}000}\,g}{1g} = \frac{1}{1{,}000{,}000}\,g$$

$$(2)\quad 1\ \mu l/l = \frac{\frac{1}{1{,}000{,}000}\,l}{1\ l} = \frac{1}{1{,}000{,}000}\ l$$

(3) 1 mg/l:

$$1\ mg = g,\ \frac{1}{1{,}000}\quad 1l = 1{,}000\ ml$$

Therefore:

$$1\ mg/l = \frac{\frac{1}{1{,}000}\,g}{1{,}000\ ml\ H_2O} = \frac{1}{1{,}000{,}000}\,g$$

since 1 ml of H_2O weighs 1 gram (g)

3.5.1 Atomic and Molecular Weights

Atomic and molecular weights of elements and compounds respectively must be used in calculating nutrient formulation concentration requirements. Atomic weights indicate the relative weights of different atoms, that is, how the weight of one atom compares with that of another. Tables of atomic weights for every atom have been drawn up by establishing a relative scale of atomic weights. In doing so, one element is chosen as a standard and all other elements are compared with it. Oxygen (O) has in the past been assigned the atomic weight of exactly 16, and all other elements are related to it. Table 3.5 lists the atomic weights of elements commonly used in hydroponics.

When a number of atoms combine they form a molecule. These are expressed as molecular formulae. For example water is H_2O, consisting of two atoms of hydrogen (H) and one atom of oxygen (O). The weight of any molecular formula is the molecular weight. It is simply the sum of the atomic weights within the molecule. The molecular weight of water is 18 (there are two atoms of hydrogen, each having atomic weight of 1.00, and one atom of oxygen having an atomic weight of 16.0). Molecular weights of commonly used fertilizers for hydroponics are given in Table 3.1. The atomic weights of all known elements are given in tables and periodic charts of chemistry texts.

TABLE 3.5 Atomic Weights of Elements Commonly Used in Hydroponics

Name	*Symbol*	*Atomic Wt.*
Aluminum	Al	26.98
Boron	B	10.81
Calcium	Ca	40.08
Carbon	C	12.01
Chlorine	Cl	35.45
Copper	Cu	63.54
Hydrogen	H	1.008
Iron	Fe	55.85
Magnesium	Mg	24.31
Manganese	Mn	54.94
Molybdenum	Mo	95.94
Nitrogen	N	14.01
Oxygen	O	16.00
Phosphorus	P	30.97
Potassium	K	39.10
Selenium	Se	78.96
Silicon	Si	28.09
Sodium	Na	22.99
Sulfur	S	32.06
Zinc	Zn	65.37

The following examples will clarify the use of atomic and molecular weights in their use with nutrient formulation calculations:

Calcium nitrate – $Ca(NO_3)_2$

Atomic weight:			*Molecular weight:*		
Ca	=	40.08	Ca	=	40.08
N	=	14.008	2N	=	28.016
O	=	16.000	6O	=	96.0
					164.096

Note: There are two nitrogen atoms and six oxygen atoms in calcium nitrate.

3.5.2 Calculations of Nutrient Formulations

If a nutrient formulation calls for 200 ppm of calcium (200 mg /l) we need 200 mg of calcium in every liter of water. In 164 mg of $Ca(NO_3)_2$ we have 40 mg of Ca (using the atomic and molecular weights to determine the fraction of calcium in calcium nitrate—assuming 100 percent purity of $Ca(NO_3)_2$). The first step is to calculate how much $Ca(NO_3)_2$ is required to obtain 200 mg of Ca. This is done by setting up a ratio as follows:

164 mg $Ca(NO_3)_2$ yields 40 mg Ca
x mg $Ca(NO_3)_2$ yields 200 mg Ca

$$\text{Ratio of } \frac{Ca}{NO_3}: \quad \frac{40}{164} = \frac{200}{x}$$

Solve for x:

$$40\,x = 200 \times 164 \quad \text{(cross multiply)}$$

$$x = \frac{200 \times 164}{40} \quad \text{(divide by 40)} = 820$$

Therefore, 820 mg of $Ca(NO_3)_2$ will yield 200 mg Ca. If the 820 mg of $Ca(NO_3)_2$ is dissolved in 1 liter of water, the resultant solution will have a concentration of 200 ppm (200 mg/l) of Ca. This assumes, however, that the $Ca(NO_3)_2$ is 100 percent pure. If it is not—which is the usual case—it will be necessary to add more to compensate for the impurity. For example, if the $Ca(NO_3)_2$ is 90 percent pure, it will be necessary to add:

$$\frac{100}{90} \times 820 = 911 \text{ mg } Ca(NO_3)_2$$

The 911 mg $Ca(NO_3)_2$ in 1 liter of water will give 200 ppm of Ca.

Of course, in most cases a larger volume of nutrient solution than 1 liter will be required. The second step, then, is to calculate the amount of fertilizer required for a given volume of nutrients.

Initially calculate the amount of a compound needed using the metric system of mg/l, then convert to pounds per gallon, if necessary. First convert the volume of the nutrient tank in gallons to liters. To do this, be aware that there is a difference between Imperial gallons (British system) and U.S. gallons as follows:

1 U.S. gallon (gal) = 3.785 liters (l)
1 Imp. gallon (gal) = 4.5459 liters (l)

For example: 100 U.S. gal = 378.5 l
100 Imp. gal = 454.6 l

The solution to the following problem will demonstrate the use of these conversions. Assume that a 200 ppm concentration of Ca is required in a 300 U.S.-gallon nutrient tank.

$$300 \text{ U.S. gal} = 300 \times 3.785 \text{ l} = 1{,}135.5 \text{ l}$$

If 911 mg $Ca(NO_3)_2$ is required per liter of water, the amount needed for 300 U.S. gallons would be:

$$911 \text{ mg} \times 1{,}135.5 \text{ l} = 1{,}034{,}440 \text{ mg}$$

To convert to grams (g), divide by 1,000 as 1,000 mg = 1 g:

$$\frac{1{,}034{,}440 \text{ mg}}{1{,}000 \text{ mg}} = 1{,}034.4 \text{ g}$$

Now, convert to kilograms by dividing by 1,000 as 1,000 g = 1 kg:

$$\frac{1{,}034.4 \text{ g}}{1{,}000 \text{ g}} = 1.034 \text{ kg}$$

To calculate pounds, use either: 1 lb = 454 g; or 1 kg = 2.2046 lb

Therefore: $$\frac{1{,}034.4 \text{ g}}{454 \text{ g}} = 2.278 \text{ lb} = 2.28 \text{ lb}$$

or: $$1.034 \text{ kg} \times 2.2046 \text{ lb} = 2.28 \text{ lb}$$

Then, convert the fraction of pounds to ounces: 1 lb = 16 oz

$$2 \text{ lb} + (0.28 \text{ lb} \times 16 \text{ oz} = 4.5 \text{ oz})$$

That is: 2 lb 4.5 oz

While this conversion to pounds is precise enough for larger weights, anything under a pound should be weighed in grams for more accuracy. A suitable gram scale for most weighing under one pound is a triple-beam balance accurate to 0.1 grams.

If the compound used contains more than one essential element—this is the usual case—the third step is to determine how much of each of the other elements was added when satisfying the needs of the first essential element. Calcium nitrate contains both calcium and nitrogen. Therefore, the third step is to calculate the amount of nitrogen added while satisfying the calcium needs.

This should be done using the ppm concepts so that adjustments can be made for this element. The calculation is made using the fraction of nitrogen in calcium nitrate and multiplying the amount of calcium nitrate used by this fraction. Use the weight of calcium nitrate before adjustments for impurities, that is, the 820 mg not the 911 mg, as follows:

$$\frac{2\,(14)}{164} \times 820 \text{ mg/l} = 140 \text{ mg/l (ppm)}$$

In summary, 820 mg/l of calcium nitrate will yield 200 mg/l of calcium and 140 mg/l of nitrogen (assuming 100% purity). Further, using the example for a 300 U.S.-gallon tank, 2 lb 4.5 oz (1,034 g) of calcium nitrate provides 200 ppm of calcium and 140 ppm of nitrogen in the tank of solution.

By use of the conversion factors in Table 3.2, the calculations may be simplified. The atomic and molecular weights and their fractions need not be used as that is how the conversion factors of Table 3.2 were derived as explained earlier.

Going back to step one to determine the amount of calcium nitrate needed to supply 200 ppm (mg/l) of calcium, use the conversion factor in Table 3.2. We want 200 ppm of Ca from the source $Ca(NO_3)_2$. Use the factor A to B of 4.094:

$$200 \times 4.094 = 819 \text{ mg/l of } Ca(NO_3)_2$$

Note that there is a slight difference between this value and the previous one (819 mg/l vs. 820 mg/l). This small difference is insignificant for our purposes.

Similarly, the amount of nitrogen in 819 mg/l of calcium nitrate may be calculated using conversion factor B to A of 0.171 in the upper part of Table 3.2:

$$0.171 \times 819 \text{ mg/l} = 140 \text{ mg/l (ppm) of N}$$

The fourth step is to calculate the additional amount of the second element needed from another source. For example, if the nutrient formulation asked for 150 ppm of N, the additional requirement would be:

$$150 - 140 = 10 \text{ ppm of N}$$

This could be obtained from KNO_3. Then, the amount of KNO_3 needed to supply 10 ppm of N using the Table 3.2 conversion factor A to B of 7.221 would be:

$$7.221 \times 10 = 72.21 \text{ mg/l of } KNO_3$$

Since KNO_3 also contains potassium (K), we must calculate the amount present using conversion factor B to A of 0.387 (middle of Table 3.2):

$$0.387 \times 72.21 \text{ mg/l} = 28 \text{ mg/l (ppm) of K}$$

To determine the amount of KNO_3 needed for a 300 U.S.-gallon nutrient tank to provide 10 ppm of N:

(1) Adjust for impurity from Table 3.3 (95% pure):

$$\frac{100}{95} \times 72.21 = 76 \text{ mg/l}$$

(2) For a 300 U.S.-gallon tank:

$$300 \times 3.785 \text{ l} = 1{,}135.5 \text{ l}$$

(3) Amount of KNO_3 needed:

$$1{,}135.5 \text{ l} \times \frac{76 \text{ mg}}{1{,}000 \text{ mg}} = 86.3 \text{ g}$$

This weight may be kept in grams as it is under one pound, or it may be converted to ounces as follows:

$$1 \text{ oz} = 28.35 \text{ g}$$

(4) Conversion to ounces:

$$\frac{86.3 \text{ g}}{28.35 \text{ g}} = 3.0 \text{ oz}$$

Note that gram measurements are more accurate than ounces.

These calculations may be continued for all the essential elements. The types and quantities of the various fertilizer salts must be manipulated until the desired formulation is achieved.

In some cases a problem may arise if the requirements of one element are satisfied by use of a compound which contains two or more essential elements, but the concentration of another element exceeds the level required.

For instance, if the nutrient formulation called for 300 ppm Ca and 150 ppm N, the calcium would be supplied by calcium nitrate as follows:

(1) Weight of $Ca(NO_3)_2$ needed (use Table 3.2 conversion factors):

$$300 \text{ ppm (mg/l)} \times 4.094 = 1{,}228 \text{ mg/l}$$

(2) Amount of N added:

$$1{,}228 \text{ mg/l} \times 0.171 = 210 \text{ mg/l (ppm) N}$$

This gives an excess of 60 ppm of N over the recommended 150 ppm N given in the formulation. Therefore, the level of N will govern how much $Ca(NO_3)_2$ can be used as a source of Ca. The previous steps must be recalculated using the limit of 150 ppm N:

(1) Weight of $Ca(NO_3)_2$ needed (use factors of Table 3.2):

$$150 \text{ ppm (mg/l)} \times 5.861 = 879 \text{ mg/l}$$

(2) Amount of Ca added:

$$879 \text{ mg/l} \times 0.244 = 214 \text{ mg/l (ppm) Ca}$$

If the recommended level of Ca is 300 ppm, then (300 – 214) = 86 ppm Ca must be supplied from other sources than $Ca(NO_3)_2$. Since calcium sulfate ($CaSO_4$) is very insoluble, the only alternative is to use calcium chloride ($CaCl_2 \cdot 6H_2O$). Note the water of crystallization on calcium chloride. This has been taken into account in calculating the conversion factors in Table 3.2.

(3) Weight of $CaCl_2 \cdot 6H_20$ needed:

86 ppm (mg/l) × 5.467 = 470 mg/l

(4) Amount of Cl added (Table 3.2 has no factor for the conversion from calcium chloride to chlorine, so the atomic-weight fraction must be used):

$$\frac{\text{Atomic weight of chlorine}}{\text{Molecular weight of calcium chloride}} \times 470 \text{ mg/l}$$

$$\frac{2(35.5)}{219} \times 470 = 152 \text{ mg/l (ppm) Cl}$$

This level of chlorine is tolerable to the plants as long as the level of sodium in the raw water and other fertilizers used is negligible.

Once the weight for each compound in the nutrient formulation has been determined for a given tank volume, changes can be easily calculated by use of ratios. These changes using ratios allow you to adjust tank volumes or concentrations for any of the elements. For example, to calculate the amount of calcium nitrate needed to provide 200 ppm Ca in a 500 U.S.-gallon tank, instead of the original 300 U.S.-gallon tank used in the example earlier, simply use the ratio:

$$\frac{500 \text{ U.S. gal}}{300 \text{ U.S. gal}} \times \underset{Ca(NO_3)_2}{1{,}034.4 \text{ g}} = \underset{Ca(NO_3)_2}{1{,}724 \text{ g}}$$

or: $$\frac{500}{300} \times 2.28 \text{ lb} = 3.8 \text{ lb}$$

That is: 3 lb + (0.8 lb × 16 oz = 12.8 oz)

3 lb 12.8 oz.

Often it is necessary to change the levels of individual elements during weather changes, plant growth stages, or the presence of deficiencies or toxicities revealed by symptoms on the plants and/or nutrient and tissue analyses. For example, to change the concentration of Ca from 200 ppm to 175 ppm in the same tank volume (300 U.S. gallons):

$$\frac{175 \text{ ppm Ca}}{200 \text{ ppm Ca}} \times 1{,}034.4 \text{ g} = 905 \text{ g of calcium nitrate}$$

or: $$\frac{175}{200} \times 2.28 \text{ lb} = 1.995 \text{ lb; round off to 2.0 lb}$$

Remember, this also affects the level of the other elements in compounds containing more than one essential element. The change in the nitrogen level is:

$$\frac{175}{200} \times 140 \text{ ppm N} = 122.5 \text{ ppm N}$$

3.5.3 Calculations for Chemical Substitutions for Fertilizers

In some areas of the world a number of basic fertilizers required may not be available. In that case it becomes necessary to substitute other chemicals that are available to provide the needed amount of essential elements. Calculations for such substitutions are as follows:

1. Substitute potassium hydroxide (KOH) and phosphoric acid (H_3PO_4) for diammonium phosphate ($(NH_4)_2HPO_4$) or monopotassium phosphate (KH_2PO_4) to supply phosphorus (P) and some potassium (K).

Note that KOH must be used to neutralize the strong acid of H_3PO_4. The reaction is as follows:

$$H_3PO_4 + KOH \rightarrow K^+ + OH^- + 3H^+ + PO_4^{\equiv}$$
$$\rightarrow K^+ + PO_4^{\equiv} + H_2O + 2H^+$$

H_3PO_4: $M_{H_3PO_4} = 97.99$ (Molecular weight of H_3PO_4)

Need: 60 ppm of P from H_3PO_4:
60×3.164 (from Table 3.2) $= 189.8$ mg/l

KOH: $M_{KOH} = 56.108$

Need:
$$\begin{matrix} 189.8 & & x \\ H_3PO_4 & + & KOH, \\ 97.99 & & 56.108 \end{matrix} \quad \text{therefore } x = \frac{189.8 \times 56.108}{97.99} = 108.7 \text{ mg/l}$$

Amount of K: $\frac{39.1}{56.108} \times 108.7 = 0.6969 \times 108.7 = 75.7$ mg/l

But, phosphoric acid is a liquid, therefore the weight required must be converted to volume measure.

To do this the specific gravity or density (D) must be used. Density is the ratio of weight to volume ($D = W/V$). Density of phosphoric acid is 1.834 (see Appendix 4).

To find the volume: $D = \frac{W}{V}$ or; $V = \frac{W}{D}$

$$\text{that is, } V = \frac{189.8}{1.834} = 103.5\ \mu l/l$$

2. Substitute nitric acid (HNO_3) and calcium carbonate ($CaCO_3$) for calcium nitrate ($Ca(NO_3)_2 \cdot 4H_2O$) to supply calcium (Ca) and nitrogen (N). Note that $CaCO_3$ must be used to neutralize the strong acid of HNO_3. The reaction is as follows:

$$CaCO_3 + HNO_3 \rightarrow Ca^{++} + NO_3^- + H^+ + CO_3^=$$
$$\rightarrow Ca^{++} + NO_3^- + HCO_3^-$$

$CaCO_3$ $M_{CaCO_3} = 100.1$

Need: 150 ppm of Ca from $CaCO_3$:

$$150 \times \frac{100.1}{40.08} = 374.6\ mg/l$$

HNO_3: $M_{HNO_3} = 63.016$

$$\begin{matrix} 374.6 & x \\ CaCO_3 + & HNO_3, \\ 100.1 & 63.016 \end{matrix} \quad \text{therefore; } x = \frac{374.6 \times 63.016}{100.1} = 235.8\ mg/l$$

$$\text{Amount of N: } 235.8 \times \frac{14}{63.016} = 52\ mg/l$$

Since HNO_3 is a liquid it must be converted to volume:

$D = 1.5027$

$$D = \frac{W}{V} \text{ or: } V = \frac{W}{D} = \frac{235.8}{1.5027} = 156.9\ \mu l/l$$

Often HNO_3 is not 100% pure, therefore it must be adjusted by the percentage purity.

In summary, 375 mg/l of $CaCO_3$ and 157 $\mu l/l$ of HNO_3 will provide 150 ppm Ca and 52 ppm N.

3. Substitute nitric acid (HNO_3) and potassium hydroxide (KOH) for potassium nitrate (KNO_3). Note that KOH must be used to neutralize the strong acid of HNO_3. The reaction is as follows:

$$KOH + HNO_3 \rightarrow K^+ + NO_3^- + H_2O$$

Amount of K needed in formulation is 150 ppm. From KOH used with H_3PO_4 for P source, there is 76 ppm of K, therefore need:
150 – 76 = 74 ppm K.

$$\text{KOH: } 74 \times \frac{56.108}{39.1} = 106 \text{ mg/l}$$

HNO_3:

$$\text{Need: } \underset{56.108}{\overset{106}{KOH}} + \underset{63.016}{\overset{x}{HNO_3}}, \quad \text{therefore, } x = 106 \times \frac{63.016}{56.108} = 119 \text{ mg/l}$$

$$\text{Amount of N: } 119 \times \frac{14}{63.016} = 26.4 \text{ mg/l}$$

Since HNO_3 is a liquid, it must be converted to volume:

D = 1.5027

$$D = \frac{W}{V} \quad \text{or: } V = \frac{W}{D} = \frac{119}{1.5027} = 79.2\ \mu\text{l/l}$$

That is, 106 mg/l of KOH and 79.2 μl/l of HNO_3 will provide 74 ppm of K and 26 ppm of N.

4. To make FeEDTA chelate (ethylene-diaminetetra acetatoferrate):

The objective is to make a 200 kg stock solution containing 10,000 mg/l (ppm) (1% iron) of chelated iron.

(1) Dissolve 10.4 kg EDTA (acid) in a solution of 16 kg of KOH in 114 l of water. Adjust the weight of KOH used accordingly if the KOH is not 100% pure. Do not add all of the KOH to the solution initially in order to maintain the *p*H at 5.5. Should the *p*H exceed 5.5, reduce it by addition of a 10% nitric acid (HNO_3) solution. If the *p*H is substantially less than 5.5, add KOH dissolved in water to the solution, slowly stirring until the *p*H reaches 5.5.

(2) Separately dissolve 10 kg of ferrous sulfate ($FeSO_4$) in 64 l of hot water. Slowly add the ferrous sulfate solution, while stirring, to the EDTA/KOH solution of *p*H 5.5. If the *p*H goes below 5.0, add some of the KOH

solution while stirring vigorously. With each addition of KOH solution, precipitation of ferrous hydroxide ($Fe(OH)_2$) will occur. As the *p*H adjusts itself this will redissolve, but the redissolution will become slower as the *p*H of the stock solution rises closer to *p*H 5.5.

(3) After all of the ferrous sulfate and KOH solutions have been added to the EDTA/KOH solution, weigh the final solution and adjust the volume with addition of water until a final solution weight of 200 kg is achieved.

Example:

In order to obtain 5 ppm of iron in 30,000 liters of water:

1. The stock solution contains 10,000 mg/l of iron.
2. We need 5 mg/l (ppm) of iron in the nutrient solution.
3. In 30,000 liters of water we need: $30{,}000 \times 5 = 150{,}000$ mg or 150 g of iron.
4. If the FeEDTA stock solution contains 10,000 mg/l or 10 g/l of iron, we need:

   For 150 g of iron: $\dfrac{150}{10} = 15$ l of FeEDTA stock solution.

Note: The density of a compound should be obtained from the manufacturer as variations will occur from one source to another. A general table of solubilities and densities is given in the Appendix.

3.5.4 Nutrient Formulation Adjustments

Since the nutrient formulation will have to be adjusted frequently during the growing of any crop, it is essential to understand the calculations and manipulations outlined in the previous section. Many claims have been made as to the derivation of "optimum formulations" for a particular crop. Too often, however, these claims are not substantiated and cannot be supported since the optimum formulation depends on too many variables which cannot be controlled. An optimum formulation would depend on the following variables:

1. Plant species and variety.
2. Stage of plant growth.
3. Part of the plant representing the harvested crop (root, stem, leaf, fruit).
4. Season of year—day length.
5. Weather—temperature, light intensity, sunshine hours.

Different varieties and plant species have different nutrient requirements, particularly nitrogen, phosphorus and potassium. For example, lettuce and other leafy vegetables can be given higher rates of nitrogen than tomatoes or cucumbers. The latter two require higher rates of phosphorus, potassium and calcium than do leafy plants.

Ulises Durany (1982) states that the nitrogen (N) level should remain lower (N = 80 to 90 ppm) for species that produce fruit than those species producing leaves (N = 140 ppm). For species that are grown for roots, potassium (K) should be higher (K = 300 ppm). For lettuce, on the other hand, relatively low levels of potassium (K = 150 ppm) favor the closing of the heads and therefore result in greater weights.

The proportions among the various elements must vary according to the species of plant, the growth cycle and development of the plant, and climatic conditions, particularly light intensity and duration.

Nutrient formulations are often composed of several different levels to be used at the different stages of plant growth. Formulations specifically for tomatoes usually consist of three levels—A, B and C. These levels apply only to the macroelements; the microelements are kept the same for all three levels. The A formulation is approximately one-third of C, and B is approximately two-thirds of C. Nonetheless, individual elements are often adjusted independently. The A formulation is used for seedlings from first true-leaf stage (10 to 14 days old) until they are 14 to 16 inches tall. The B formulation is used from 14 to 16 days until the plants are 24 inches tall, when initial fruit is about 1/4 to 1/2 inch in diameter. After that the C formulation is used.

Cucumbers use only A and B formulations, where A is approximately one-half of the B level. The A level is used until the first cucumbers have set. Similarly, leafy vegetables also use a two-level formulation. The first level (lower) is used until the plants are about 3 weeks old and the second level thereafter.

In general, plants harvested for their leaves can tolerate higher N levels since nitrogen promotes vegetative growth. However, plants grown for fruit production should have lower N and higher P, K, and Ca levels. Under high light conditions, plants will use more nitrogen than under poor light. High potash (K) levels during the fall and early winter will improve fruit quality. This potassium/nitrogen ratio is more important and should be varied with the climate. During the longer sunny summer days, the plant needs more nitrogen and less potassium than during the shorter, darker winter days. It is common practice therefore to double the ratio K/N during the winter. This will make for harder growth in winter than would take place on a summer formulation. Some typical nutrient formulations derived over the past by various researchers and commercial growers are given in Table 3.6.

TABLE 3.6 Composition of Nutrient Solutions (ppm)

Reference	*pH*	Ca^{++}	Mg^{++}	Na^{+}	K^{+}	*N as* NH_4^+	*N as* NO_3^-	*P as* $PO_4^{\equiv}$	*S as* $SO_4^{=}$	Cl^-	*Fe*	*Mn*	*Cu*	*Zn*	*B*	*Mo*
Knopp (1865)		244	24		168		206	57	32		Trace					
Shive (1915)		208	484		562		148	448	640		Trace					
Hoagland (1919)	6.8	200	99	12	284		158	44	125	18	As reqd.					
Jones & Shive (1921)		292	172		102	39	204	65	227		0.83					
*Rothamsted	6.2	116	48		593		139	117	157	17	8	0.25			0.2	
Hoagland & Snyder (1933, 1938)		200	48		234		210	31	64	~1	As reqd.	0.1	0.014	0.01	0.1	0.016
Hoagland & Arnon (1938)		160	48		234	14	196	31	64		0.6 3× weekly	0.5	0.02	0.05	0.5	0.01
Long Ashton Soln	5.5–6.0	134–300	36	30	130–295		140–284	41	48	3.5	5.6 or 2.8	0.55	0.064	0.065	0.5	0.05
Eaton (1931)		240	72		117		168	93	96		0.8	0.5			1	
Shive & Robbins (1942)		60	53	92	117		56	46	70	107	As reqd.	0.15		0.15	~0.1	
Robbins (1946)		200	48		195		196	31	64		0.5	0.25	0.02	0.25	0.25	0.01
White (1943)	4.8	50	72	70	65		47	4	140	31	1.0	1.67	0.005	0.59	0.26	0.001
Duclos (1957)	5–6	136	72		234		210	27	32		3	0.25	0.15	0.25	0.4	2.5
Tumanov (1960)	6–7	300–500	50		150		100–150	80–100	64	4	2	0.5	0.05	0.1	0.5	0.02
A. J. Abbott	6.5	210	50		200		150	60	147		5.6	0.55	0.064	0.065	0.5	0.05
E. B. Kidson	5.5	340	54	35	234		208	57	114	75	2	0.25	0.05	0.05	0.5	0.1
Purdue (1948) A		200	96		390	28	70	63	607		20	0.3	0.02	0.05	0.5	
B		200	96		390	28	140	63	447		1.0	0.3	0.02	0.05	0.5	
C		120	96		390	14	224	63	64		1.0	0.3	0.02	0.05	0.5	

Schwartz	124	43		312		*98	93	160							
(Israel)															
California	160	48		234	15	196	31	64							
New Jersey	180	55		90	20.5	126	71	96							
South Africa	320	50		300		200	65								
CDA A	131	22		209	33	93	36.7	29.5	188	1.7	0.8	0.035	0.094	0.46	0.027
Saanichton B	146	22		209	33	135	36.7	29.5	108	1.7	0.8	0.035	0.094	0.46	0.027
B.C., Canada C	146	22		209	33	177	36.7	29.5	—	1.7	0.8	0.035	0.094	0.46	0.027
Dr. Pilgrim C	272	54.3	—	400	—	143.4	93.0	237.5	—		—	—		—	—
Elizabeth B	204	40.7	—	300	—	107.6	69.75	178.1	—		—	—		—	—
N.C., USA A	136	27.15	—	200	—	71.7	46.5	118.75	—		—	—		—	—
Dr. H.M. Resh C	197	44	—	400	30	145	65	197.5	—	2	0.5	0.03	0.05	0.5	0.02
Univ. of B.C. B	148	33	—	300	20	110	55	144.3	—	2	0.5	0.03	0.05	0.5	0.02
Vancouver, A	98.5	22	—	200	10	80	40	83.2	—	2	0.5	0.03	0.05	0.5	0.02
B.C. Canada (1971)															
Dr. H.M. Resh, Dry Season	250	36	—	200	53	177	60	129	—	5	0.5	0.03	0.05	0.5	0.02
Tropical Wet Season	150	50	—	150	32	115	50	52	—	5	0.5	0.03	0.05	0.5	0.02
Formulations															
(lettuce) (1984)															
Dr. H. M. Resh,	200	40	—	210	25	165	50	113	—	5	0.5	0.1	0.1	0.5	0.05
Lettuce Formulations,															
Florida, California (1989) (1993)															
Dr. H. M. Resh,															
Cucumbers,															
Florida (1990)															
Seedlings (0–10 days)	100	20	—	175	3	128	27	26	—	2	0.8	0.07	0.1	0.3	0.03
10 days to 1st fruit swell	220	40	—	350	7	267	55	53	—	3	0.8	0.07	0.1	0.3	0.03
Maturity, after 1st fruit swell	200	45	—	400	7	255	55	82	—	2	0.8	0.1	0.33	0.4	0.05

* Add additional 1 m M (14 ppm) per week to a total of about 12 m M, (128 ppm), for the first 6 to 8 weeks.

Ulises Durany (1982) recommends that for the development of tomatoes during the initial vegetative phase the N:K proportion should be 1:5 (e.g.: 80 ppm N:400 ppm K); the intermediate phase during blossoming and fruit-set, the N:K ratio should be 1:3 (e.g.: 110 ppm N:330 ppm K); and the mature stage with ripening fruit should have a N:K ratio of 1:1.5 (e.g.: 140 ppm N:210 ppm K). This can be achieved through the use of potassium nitrate and calcium nitrate with potassium sulfate.

Schwarz (1968) lists a number of ratios for N:P:K to be used during the summer and winter seasons for various crops grown in European, Mediterranean and subtropical climates (Table 3.7).

TABLE 3.7 Ratios of N:P:K Recommended for Summer and Winter Seasons in Several Climate Regions

Crop, Climate, Season		*N*	*P*	*K*
Tomato (mature stage)				
Middle European climate	— summer	1	0.2–0.3	1.0–1.5
	— winter	1	0.3–0.5	2–4
Mediterranean and subtropical climate	— summer	1	0.2	1
	— winter	1	0.3	1.5–2.0
Lettuce and other leafy vegetables				
	— summer	1	0.2	1
	— winter	2	0.3	2
Ammonium:Nitrate ratio is:				
(NH_4:NO_3)	— summer	1:3–4		
	— winter	1:4–8		

Modified from Schwarz (1968) p.32.

3.6 Nutrient Stock Solutions

3.6.1 Injector or Proportioner System

Fertilizer injection systems have become very popular with commercial growers since they save time by reducing the number of nutrient solution preparations. They also work well in the automation of nutrient solution adjustment using computer monitoring and injection of stock solutions. Therefore, more accurate, stabilized solutions can be maintained. Injector systems are used with both open and re-circulating hydroponic designs. With solution and tissue analyses, appropriate adjustments in the formulation can be made by altering the settings on the injector heads.

A fertilizer injector or proportioner automatically makes up the nutrient solution by injecting preset amounts of concentrated stock solutions into the irrigation lines. In an "open" hydroponic system (nutrient solution is not recycled) a new nutrient solution is prepared automatically with each watering cycle (Fig. 3.1). Changing of the nutrient solution is eliminated. Additional stock solutions are simply made up every week or so.

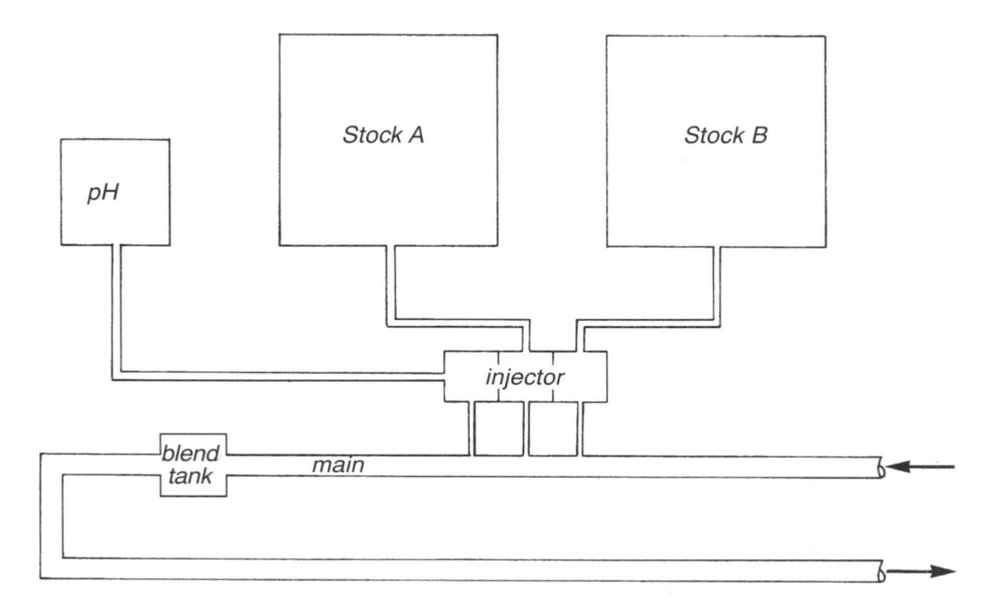

Figure 3.1. Layout of a basic injector system.

Injectors may also be used with recycle systems to automatically adjust the returning nutrient solution. Analysis of the returning solution indicates the modifications to be made to the stock solution to bring the nutrients to optimum levels before once again irrigating the plants. While nutrient solution analysis must be done by a laboratory to determine the levels of all essential elements, the general total salt levels can be determined by an electrical conductivity (EC) meter. The EC meter and a *p*H meter act as sensors for the computer in monitoring the current status of the returning and outgoing solutions. The computer then can activate the injector to adjust the nutrient solution according to preset levels stored by the computer (Fig. 3.2). The formulations of the nutrient stock solutions and the settings of the injector heads allow the operator to make changes in the outgoing nutrient solution to achieve optimum nutrient levels of each ion.

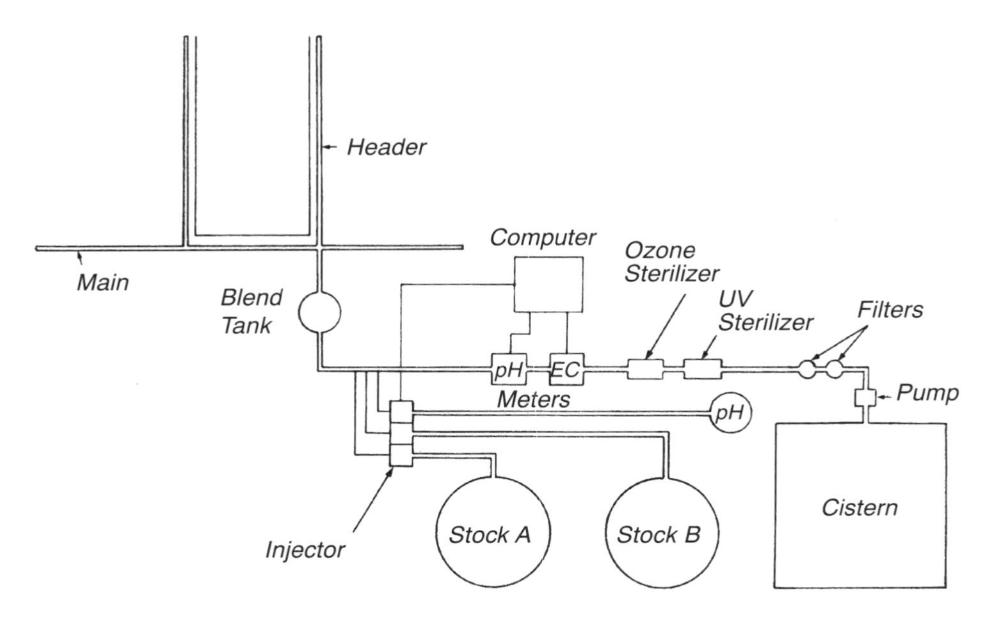

Figure 3.2. Layout of injector system for recirculating systems. Note the sterilization of the nutrient solution before it returns to the plants.

There are a number of different manufacturers of injectors (see Appendix 5). The choice of a particular make will depend upon the volume of nutrient solution to be injected at any given time in gallons per minute (gpm), the accuracy required for the system and the ability to expand the system. Some of the better makes of injectors allow you to add injector heads with the expansion of the hydroponic system, so you need not purchase a new injector.

For example, a 3-acre (1.2-hectare) hydroponic herb operation used an Anderson injector with 5 heads as shown in Figure 3.3. The system consisted of a 3-inch (7.6-cm) diameter water main forming a loop with the injector, a blending tank and a filter before going into the irrigation system of the hydroponic beds. The water flows from right to left in the system illustrated in Figure 3.3.

A paddle-wheel sensor upstream from the injector (extreme right of Figure 3.3) monitors the flow of water. For every 4 gallons (15 l) of water that passes, the sensor sends impulses to the controller (the grey box on the white panel shown in Figure 3.4). The controller activates one stroke of the injector for every 4 gallons (15 l) of water going through the main loop. A stroke occurs when the controller sends an electric current to the solenoid valve on the injector, opening the valve and allowing the pressure of the raw water to drive the diaphragms of the injector heads. The raw water passes through a 200-micron household filter prior to entering

Figure 3.3. Anderson injector with five heads (on the left), paddle-wheel sensor on main line (grey area on pipe before inlet lines from injector). Water flows from back to front. Clear flexible hoses attach from the stock tank pipes on the floor to the injector heads and from the injector heads to the inlet back-flow preventer valves on 1-inch diameter pipes attached to the main line.

the solenoid to prevent any silt from damaging the valve. As the pressurized water (minimum of 15–20 psi or 103.5–138 kPa) enters the back of the diaphragm of each head, it pushes the diaphragm forward causing the displacement of the stock solution in front of it. The stock solution flows to the main loop via tubing and a header. Backflow preventer valves on the heads of the injector and on the inlets to the stock solution header on the main loop prevent solution from backing up on the return stroke of the diaphragms (Figures 3.3). This injector system has 5 heads: two for stock A, two for stock B and a smaller acid head (on the left of Figure 3.3).

Each stroke of the injector displaces 40 ml of stock solution per head at the "10" setting on the head dial. At optimum injector operation of 32 strokes per minute, the maximum flow through the system is: 32×4 gal $= 128$ gal/min (484.5 l/min). The ratio of fertilizer stock to water achieved per head is: 40 ml : 15,141 ml (4 U.S. gal), or 1:378. For two heads: 80 ml:15,141 ml, or 1:189. To obtain a 1:200 dilution, set the dial on each head of stocks A and B to 9.5. This setting gives: $9.5/10 \times 40$ ml $= 38$ ml of $200\times$ concentration stock solution. The ratio for the two heads of each stock solution is: 76 ml: 15,141 ml, or 1:200 (fertilizer to water).

For a larger system, the impulse setting from the paddle-wheel sensor is increased at the controller. For example, if the setting is increased to one stroke per 5 gallons (19 l), the optimum operating would allow a flow of 5 gal × 32 strokes/min = 160 gal/min (606 l/min). However, if a 1:200 proportion was still required, an additional injector head would have to be installed. Each head is only capable of displacing 40 ml of stock solution per stroke. The maximum dilution of a two-head injector on a 5 gal/stroke basis is: 80 ml:18,927 ml (5 U.S. gal), or 1:236.

Alternatively, the strength of the stock solutions could be increased above 236 times. Following these principles, the injector system is expandable with additional heads to operate with higher volumes of water flow.

The entrance of stock solutions A and B into the main loop is separated by at least 18 inches (46 cm), so that adequate mixing with raw water occurs before they come in contact with each other. As the water-nutrient passes downstream, it is mixed further in an 80-gallon (303-liter) blending tank as shown in Figure 3.4. The acid enters the main loop several feet (60 cm) downstream from the stock solutions (small black tubing entering underneath main loop in Figure 3.3).

Figure 3.4. Stock tanks A and B, 30-gallon acid tank next to the blending tank (galvanized steel upright tank on left), injector and main line loop.

The stock solutions and acid enter the injector heads with flexible tubing (Figure 3.3) from the header lines connected to the 1,500-gallon (5,678-liter) stock tanks A and B and a 30-gallon (114-liter) acid tank (Figure 3.4). The main loop carries the mixed solution from the blending tank to the hydroponic growing system (Figure 3.4). A 200-mesh filter installed downstream removes any particulate matter before the solution enters the drip irrigation system.

3.6.2 Stock Solutions

Stock solutions are concentrated nutrient solutions. Depending upon the ability of the injector, the stock solutions may be prepared as 50, 100 or 200 times normal strength. A second factor that may limit the degree of concentration of the stock solution is that of fertilizer solubility. The least soluble fertilizer will be the limiting factor for the entire stock solution. When determining the strength of the stock solution, refer to the information on "physical constants of inorganic compounds" listed in Appendix 4.

Two separate stock solutions plus an acid solution must be prepared in individual tanks. These are generally called "stock A," "stock B" and "acid." The reason for separate solutions is that precipitation will occur among the sulfates and nitrates of some compounds if mixed together at high concentrations. For example, the sulfates of potassium sulfate or magnesium sulfate will precipitate with calcium from calcium nitrate.

Stock A may contain half of the total potassium nitrate requirement, all of calcium nitrate, ammonium nitrate, nitric acid (to lower the *p*H of the stock solution under 5.0) and iron chelate. Stock B would consist of the other half of potassium nitrate, all of potassium sulfate, monopotassium phosphate, phosphoric acid, magnesium sulfate, and the remaining micronutrients apart from iron. Acid stock solution is diluted to about 15–20% of the concentrated liquid form available from suppliers. Always be careful with such strong acids as they cause severe injury to a person. Add acid to water, never the opposite!

Some of the acids used are: nitric acid (HNO_3) (42%) (produces harmful vapors and burns the skin), sulfuric acid (H_2SO_4) (66%) (burns the skin and produces holes in clothing), phosphoric acid (H_3PO_4) (75%), and hydrochloric acid (HCl) (muriatic acid). With these strong acids, wear protective plastic/rubber gloves, apron, eye goggles and an approved respirator. Particular care must be taken with nitric acid as it gives off toxic fumes when it comes in contact with the air.

To determine the upper limit of concentration for a stock solution, use the solubility products listed in Appendix 4. The solubility is given as the amount of grams of a specific fertilizer that may be dissolved in 100 ml of cold or hot water. Since the stock solution probably will not be heated, the cold water figure should be used. The following example will demonstrate the use of these solubility factors.

Stock A:

	Compound	*Solubility (g/100ml cold water)*
*	potassium nitrate	13.3
	calcium nitrate	121.2
	ammonium nitrate	118.3
	nitric acid	no limit
	iron chelate	not in table, but is very soluble

Stock B:

	Compound	*Solubility (g/100ml cold water)*
	potassium nitrate	13.3
*	potassium sulfate	12.0
	monopotassium phosphate (potassium dihydrogen phosphate)	33.0
	phosphoric acid	548
	magnesium sulfate	71

* Note: These are the least soluble compounds and possibly the limiting ones depending upon the weights of each required. Although micronutrients will be included in stock solution B, they have not been included in this list of solubilities since they will be required in very small amounts which will not exceed their solubilities at the 200 times concentration.

To demonstrate the calculations of a stock solution, the following nutrient formulation will be used:

N – 200 ppm	P – 50 ppm	K – 300 ppm
Ca – 200 ppm	Mg – 40 ppm	Fe – 5 ppm
Mn – 0.8 ppm	Cu – 0.07 ppm	Zn – 0.1 ppm
B – 0.3 ppm	Mo – 0.03 ppm	

Assume that the raw water contains 30 ppm of Ca and 20 ppm of Mg. The adjustments to the formulation then are:

1. The amount of Ca to be added will be 170 ppm (200 – 30).
2. The amount of Mg to add is 20 ppm (40 – 20).

Using 1,200 U.S.-gallon tanks to store each stock solution, and a 200 times concentration of solution, the procedure is:

1. Determine the amount of each compound to satisfy the macronutrients and adjust for impurities.

A. Ca: 170 ppm (mg/l)

(i) Weight of $Ca(NO_3)_2$ (use Table 3.2 conversion factors):
$170 \times 4.094 = 696$ mg/l

(ii) Adjust for impurity from Table 3.3 (90% pure):

$$\frac{100}{90} \times 696 = 773 \text{ mg/l}$$

2. Calculate the amount of the compound for the stock solution concentration (200 times in this case).

200×773 mg/l = 154,600 mg/l or: 154.6 g/l
(1000 mg = 1 g)

3. Compare this amount with the solubility given in g/100 ml of cold water.

(i) Convert to g/100 ml:
154.6 g/l 1 l = 1,000 ml
That is, 154.6 g/1000 ml
or: 15.46 g/100 ml (divide by 10 to get 100 ml)

(ii) Compare with the solubility:
15.46 g/100 ml vs. 121.2 g/100 ml

Therefore, this amount is well within the solubility limits of calcium nitrate.

4. Continue with the calculations for all macronutrient compounds.

B. N: We require a total of 200 ppm from all sources. Assume that we wish to use some ammonium nitrate to supply 10 ppm of N from NH_4.

(i) Amount of N in $Ca(NO_3)_2$ added:
696 mg/l of $Ca(NO_3)_2$ (before adjustment for purity)
696 mg/l $\times$ 0.171 (Table 3.2 factor) = 119 mg/l (ppm)

(ii) Balance needed from sources other than calcium nitrate:
$200 - 119 = 81$ ppm (mg/l)

(iii) Amount of NH_4NO_3 needed to obtain 10 ppm of $NH_4 - N$ and 10 ppm of $NO_3 - N$ (that is a total of 20 ppm N from NH_4NO_3 source):

20 mg/l × 2.857 (Table 3.2 factor) = 57 mg/l

(iv) Adjust for purity (Table 3.3 – 98%):

$$\frac{100}{98} \times 57 \text{ mg/l} = 58 \text{ mg/l}$$

(v) For 200 × concentration:

200 × 58 mg/l = 11,600 mg/l or: 11.6 g/l

(vi) Express in g/100 ml and compare with solubility:

11.6 g/l = 1.16 g/100 ml (divide by 10 as 1 l = 1000 ml)

1.16 g/100 ml vs. 118.3 g/100 ml

Therefore, the level of ammonium nitrate is within the solubility range.

(vii) Balance of N required from sources other than calcium nitrate and ammonium nitrate:

200 – (119 + 20) = 61 ppm (mg/l)

(viii) Final source of N is from KNO_3:

61 mg/l × 7.221 (Table 3.2 factor) = 440.5 mg/l

(ix) Adjust for purity (Table 3.3 – 95%):

$$\frac{100}{95} \times 440.5 \text{ mg/l} = 464 \text{ mg/l}$$

(x) For 200 × concentration:

200 × 464 mg/l = 92,800 mg/l or: 92.8 g/l

(xi) Express in g/100 ml and compare to solubility:

92.8 g/l = 9.28 g/100 ml

9.28 g/100 ml vs. 13.3 g/100 ml

This is within the solubility limit. Note that half of the potassium nitrate is added to stock A and the other half to stock B.

Therefore, actually 9.28 ÷ 2 = 4.62 g/100 ml concentration is added to each tank. Often a higher level of potassium nitrate is used, and therefore splitting it between the two stock solutions will keep it within the solubility limit.

C. K: 300 ppm (mg/l)

(i) Amount of K in KNO_3 used:
440.5 mg/l × 0.387 (Table 3.2 factor) = 170.5 mg/l

(ii) Balance of K needed: 300 – 170 = 130 mg/l

(iii) Other sources: KH_2PO_4 and K_2SO_4

First calculate the amount of KH_2PO_4 to use to get 50 ppm of P.

D. P: 50 ppm (mg/l) from KH_2PO_4

(i) 50 mg/l × 4.394 (Table 3.2 factor) = 220 mg/l

(ii) Adjust for purity (Table 3.3 – 98%)

$$\frac{100}{98} \times 220 \text{ mg/l} = 224 \text{ mg/l}$$

(iii) For 200 × concentration :

$$200 \times \frac{224 \text{ mg}}{1{,}000 \text{ mg}} = 44.8 \text{ g/l} \quad \text{or: } 4.5 \text{ g/100 ml}$$

(iv) Compare with solubility:
4.5 g/100 ml vs. 33.0 g/100 ml

This is well below the solubility limit.
Now go back and calculate the K requirement from other sources (see **C.** above).

K: 300 ppm (mg/l)

(v) Amount of K in 220 mg/l of KH_2PO_4
220 mg/l × 0.287 (Table 3.2 factor) = 63 ppm

(vi) Balance of K needed: 300 – (170 + 63) = 67 ppm (mg/l)

(vii) Amount of K_2SO_4:
67 mg/l × 2.229 = 149 mg/l

(viii) Adjust for purity (90%):

$$\frac{100}{90} \times 149 \text{ mg/l} = 166 \text{ mg/l}$$

(ix) For 200 × concentration:

$$200 \times \frac{166 \text{ mg}}{1{,}000 \text{ mg}} = 33.2 \text{ g/l} \quad \text{or: } 3.32 \text{ g/100 ml}$$

(x) Compare to solubility:
3.32 g/100 ml vs. 12.0 g/100 ml

This level is within the solubility range.

E. Mg: 20 ppm (mg/l)

(i) Amount of $MgSO_4$ required:
20 mg/l × 10.14 = 203 mg/l

(ii) Adjust for purity (98%):

$$\frac{100}{98} \times 203 \text{ mg/l} = 207 \text{ mg/l}$$

(iii) For 200 × concentration:

$$200 \times \frac{207 \text{ mg}}{1{,}000 \text{ mg}} = 41.4 \text{ g/l} \quad \text{or: } 4.14 \text{ g/100 ml}$$

(iv) Compare to solubility:
4.14 g/100 ml vs. 71 g/100 ml

This level is soluble.

5. Convert all compound weights to the amounts for 1,200 U.S.-gallon stock tanks.

A. Convert 1,200 U.S. gallons to liters:
1,200 × 3.785 l = 4,542 l

B. Calculate the weights of each compound for this volume.

(i) $Ca(NO_3)_2$: 154.6 g/l
154.6 × 4,542 = 702,193 g or: 702.2 kg

That is; 702.2 kg × 2.2 lb = 1,545 lb

(ii) NH_4NO_3: 11.6 g/l
11.6 × 4,542 = 52,687 g or: 52.7 kg

That is; 52.7 kg × 2.2 lb = 116 lb

(iii) KNO_3: 92.8 g/l
92.8 × 4,542 = 421,498 g or: 421.5 kg

That is; 421.5 kg × 2.2 lb = 927 lb

Note: This total of 927 lb will be divided into two equal portions, one in each stock tank. Therefore, 927×2 = 463.5 lb should be added to each of stock A and B tanks.

(iv) KH_2PO_4: 44.8 g/l
44.8 × 4,542 = 203,482 g or: 203.5 kg
That is; 203.5 kg × 2.2 lb = 448 lb

(v) K_2SO_4: 33.2 g/l
33.2 × 4,542 = 150,794 g or: 150.8 kg
That is; 150.8 kg × 2.2 lb = 332 lb

(vi) $MgSO_4$: 41.4 g/l
41.4 × 4,542 = 188,039 g or: 188 kg
That is; 188 kg × 2.2 lb = 414 lb

7. Now calculate the weights for each of the micronutrient compounds.

A. Fe: 5ppm (mg/l)

(i) Source: FeEDTA (10% Fe)
5 mg/l × 10.0 (Table 3.2 factor) = 50 mg/l

(ii) For 200 × concentration:
200 × 50 mg/l = 10,000 mg/l or: 10 g/l

(iii) For 1,200 U.S. gallons (4,542 l):
10 × 4,542 = 45,420 g or: 45.4 kg
That is; 45.4 kg × 2.2 lb = 100 lb

B. Mn: 0.8 ppm (mg/l)

(i) Amount of $MnSO_4$
0.8 mg/l × 4.061 (Table 3.2 factor) = 3.25 mg/l

(ii) Adjust for percentage purity. The purity of compounds, especially those of micronutrients, differ with the manufacturer. The grower should obtain the precise percent purity for any given product from the fertilizer dealer. For calculation purposes, a purity of 90% will be used. Adjustments may be made after a solution analysis has been carried out for the new solution. The actual level of each element can be compared to the theoretical expected level of each.

$$\frac{100}{90} \times 3.25 \text{ mg/l} = 3.6 \text{ mg/l}$$

(iii) For 200 × concentration:
200×3.6 mg/l = 720 mg/l or: 0.72 g/l

(iv) For 1,200 U.S. gallons (4,542 l):
$0.72 \times 4{,}542 = 3{,}270$ g or: 3.27 kg

That is; 3.27 kg × 2.2 lb = 7.2 lb
or: 7 lb 3 oz (1 lb = 16 oz; 0.2 × 16 = 3 oz)

C. Cu: 0.07 ppm (mg/l)

(i) Source: $CuSO_4$
0.07 mg/l × 3.93 (Table 3.2 factor) = 0.275 mg/l

(ii) Percent purity (98%):

$$\frac{100}{98} \times 0.275 \text{ mg/l} = 0.281 \text{ mg/l}$$

(iii) For 200 × concentration:
200×0.281 mg/l = 56 mg/l or: 0.056 g/l

(iv) For 1,200 U.S. gallons:
$0.056 \times 4{,}542 = 254$ g

Since this is less than one pound, it is more accurate to weigh it using a gram scale.

D. Zn: 0.1 ppm (mg/l)

(i) Source: ZnEDTA (14% powder)
0.1 mg/l × 7.143 = 0.7143 mg/l

(ii) For 200 × concentration:
200×0.7143 mg/l = 143 mg/l or: 0.143 g/l

(iii) For 1,200 U.S. gallons:
$0.143 \times 4{,}542 = 650$ g or 1.43 lb, which is 1 lb 7 oz

E. B: 0.3 ppm (mg/l)

(i) Source: H_3BO_3
0.3 mg/l × 5.717 = 1.715 mg/l

(ii) Percent purity (about 95%):

$$\frac{100}{95} \times 1.715 \text{ mg/l} = 1.805 \text{ mg/l}$$

(iii) For 200 × concentration:
$200 \times 1.805 \text{ mg/l} = 361 \text{ mg/l}$ or: 0.361 g/l

(iv) For 1,200 U.S. gallons:
$0.361 \times 4{,}542 = 1{,}640$ g or: 3.61 lb

That is; 3 lb 10 oz

F. Mo: 0.03 ppm (mg/l)

(i) Source: ammonium molybdate
$0.03 \text{ mg/l} \times 1.733 = 0.052 \text{ mg/l}$

(ii) Percent purity is about 95%:

$$\frac{100}{95} \times 0.052 \text{ mg/l} = 0.055 \text{ mg/l}$$

(iii) For 200 × concentration:
$200 \times 0.055 \text{ mg/l} = 11 \text{ mg/l}$

(iv) For 1,200 U.S. gallons:
$11 \times 4{,}542 = 49{,}962$ mg or: 50 g

8. Construct a table to summarize all of the information.

	Compound	*Weight (lb)*	*Stock Sol. (g/100 ml)*	*Max. Sol. (g/100 ml)*	*Nutrient Sol. Elements (ppm) After Injection*
STOCK A: (200 ×)	KNO_3	463.5	4.64	13.3	K – 85 N – 30.5
(1,200-gal tank)	$Ca(NO_3)_2$	1,545	15.46	121.2	Ca – 170 N – 119
	NH_4NO_3	116	1.16	118.3	N – 20
	FeEDTA	100	—	very sol.	Fe – 5
	*HNO_3	—	—	no limit	

	Compound	Weight (lb)	Stock Sol. (g/100 ml)	Max. Sol. (g/100 ml)	Nutrient Sol. Elements (ppm) After Injection
STOCK B: (200 ×)	KNO_3	463.5	4.64	13.3	K – 85 N – 30.5
(1,200-gal tank)	K_2SO_4	332	3.32	12.0	K – 67
	KH_2PO_4	448	4.5	33.0	K – 63 P – 50
	$MgSO_4$	414	4.14	71	Mg – 20
	* H_3PO_4	—	—	548	
	$MnSO_4$	7 lb 3 oz	—	105.3	Mn – 0.8
	$CuSO_4$	254 g	—	31.6	Cu – 0.07
	ZnEDTA	1 lb 7 oz	—	very sol.	Zn – 0.1
	H_3BO_3	3 lb 10 oz	—	6.35	B – 0.3
	NH_4 – Mo	50 g	—	43	Mo – 0.03

* These acids are added in sufficient quantity to lower the *p*H to 5.5. Phosphoric acid (H_3PO_4) may be substituted for KH_2PO_4 as a source of P. Then adjust the remaining compounds containing K, especially K_2SO_4, as no K will be obtained from H_3PO_4.

Nutrient solution totals: NO_3 – N: 190 ppm; NH_4 – N: 10 ppm;
P: 50 ppm; K: 300 ppm

The raw water in this example contained 30 ppm of Ca and 20 ppm of Mg. This brings the total Ca to 200 ppm and Mg to 40 ppm in the final nutrient solution after injection.

Stock solutions require continuous agitation to prevent settling out of some of the fertilizer components. This may be accomplished in several ways. A motor having a long shaft with a propeller-type blade on the end may be mounted above the opening of the tank. The blade and shaft should be submersed to a position very close to the bottom of the tank. It is very important that the shaft and blade is of stainless steel to resist the corrosive nature of the nutrient solution. A submersible pump may be placed at the bottom of each tank to circulate the solution. But it must be well-sealed; have either a plastic or stainless-steel impeller and stainless-steel screws holding the pump components together. If there are any regular steel or galvanized screws in contact with the nutrient solution, they will be dissolved by the electrolitic nature of the solution within several weeks. This would then expose the motor to the solution, and the operator could easily receive an electric shock when working with the injection system.

An alternative to these methods is to use a circulating pump which has components resistant to corrosive solutions, such as a swimming-pool circulation pump. The pump would be located outside and near each tank (one pump per tank). Plastic PVC piping of at least 1½-inch diameter would connect the pump inlet and outlets to the bottom of the stock-solution tank as shown in Figure 3.5. Place an elbow at the end of the outlet line in the tank to deflect the solution around the tank. The inlet end requires a check valve at its end near the base of the tank to keep the pump from losing prime. The pump should be anchored to a base of concrete or heavy wood support with anchor bolts to prevent it from shifting in position. The pump must run continuously.

Figure 3.5. Circulation pump with plumbing to agitate the stock solution.

If a tee, plastic ball-valve, followed by a hose bibb is attached to the return line at a convenient height, this will assist in the cleaning of the tank between solution makeups. There will always be some sediment by the end of the period when the solution needs to be made up. When rinsing the tank, use the waste on some nearby plants around the greenhouse. Final cleaning and removal of the remaining solution below the level of the intake line may be done using a submersible pump. If this submersible pump is rinsed with raw water after its use, no corrosive wear will result, even if some of the components are not entirely of stainless steel or plastic.

The acid stock solution does not require agitation since acids dissolve well in water and do not form any precipitates.

To facilitate the filling of the stock tanks, install a 1-inch diameter raw-water line above the tanks. At the inlet to each tank, attach a gate valve (Figure 3.6).

Figure 3.6. Stock tanks (1,500 gal) A and B with circulation pumps and overhead raw water line to fill tanks.

When using small injector system stock tanks of less than 100 U.S. gallons (378 l), the weights of individual micronutrient compounds become very small. For example, the amount of ammonium molybdate for a 200 times concentration stock solution for a 100 U.S.-gallon (378 l) stock tank would be 4.17 grams. Such small weights are difficult to measure on a triple-beam balance which is accurate to 1.0 gram. Also, if a nutrient solution is prepared in a storage tank at its final plant formulation without the use of stock solutions, the quantities required for micronutrient compounds are very small and therefore difficult to weigh accurately.

In these cases, it is better to use a highly concentrated stock solution for the micronutrients and store it in an opaque container. Make up a 10- to 20-gallon (38- to 76-liter) quantity. Do not make up more than what you expect to use in about one month, as changes may have to be made in the formulation to better suit the plants during different stages of growth. Some sedimentation may occur over this period of time, so agitate the solution with a small motor with shaft and impeller blade as described above. With a small tank, a paint mixer can be used. These products should be available at hardware, irrigation and greenhouse suppliers.

To illustrate the derivation of a micronutrient stock solution, the same micronutrient formulation will be used as in the above example. That is: Mn – 0.8 ppm; Cu – 0.07 ppm; Zn – 0.1 ppm; B – 0.3 ppm; Mo – 0.03 ppm. Note: Iron (Fe) is not included in this micronutrient stock solution as sufficient weight of it is required such that it can be accurately measured on a balance.

The following situation will exemplify the calculations involved in determining the weights of micronutrient compounds and their comparisons to the solubilities for a concentrated stock solution. Prepare a nutrient solution at normal plant strength in a 1,000 U.S.-gallon nutrient tank (not a stock solution) and use a 600 times strength micronutrient stock solution to supply the micronutrients to this tank. The three-step procedure is as follows:

Step 1. Determine the quantities of each compound needed for the 1,000 U.S.-gallon nutrient tank. At this time, only the calculations for the micronutrients will be demonstrated as those for the macronutrients shown in Section 3.5.2.

A. Mn: 0.8 ppm (mg/l)

(i) Amount of $MnSO_4$
0.8 mg/l × 4.061 (Table 3.2 factor) = 3.25 mg/l

(ii) Adjust for purity (90%)

$$\frac{100}{90} \times 3.25 \text{ mg/l} = 3.6 \text{ mg/l}$$

(iii) For 1,000 U.S.-gallon tank:
1,000 × 3.785 l = 3,785 liters
3,785 × 3.6 mg/l = 13,626 mg or: 13.6 g

B. Cu: 0.07 ppm (mg/l)

(i) Amount of $CuSO_4$
0.07 mg/l × 3.93 = 0.275 mg/l

(ii) Adjust for purity (98%) :

$$\frac{100}{98} \times 0.275 \text{ mg/l} = 0.281 \text{ mg/l}$$

(iii) For 1,000 U.S.-gallon tank:
$3{,}785 \times 0.281$ mg/l = 1,064 mg or: 1.064 g

C. Zn: 0.1 ppm (mg/l)

(i) Amount of $ZnSO_4$
0.1 mg/l × 4.40 = 0.44 mg/l

(ii) Adjust for purity (90%) :

$$\frac{100}{90} \times 0.44 \text{ mg/l} = 0.4889 \text{ mg/l}$$

(iii) For 1,000 U.S.-gallon tank:
$3{,}785 \times 0.4889$ mg/l = 1,850 mg or: 1.85 g

D. B: 0.3 ppm (mg/l)

(i) Source: H_3BO_3
0.3 mg/l × 5.717 = 1.715 mg/l

(ii) Percent purity (95%):

$$\frac{100}{95} \times 1.715 \text{ mg/l} = 1.805 \text{ mg/l}$$

(iii) For 1,000 U.S.-gallon tank:
$3{,}785 \times 1.805$ mg/l = 6,832 mg or: 6.83 g

E. Mo: 0.03 ppm (mg/l)

(i) Source: ammonium molybdate
0.03 mg/l × 1.733 = 0.052 mg/l

(ii) Percent purity (95%):

$$\frac{100}{95} \times 0.052 \text{ mg/l} = 0.055 \text{ mg/l}$$

(iii) For 1,000 U.S.-gallon tank:
$3{,}785 \times 0.055$ mg/l = 208.2 mg or: 0.208 g

It is clear from these calculations that the amounts are too small to be weighed accurately by a triple-beam balance.

Step 2. Calculate the quantities of each micronutrient compound for a 600 times concentration stock solution (use a 10 U.S.-gallon stock tank).

A. Mn: For 0.8 ppm (mg/l), use 3.6 mg/l of $MnSO_4$ (as calculated in 1. A. (ii) above).

(i) For 10 U.S.-gallon stock tank:
$10 \times 3.785 \text{ l} = 37.85 \text{ l}$
$37.85 \times 3.6 \text{ mg/l} = 136.26 \text{ mg}$

(ii) 600 × concentration:
$600 \times 136.26 \text{ mg} = 81{,}756 \text{ mg}$ or: 81.8 g

(iii) Convert 81.8 g in 37.85 l to g/100 ml and compare it to the solubility limit:

$$\frac{81.8 \text{ g}}{37.85 \text{ l}} = 2.16 \text{ g/l} \text{ or: } 0.216 \text{ g/100 ml}$$

0.216 g/100 ml vs. 105.3 g/100 ml

This concentration is well within the solubility range.

B. Cu: For 0.07 ppm (mg/l) use 0.281 mg/l of $CuSO_4$.

(i) For 10 U.S.-gallon stock tank:
$37.85 \times 0.281 \text{ mg/l} = 10.64 \text{ mg}$

(ii) 600 × concentration:
$600 \times 10.64 \text{ mg} = 6{,}384 \text{ mg}$ or: 6.4 g

(iii) Convert to g/100 ml:

$$\frac{6.4 \text{ g}}{37.85 \text{ l}} = 0.169 \text{ g/l} \text{ or: } 0.0169 \text{ g/100 ml}$$

0.0169 g/100 ml vs. 31.6 g/100 ml

This is within the solubility limit.

C. Zn: For 0.1 ppm (mg/l) use 0.4889 mg/l of $ZnSO_4$.

(i) For 10 U.S.-gallon stock tank:
$37.85 \times 0.4889 \text{ mg/l} = 18.505 \text{ mg}$

(ii) 600 × concentration:
$600 \times 18.505 \text{ mg} = 11{,}103 \text{ mg}$ or: 11.1 g

(iii) Convert to g/100 ml:

$$\frac{11.1\ g}{37.85\ l} = 0.2932\ g/l \quad \text{or:}\ 0.0293\ g/100\ ml$$

0.0293 g/100 ml vs. 96.5 g/100 ml (from Appendix 4)

This concentration is well below the solubility limit.

D. B: For 0.3 ppm (mg/l), use 1.805 mg/l of H_3BO_3.

(i) For 10 U.S.-gallon stock tank:
37.85 × 1.805 mg/l = 68.32 mg

(ii) 600 × concentration:
600 × 68.32 mg = 40,992 mg or: 41 g

(iii) Convert to g/100 ml:

$$\frac{41\ g}{37.85\ l} = 1.083\ g/l \quad \text{or:}\ 0.108\ g/100\ ml$$

0.108 g/100 ml vs. 6.35 g/100 ml

This level is within the solubility range.

E. Mo: For 0.03 ppm (mg/l), use 0.055 mg/l of ammonium molybdate.

(i) For 10 U.S.-gallon stock tank:
37.85 × 0.055 mg/l = 2.082 mg

(ii) 600 × concentration:
600 × 2.082 mg = 1,249 mg or: 1.249 g

(iii) Convert to g/100 ml:

$$\frac{1.249\ g}{37.85\ l} = 0.033\ g/l \quad \text{or:}\ 0.0033\ g/100\ ml$$

0.0033 g/100 ml vs. 43 g/100 ml

This is well below the solubility limit.

Step 3. The final step is to calculate the volume of a 600 × concentration stock solution in a 10 U.S.-gallon tank to add to a 1,000 U.S.-gallon nutrient solution to obtain 1 × concentration. It must be diluted to a ratio of 1 part of stock solution to 600 parts of water, that is, 600:1. Therefore, for 1 U.S. gallon of nutrient solution, 1/600 U.S. gallons of 600 × stock solution would be required. It is easiest to set up a ratio with one unknown (x) as follows:

$$\frac{1}{1/600} = \frac{1{,}000}{x} \qquad \frac{(1 \times \text{ nutrient solution})}{(600 \times \text{ stock solution})}$$

$$x = 1{,}000 \times 1/600 \quad \text{(cross multiply)}$$
$$x = 1{,}000/600 = 1\tfrac{2}{3} \text{ U.S. gallons}$$

Convert to liters:

$$1\tfrac{2}{3} \text{ U.S. gallons} \times 3.785 \text{ l} = 6.308 \text{ l} \text{ or: } 6{,}308 \text{ ml}$$

In summary, 6.308 l of 600 × concentration stock solution would be added to the 1,000 U.S.-gallon nutrient tank of 1 × strength solution. If the nutrient solution had to be prepared once a week, the 10 U.S. gallons of 600 × stock solution would last for 6 weeks. Comprise a summary table as follows:

Micronutrient Stock Solution 600 × in 10 U.S.-gallon Tank

Compound	*Weight (g)*	*Element (ppm)*
$MnSO_4$	81.8	Mn – 0.8
$CuSO_4$	6.4	Cu – 0.07
$ZnSO_4$	11.1	Zn – 0.1
H_3BO_3	41.0	B – 0.3
NH_4 – Mo	1.25	Mo – 0.03

3.7 Preparing the Nutrient Solution

The preparation of the nutrient solution will vary depending upon the nutrient-solution tank volume and whether a normal strength or a stock solution is used.

3.7.1 Preparing Normal Strength Solutions

For small-volume tanks (less than 2,000 gallons), individual fertilizer compounds may be weighed in advance and placed in plastic bags. Use a felt pen to write the compound formula on each bag to avoid any confusion in its future use. Use a triple-beam balance to weigh the smaller compounds in grams. The micronutrients may be placed in one bag together, with the exception of iron, which should be in a separate bag. The macronutrient compounds should be weighed using a scale capable of 20 to 30 pounds. As mentioned earlier, any compound that is required in an amount less than a pound should be weighed using the gram balance. Generally, it is faster if at least 5 batches are weighed when making up the formulations. Do not make too many as it may be necessary to change the formulation due to plant growth or weather conditions.

With large-volume tanks, weigh only those fertilizers needed for that batch. Weigh each fertilizer salt separately, arranging them in piles on polyethylene sheets or in buckets so there is no loss. This should be done accurately to within plus or minus 5 percent using either a gram scale or a pound scale depending upon the weights required of each.

Continue using the following procedure:

1. Fill the nutrient-solution storage tank with water to about one-third full.

2. Dissolve each fertilizer salt individually in a 5-gallon bucket of water. Add the water to the fertilizer and stir vigorously. Use a hose and nozzle to assist in mixing. Usually the entire amount of fertilizer will not dissolve on the first addition of water. Pour off the dissolved liquid portion into the storage tank and repeat adding water, stirring and canting off the dissolved portion until the entire salt has entered solution. Use hot water with salts that are hard to dissolve.

3. Dissolve the macronutrients first, then the micronutrients.

4. In small systems such as backyard greenhouses, all the sulfates can be mixed together in dry form before dissolving; e.g.: K_2SO_4, $MgSO_4$. Then the nitrates and phosphates can be mixed in dry form before dissolving; e.g.: KNO_3, KH_2PO_4. Add $Ca(NO_3)_2$ last.

5. With larger systems, add potassium sulfate first. For better mixing, operate the irrigation pump of the nutrient solution system with the bypass valve fully open and the valve to the greenhouses closed. Keep the pump circulating the nutrient solution in the storage tank until all of the solution is prepared.

6. Fill the tank to at least half, but no more than two-thirds, then add potassium nitrate.

7. Fill the tank to three-quarters, then add magnesium sulfate and monopotassium phosphate.

8. Add calcium nitrate slowly, while circulating the solution.

9. Add the micronutrients with the exception of iron chelate.

10. Check the *p*H of the nutrient solution and adjust it if necessary, with either sulfuric acid (H_2SO_4) or potassium hydroxide (KOH). High *p*H (more than 7.0) causes precipitation of Fe^{++}, Mn^{++}, $PO_4^{\equiv}$, Ca^{++} and Mg^{++} to insoluble and unavailable salts.

11. Add the iron chelate (FeEDTA) and top-up the tank to the final volume.

12. Check the *p*H and adjust it between 5.8 and 6.4 depending upon the crop optimum *p*H level.

13. If a closed recirculating hydroponic system is being used, circulate the nutrient solution through the system for 5 to 10 minutes, then check the *p*H again and adjust it again if necessary.

3.7.2 Preparing Stock Solutions

When preparing stock solutions, larger weights of each compound will be required so prepare only one batch of solution. Weigh each compound separately. Often whole sacks (50 lb or 100 lb) will be used, so only weigh the additional amount needed. For example, if 414 pounds of magnesium sulfate is one component, use eight 50-lb or four 100-lb bags and then weigh out the additional 14 lb on a scale.

If 5-gallon buckets are being used to dissolve the fertilizer salts, fill the buckets only to one-third capacity to allow sufficient water to dissolve the compound. By using 15 to 20 buckets, one person can steadily be dissolving the fertilizer while another pours off the solution into the storage tank. Another method, other than using buckets, is to have a mixing tank with a circulation pump to mix each fertilizer separately, then once dissolved, pump the solution into the stock-solution storage tank. Be careful not to use more water in the mixing process than the stock tank can hold.

When making up stock solutions, do not add a lot of water to the stock tank before beginning the mixing process, or the final volume may be exceeded in the dissolution process alone.

Add only half of the potassium nitrate to each stock tank.

For stock A, the sequence of adding the fertilizers is: Half of the potassium nitrate, calcium nitrate, ammonium nitrate and iron chelate after adjusting the *p*H to about 5.5. Top-up the tank with water to within 10 gallons of the final solution level before adjusting the *p*H and adding the iron chelate.

For tank B, the sequence is: The other half of the potassium nitrate, potassium sulfate. Fill the tank to three-quarters, then continue adding magnesium sulfate and monopotassium phosphate. Adjust the *p*H to about 5.5 after topping-up the tank with water to within 30 gallons of final level, then add the micronutrients. Finally, check the *p*H again and adjust it if necessary.

The acid, tank C, is made up last using one of the acids described earlier. Fill the acid tank to three-quarters with water before adding the acid to the water. Then after stirring with a plastic pipe, top-up the tank slowly with water to the final level.

3.8 Plant Relations and Cause of Nutrient Solution Changes

In a cyclic (closed) system in which the nutrient solution is drained back to the reservoir after use, the life of the nutrient solution is 2–3 weeks, depending upon the season of the year and the stage of plant growth. During the summer months, under mature high-yielding plants, the nutrient solution may have to be changed as often as every week. The reason for changing the solution is that plants differentially absorb the various elements. As a result, some elements become in short supply before others. Just how deficient they are at any point can be determined only by atomic absorption analyses of the nutrient solution. Such analyses can be done only in costly laboratory facilities. Consequently, many people are unable to carry out such analyses. The only safeguard against a nutrient disorder then is to change the solution periodically. In some cases it is possible to add partial formulations between changes, but this is only by trial and error and can result in excess salt buildup of nutrients which are taken up by the plant at relatively low rates.

The relative uptake of the various mineral elements by the plant is affected by:

a. Environmental conditions: temperature, humidity, light intensity.
b. Nature of the crop.
c. Stage of plant development.

As a result of the differential uptake of the various elements, the composition of the nutrient solution is changing constantly. Some of the elements are being depleted more rapidly than others, and the concentration is being increased by the plants' relatively greater absorption of water than of salts. In addition to changes in salt composition, the *p*H is also changing as a result of reactions with the aggregate and the unbalanced absorption of the anions and cations from the solutions.

3.8.1 Nutrient Analysis

Before replacing mineral elements, it is necessary to determine their concentration by chemical methods of analysis in order to determine the quantities absorbed by the plant. The difference in concentration of the mineral elements from the time of first mixing the nutrient solution to that at the time of analysis tells how much of each element must be added to bring the concentration to its original level.

Besides testing the solutions for depletions, it is necessary to test for the accumulation of unused ions such as sodium, sulfate or chloride, or for the presence of excesses of toxic elements as copper or zinc.

3.8.2 Plant Tissue Analysis

By conducting both plant-tissue and nutrient-solution analyses, we can compare and relate plant physiological upsets to imbalances of various mineral elements in the nutrient solution. It is possible to control the changes in the nutrient solution once a precise relationship is established between fluctuations in mineral elements in the plant tissue and those in the nutrient solution. Then the nutrient solution should be adjusted before visual symptoms appear in the plant tissue. This would prevent any mineral stress from occurring within the plant and thus increase yields by allowing the plant to grow under optimum mineral nutrition conditions.

An advantage of tissue analysis over nutrient analysis is that tissue analysis indicates what has been or is being absorbed from the nutrient solution by the plant, whereas nutrient analysis indicates only the relative availability of nutrients to the plant. Actual uptake of essential elements may be restricted by conditions of the medium, solution, environmental factors or the plant itself. For instance, if the medium is not inert and reacts with the nutrient solution, ions may be retained by the particles and unavailable to the plant. Imbalances within the nutrient solution or fluctuating *p*H levels will reduce uptake by the plants. Diseases or nematodes in the roots of plants would reduce their absorption capacity for nutrients. Environmental factors such as insufficient light, extremes in temperatures, and insufficient carbon dioxide levels will prevent plants from efficiently utilizing the available nutrients in the solution. Plant-tissue analysis measures the effects of these conditions on nutrient uptake.

The use of tissue analysis in relating the nutrient status of a plant is based upon the fact that normal, healthy growth is associated with specific levels of each nutrient in certain plant tissues. Since these normal levels are not the same for all tissues on any plant or for all species, it is necessary to select an indicator tissue which is representative of a definite stage of growth. Generally, a young, vigorous leaf near the growing point of the main stem of the plant is selected. It is more reliable to take samples at various stages of growth than to concentrate on a large number of samples at any one stage since the concentration of most nutrients decreases as the plants age and mature.

To accurately relate tissue analysis results to nutritional requirements of plants considerable data must be available on optimum levels of nutrients in specific plant species and tissues to be compared. This information is available for greenhouse lettuce, tomatoes and cucumbers (Table 3.8). The indicator tissue for tomatoes is the fifth leaf down from the growing tip of the main stem and it includes both petiole and blade tissue. At least ten samples should be taken of only one variety under one fertilizer treatment. The indicator tissue for cucumbers is a young leaf, without petiole, about 10 cm (4 inches) in diameter, usually the third visible leaf from the top of the main stem. A representative sample should consist of ten uniform replicates. The leaves should be either oven dried (70° C for 48 hours) or delivered promptly in a fresh condition to a commercial laboratory. Before drying, separate the blades from the petioles. Analysis for NO_3–N, PO_4–P, K, Ca, and Mg are usually performed on the petioles; minor elements on the blades.

To examine the results, date of planting, stage of growth, and previous fertilizer treatment must be known. A weekly series of tests will indicate clearly the trends in any nutrient level. A combination of nutrient and tissue analysis will allow the grower to anticipate any problems before they occur and make the necessary adjustments in the nutrient solution formulation.

TABLE 3.8 Range of Nutrient Levels in Tissues of Apparently Healthy Plants

Element	*Tomatoes*	*Cucumbers*	*Lettuce*
N %	4.5	5.25	4.3
	4.5–5.5	5.0–6.0	3.0–6.0
P %	0.7	0.75	1.0
	0.6–1.0	0.7–1.0	0.8–1.3
K %	4.5	4.75	5.4
	4.0–5.5	4.5–5.5	5.0–10.8
Ca %	1.5	3.0	1.5
	1.5–2.5	2.0–4.0	1.1–2.1
Mg %	0.5	0.75	0.42
	0.4–0.6	0.5–1.0	0.3–0.9
Fe (ppm)	100	125	120
	80–150	100–150	130–600
B (ppm)	50	40	32
	35–60	35–60	25–40
Mn (ppm)	70	70	70
	70–150	60–150	20–150
Zn (ppm)	30	50	45
	30–45	40–80	60–120
Cu (ppm)	5	8	14
	4–6	5–10	7–17
Mo (ppm)	2	2	2–3
	1–3	1–3	1–4
N/K ratio	1.0	1.1	
	0.9–1.2	1.0–1.5	

3.8.3 Changing of Solutions

Work by Steiner (1980) on nutrient solutions indicates that if the ratios of nutrient uptake for a certain crop under given conditions are known, ions can be applied continuously to the solution in these mutual ratios, controlled only by a conductivity meter. While generally this is true, extended use of the same nutrient solution may result in accumulation of toxic quantities of minor elements such as zinc and copper from metals in the plumbing system, fertilizer impurities, or from the water itself.

The useful life of a nutrient solution depends principally on the rate of accumulation of extraneous ions which are not utilized by the plants at rapid rates. Such an accumulation results in a high osmotic concentration of the nutrient solution. An electrical conductivity device, commercially available, should be used to determine the rate at which the nutrient solution becomes concentrated. Determinations should be made on a new nutrient solution and then repeated after each makeup with nutrient salts. As the total salt level increases it will more readily conduct an electric current. The unit of measurement used to express conductance is a *mho.* For simplicity, conductivity is often expressed as millimhos/cm, with the desired range 2.00 to 4.00. Salt levels above 4 millimhos/cm may result in wilting, suppressed growth, and fruit cracking. One mMho/cm = 1 milliSiemen/cm (mS/cm).

The total concentration of elements in a nutrient solution should be between 1,000 and 1,500 ppm so that osmotic pressure will facilitate the absorption processes by the roots. This should correspond to total salt conductivity readings between 1.5 and 3.5 millimhos (mMho). In general, the lower values (1.5–2.0 mMho) are preferred by crops such as cucumbers while higher values are better for tomatoes (2.5–3.5 mMho). One mMho/cm is approximately equal to 650 ppm of salt.

In general, no nutrient solution should be used for more than three months without complete replacement, together with a tap-water flushing of the whole system, including the beds. Two months is probably the average economical life of a nutrient solution which has been adjusted by use of analyses on a regular weekly basis. Of course, without such analyses as mentioned earlier, the life of the solution would be no more than 2 to 3 weeks.

3.8.4 Adjustment of Nutrient Solutions by Use of Electrical Conductivity

Total dissolved solutes (TDS) instruments which determine the dissolved solids in water are basically water conductivity measuring instruments. The quantity of dissolved solids in parts per million (ppm) or mg/l by weight is directly proportional to conductivity in millimhos (mMho) per unit volume. However, the electrical conductivity (EC) varies not only to the concentration of salts present, but also to the chemical composition of the nutrient solution. Some fertilizer salts conduct electric current better than others. For instance, ammonium sulfate conducts twice as much electricity as calcium nitrate and more than three times that of magnesium sulfate, whereas urea does not conduct electricity at all. Nitrate ions do not produce as close a relationship with electrical conductivity as do potassium ions (Alt, D. 1980). The higher the nitrogen to potassium, the lower will be the electrical conductivity values for the nutrient solution. Electrical conductivity measures total solutes, it does not differentiate among the various elements. For this reason, while a close theoretical relationship exists between TDS and EC, standard solutions of a nutrient formulation should be measured to determine their correlation in a given solution. For example, in Table 3.9 a 666 ppm TDS is equivalent to 1.0 mMho which is the measurement of a solution containing 490 ppm of sodium chloride or 420 ppm of calcium carbonate. That is, a 490 ppm solution of sodium chloride or a 420 ppm solution of calcium carbonate each give an electrical conductivity reading of 1.0 mMho.

TABLE 3.9 Relationship between Total Dissolved Solutes (TDS)and Electrical Conductivity (EC) for Sodium Chloride and Calcium Carbonate Solutions (Solution A)

Solution A TDS (ppm)	EC (mMho)	NaCl (ppm)	$CaCO_3$ (ppm)
10,000	15	8,400	7,250
6,660	10	5,500	4,700
5,000	7.5	4,000	3,450
4,000	6	3,200	2,700
3,000	4.5	2,350	2,000
2,000	3	1,550	1,300
1,000	1.5	750	640
750	1.125	560	475
666	1	490	420
500	0.75	365	315
400	0.6	285	250
250	0.375	175	150
100	0.15	71	60
66	0.10	47	40
50	0.075	35	30
40	0.06	28	24
25	0.0375	17.5	15
6.6	0.01	4.7	4

Table 3.9 relates concentration of sodium chloride and calcium carbonate with conductivity. A list of conductivities for 0.2% solutions (2 grams of fertilizer in 1 liter of distilled water) of various fertilizers aisgiven in Table 3.10. Conductivities for various concentrations of calcium nitrate are outlined in Table 3.11. These conductivity standards should be used to derive a theoretical relationship between conductivity and TDS. Actual conductivity measurements for fertilizers may vary somewhat from those in Tables 3.10 and 3.11 due to the solubility and purity of the particular fertilizer source. If electrical conductivity readings are not taken at the standard temperature of 25°C, a correction factor must be used (Table 3.12).

TABLE 3.10 Conductivity (EC) of 0.2% Solution in Distilled Water

Fertilizer Compound	*EC (mMho)*
$Ca(NO_3)_2$	2.0
KNO_3	2.5
NH_4NO_3	2.9
$(NH_4)_2SO_4$	3.4
K_2SO_4	2.4
$MgSO_4 \cdot 7H_2O$	1.2
$MnSO_4 \cdot 4H_2O$	1.55
NaH_2PO_4	0.9
KH_2PO_4	1.3
HNO_3	4.8
H_3PO_4	1.8

TABLE 3.11 Conductivity (EC) of Various Concentrations of Calcium Nitrate in Distilled Water

Concentration (%)	*EC (mMho)*
0.05	0.5
0.1	1.0
0.2	2.0
0.3	3.0
0.5	4.8
1.0	9.0

TABLE 3.12 Temperature Factors for Correcting Conductivity Data to Standard Temperature of 25°C*

°C	°F	Temperature Factor
5	41.0	1.613
10	50.0	1.411
15	59.0	1.247
16	60.8	1.211
17	62.6	1.189
18	64.4	1.163
19	66.2	1.136
20	68.0	1.112
21	69.8	1.087
22	71.6	1.064
23	73.4	1.043
24	75.2	1.020
25	**77.0**	**1.000**
26	78.8	0.979
27	80.6	0.960
28	82.4	0.943
29	84.2	0.925
30	86.0	0.907
31	87.8	0.890
32	89.6	0.873
33	91.4	0.858
34	93.2	0.843
35	95.0	0.829
40	104.0	0.763
45	113.0	0.705

* From "Saline and Alkali Soils," U.S. Salinity Laboratory Staff, Agricultural Handbook, No. 60, p. 90.

Some instruments such as the "Volmatic" conductivity meter, have a built-in temperature compensation which automatically makes the necessary adjustment.

Since most waters contain a number of dissolved solids and these solutes have different weights per ion, they will have a different concentration (ppm) for a given conductivity value. To adjust for these extraneous solids in a specific source of water to be used for a nutrient solution, a series of standards of known fertilizer concentrations should be made up with the water in question and their conductivities measured.

These measurements should be plotted graphically to better represent the actual relationship. When actual values for electrical conductivity for

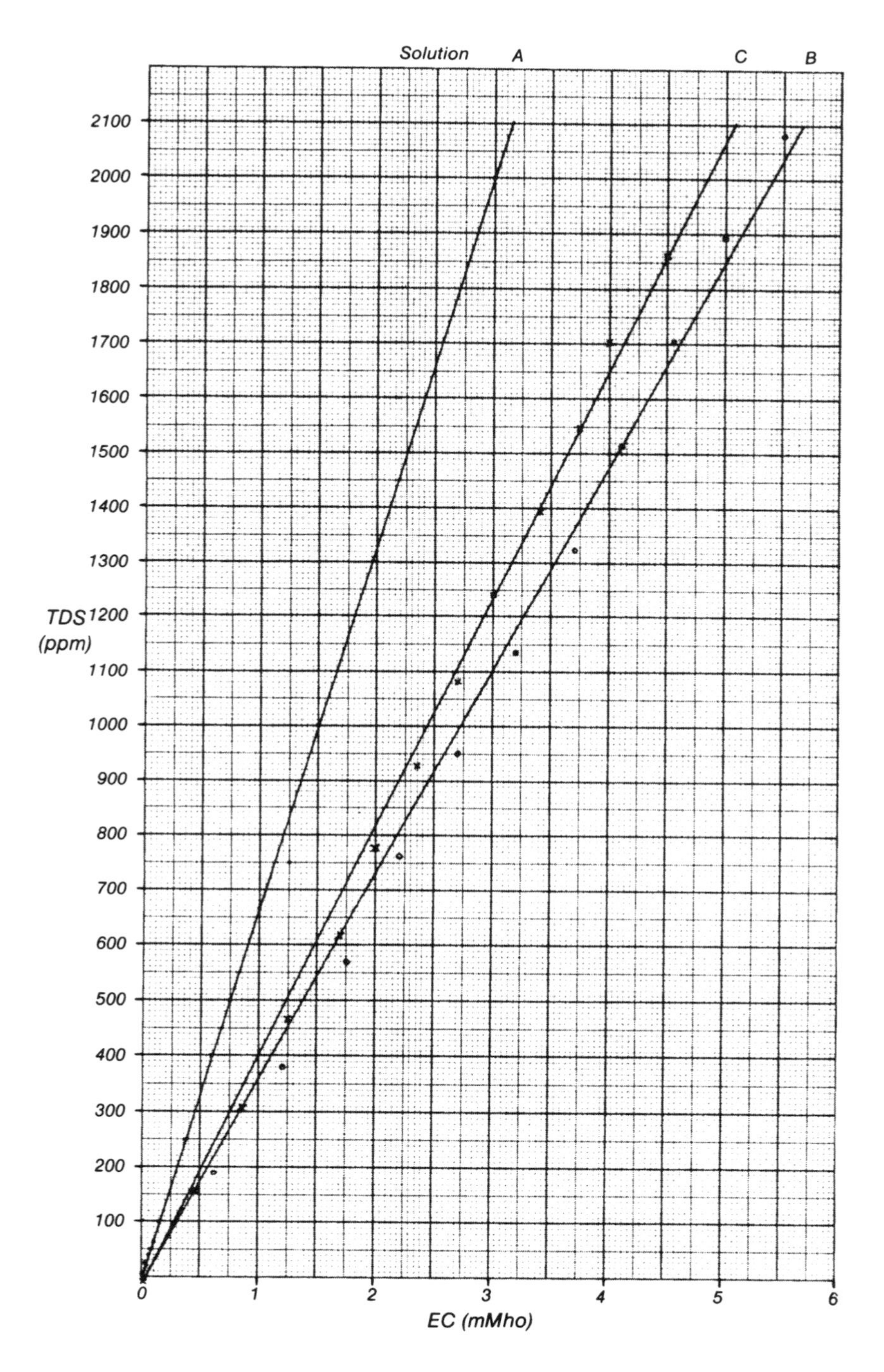

Figure 3.7. Electrical Conductivity (EC) versus Total Dissolved Solutes (TDS) for a number of nutrient solutions.

given nutrient solutions are plotted against known total dissolved solutes, a straight line direct relationship is revealed (Tables 3.13, 3.14 and Fig. 3.7). If the theoretical TDS and EC values from Table 3.9 (Solution A) are plotted on this graph, a large difference is apparent due to the different chemical compositions of the solutions (Fig. 3.7). For example, the electrical conductivities of 1.0, 2.0, 3.0, 4.0, and 5.0 mMho are equivalent to actual TDS contents of 350 ppm, 720 ppm, 1,100 ppm, 1,460 ppm, and 1,840 ppm respectively for Solution B (Fig. 3.7) while the theoretical values from Table 3.9 and Figure 3.7 are 666 ppm, 1,330 ppm, 2,000 ppm, 2,680 ppm, and 3,340 ppm respectively. If nutrient solutions differ only slightly in their composition, the correlation between EC and TDS will be similar for both solutions (Tables 3.13, 3.14 and Fig. 3.7). While a substantial difference exists between the actual and theoretical values of TDS for a given conductivity reading due to solubility differences of fertilizers, the ratios and compositions of various ions in the solution, the data is useful to determine the overall changes which occur in a specific nutrient solution.

With sufficient data collected over time to verify the statistical significance of the relationship between EC and TDS, a relatively inexpensive and simple method to monitor changes in the nutrient solution over time will indicate what adjustments should be made to keep the solution in balance for the crop being grown.

This principle can become more useful by determining for each crop the relationships among total dissolved solutes, electrical conductivity, concentration of each essential element and stage of plant growth under similar light conditions. Data can be obtained through a laboratory analysis of the nutrient solution each week. Samples should always be taken from the same nutrient tank having the same volume of solution, growing the same number of plants. If these factors change between sample periods, error will result in the interpretation, as variations would have occurred among samples, and comparisons would not have the same basis.

If these tests are taken at the same stages of growth for each crop for a number of crops during the year, the only variable remaining would be sunlight hours. Adjustments could be made in the results of the analyses for changes in weekly total sunlight hours.

Head lettuce requires about 84 days (12 weeks) to maturity. The nutrient solution should be changed completely at 6 weeks. Analysis should be taken weekly, that is, days 0, 7, 14, 21, 28, 35, and 42 for the first solution make-up and days 42, 49, 56, 63, 70, 77, and 84 for the second solution make-up. By analyzing 12 of the essential elements (N, P, K, Ca, Mg, S, Fe, Mn, B, Cu, Zn, and Mo) weekly, changes in the levels of each element could be plotted against age of the plants.

TABLE 3.13 Electrical Conductivity (EC) and Total Dissolved Solutes (TDS) of a Standard 30-Liter Nutrient Solution (Solution B)

g/30 l				TDS	EC
K_2SO_4	$Ca(NO_3)_2$	$(NH_4)_2HPO$	$MgSO_4 \cdot 7H_2O$	(ppm)	(mMho)
0	0	0	0	0	0.05
2.82	7.4	1.88	2.75	189.4	0.62
5.65	14.8	3.75	5.5	379	1.2
8.5	22.2	5.62	8.25	568	1.75
11.3	29.6	7.5	11.0	757.5	2.2
14.1	37.0	9.38	13.75	947	2.7
17.0	44.4	11.25	16.5	1,136	3.2
19.8	51.8	13.12	19.25	1,326	3.7
22.6	59.2	15.0	22.0	1,515	4.1
25.4	66.6	16.88	24.75	1,704	4.55
28.25	74.0	18.75	27.5	1,894	5.0
31.1	81.4	20.62	30.25	2,083	5.5
33.9	88.8	22.5	33.0	2,272.5	6.3

TABLE 3.14 Electrical Conductivity (EC) and Total Dissolved Solutes (TDS) of a Standard 30-Liter Nutrient Solution (Solution C)

g/30 l				TDS	EC
KNO_3	$Ca(NO_3)_2$	$(NH_4)_2HPO$	$MgSO_4 \cdot 7H_2O$	(ppm)	(mMho)
0	0	0	0	0	0.05
3.08	5.75	1.88	2.75	154.9	0.45
6.15	11.5	3.75	5.5	309.8	0.86
9.22	17.25	5.62	8.25	464.6	1.25
12.3	23.0	7.5	11.0	619.5	1.7
15.38	28.75	9.38	13.75	774.4	2.0
18.45	34.5	11.25	16.5	929.2	2.35
21.52	40.25	13.12	19.25	1,084.1	2.7
24.6	46.0	15.0	22.0	1,239	3.0
27.68	51.75	16.88	24.75	1,393.9	3.4
30.75	57.5	18.75	27.5	1,548.8	3.75
33.82	63.25	20.62	30.25	1,703.6	4.0
36.9	69.0	22.5	33.0	1,858.5	4.5

Since data is not presently available to substantiate these relationships, a hypothetical situation is used to demonstrate the expected relationships of levels of essential elements, TDS and EC with plant stage of growth for head lettuce (Fig. 3.8). When the plants are seedlings (0–21 days) little change in solution element levels would occur. As the plant leaf area increases, demand upon nutrients would increase, resulting in a depletion of existing elements within the nutrient solution to suboptimal levels. Due to differential uptake of elements by plants at various stages of their

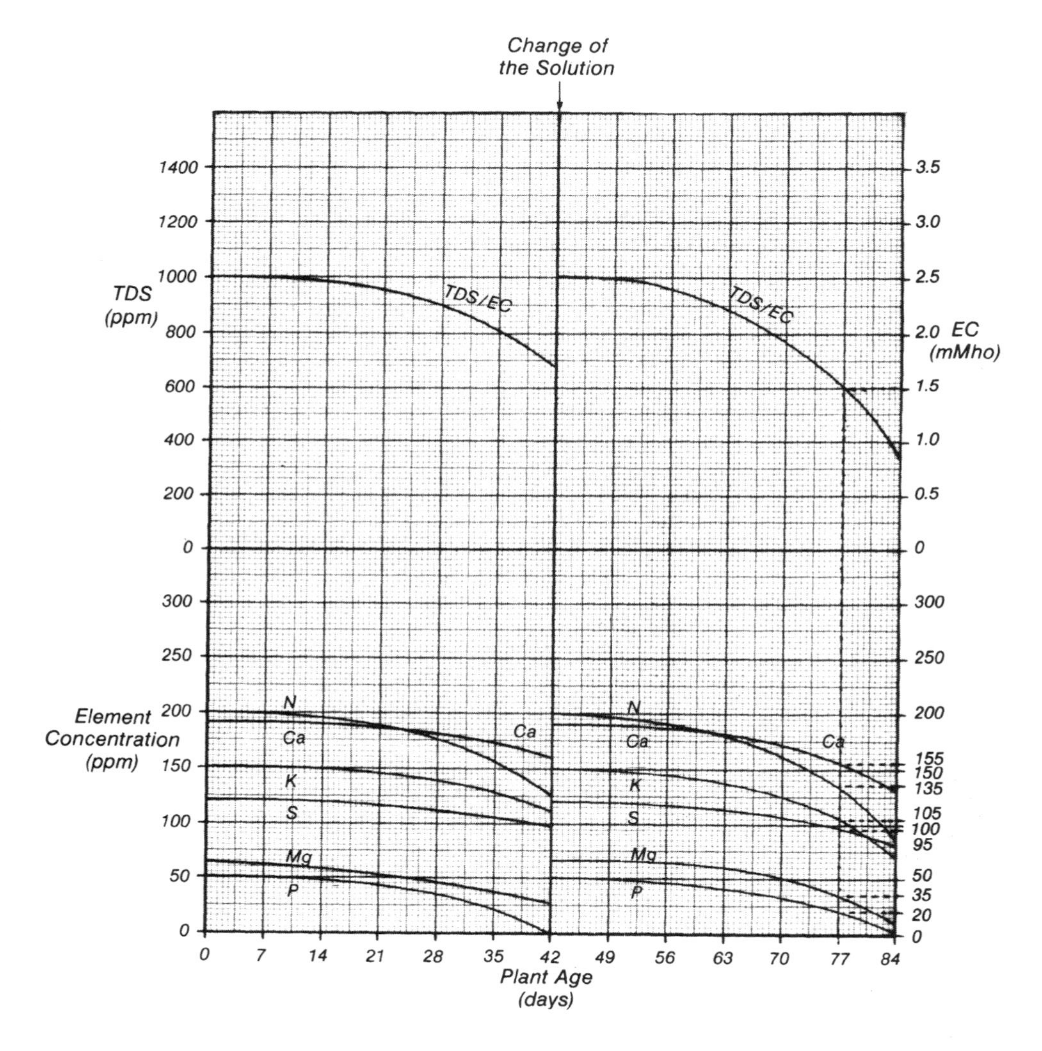

Figure 3.8. Hypothetical relationship among macroelements, Total Dissolved Solutes (TDS), Electrical Conductivity (EC) of the nutrient solution and the age of the plants. (Solution C.)

growth, the rate of decrease would differ for each element. For example, the negative slope of the nitrogen (N) curve is greater than that of K and P, indicating that nitrogen is used more rapidly by the plants than K or P. As the plants mature, the overall demand upon nutrients increases and the demand for some elements will be greater than for others, as shown by the greater negative slopes of the curves. This is more evident in the second set of curves (days 42–84) during the second nutrient solution make-up. All elements are decreasing in concentration at a faster rate, that is, the negative slope of the curves has increased.

The curves should be used to determine when to add those fertilizers whose elements are rapidly being depleted, before a deficiency causes nutritional stress in the plants and a resultant significant loss in crop productivity. In the hypothetical example above, nitrogen (N) should have been added after the 28-day period in the first solution and weekly after the 56-day period in the second solution. Phosphorus (P) and potassium (K) should have been added after the 28-day period in the first solution and weekly after the 63-day period in the second solution. Additions of fertilizers should be made to bring up the individual element to its original optimal level each time.

According to work by Steiner (1980), salts may be added to the nutrient solution during culture using electrical conductivity as a measure of the solution status. But he emphasizes that any additions of ions should be made as close as possible in the same mutual ratio as they are utilized by the plants. There is no evidence to support the idea that plants can be forced to consume ions in the ratio of the nutrient solution. Plants select their desired uptake ratio of nutrients from widely different ratios in the nutrient solution. Therefore, ions should be added in the same mutual ratio as the plants absorb them.

Based upon this knowledge, a favorable nutrient solution for any given crop can be developed and its ratios of nutrients can be adjusted to optimum levels by use of electrical conductivity of the nutrient solution, provided the relationship among the stage of crop growth, sunlight hours and total dissolved solutes has been determined. As Steiner (1980) points out, even if small deviations in the chemical composition of the nutrient solution occur, the plants themselves can select the ions in a mutual ratio favorable for their growth and development.

The same type of curves and reasoning can be made for the micronutrients, the differences being that the changes and levels are much smaller. Nonetheless, adjustments are equally important in order to maintain an optimum nutrient status without encountering deficiencies.

With total dissolved solutes (TDS) and electrical conductivity (EC) included in the above graphs (Fig.3.8, Solution C), the relationships to the age of plants and levels of individual elements are evident. An electrical conductivity reading would indicate the total dissolved solute level and the probable levels of each essential element at any given plant age. With this information supported by sufficient data for a given nutrient formulation, future EC readings on a nutrient-solution sample having the same formulation but no laboratory analysis carried out on it, could be used to determine the levels of individual elements within the solution. Accurate adjustments could be made in the nutrient solution at any given season and stage of plant growth.

For example, if the EC was taken at day 56 and was found to be 1.5, follow the TDS/EC curve (Fig. 3.8) along until the point of EC = 1.5, then read directly below it the levels of each element. Since the plant age is within the second nutrient-solution makeup, the second set of graphs should be used. As plants mature, their uptake of specific elements changes. For instance, more K and Ca would be taken up by tomatoes as they produce fruit. That is, their ratio of uptake of K and Ca increases in comparison with the other elements. During early vegetative growth, N would be more in demand. In this example, it is assumed that no fertilizers had been added earlier to adjust for plant uptake. If some were added, the EC would have been higher and the intersection point on the TDS/EC curve would have been further to the left, indicating the solution varied little from its original preparation at day 42.

The points of intersection of the individual elements corresponding to the TDS/EC curve at EC = 1.5 would be: Ca = 155 ppm, N = 135 ppm, K = 105 ppm, S = 95 ppm, Mg = 35 ppm, and P = 20 ppm. If the original formulation had Ca = 190 ppm, N = 200 ppm, K = 150 ppm, S = 120 ppm, Mg = 65 ppm, and P = 50 ppm, the differences that have to be made up by addition of fertilizers are: Ca = 35 ppm, N = 65 ppm, K = 45 ppm, S = 25 ppm, Mg = 30 ppm, and P = 30 ppm. Generally, sulfur (S) need not be added specifically as it is included in several fertilizers such as potassium sulfate and magnesium sulfate which would be used to increase potassium (K) and magnesium (Mg) levels.

If the original nutrient-solution volume was 30,000 liters (7,925 U.S. gallons) and it had to be maintained, the solution level should be topped-up before recording the electrical conductivity. If the conductivity value was taken within several weeks of crop maturity and the solution volume was at 20,000 liters (5,283 U.S. gallons) but that volume was sufficient to maintain the crop until harvest, then the conductivity should be recorded on that solution volume without topping it up since less fertilizers would be required to adjust the solution. If water needs to be added to the solution, it should be done before recording conductivity. In general, as the volume of water decreases, it concentrates the solution, unless the plants are taking up nutrients faster than utilizing water.

The amount of fertilizers needed to adjust the 30,000 liters (7,925 U.S. gallons) of solution in the above example would be as follows (using conversion factors from Table 3.2):

Ca: 35 ppm, source: $Ca(NO_3)_2$
Need: $35 \times 4.094 = 143.3$ mg/l

For 30,000 liters: $30{,}000 \times \dfrac{143.3}{1{,}000} = 4{,}299$ g or: 4.3 kg

Amount of N: $143.3 \times 0.171 = 24.5$ mg/l (ppm)

Mg: 30 ppm, source: $MgSO_4 \cdot 7H_2O$
Need: $30 \times 10.14 = 304$ mg/l or: 9.1 kg/30,000 l
Amount of S: $304 \times 0.130 = 39.5$ ppm

P: 30 ppm, source: KH_2PO_4
Need: $30 \times 4.394 = 131.8$ mg/l or: 3.95 kg/30,000 l
Amount of K: $131.8 \times 0.287 = 37.8$ ppm

K: 38 ppm, source: KH_2PO_4; Need: 45–38 = 7 ppm, source KNO_3
Need: $7 \times 2.586 = 18.1$ mg/l or: 0.54 kg/30,000 l
Amount of N: $18.1 \times 0.1385 = 2.5$ ppm

S: 39.5 ppm, source: $MgSO_4 \cdot 7H_2O$

* N: 24.5 ppm, source: $Ca(NO_3)_2$
2.5 ppm, source: KNO_3
Total: 27 ppm

* Since the amount of N that can be obtained is limited by the amount of KNO_3 used as a source of K, either additional N could be added from $Ca(NO_3)_2$, NH_4NO_3 or alternatively, the N level should be left at 27 ppm from the present levels of $Ca(NO_3)_2$ and KNO_3 required to provide Ca and K respectively.

The amount of fertilizers to be added to 30,000 liters (7,925 U.S. gallons) of solution are: 4.3 kg of calcium nitrate, 9.1 kg of magnesium sulfate, 3.95 kg of monopotassium phosphate, and 0.54 kg of potassium nitrate. The fertilizers add the following amounts of each element: N = 27 ppm, P = 30 ppm, K = 45 ppm, Ca = 35 ppm, Mg = 30 ppm, and S = 39.5 ppm. After adding the fertilizers, the *p*H and electrical conductivity should be recorded. The conductivity value should approach that when the solution was first prepared.

Using EC to adjust a nutrient solution weekly instead of changing it every few weeks could extend its life to at least 6 weeks. In this way substantial savings in fertilizers, water and labor would be realized. These savings would be particularly beneficial in countries having a scarcity of water and/or a shortage of fertilizers. Such savings would more than compensate for the initial cost of solution analyses.

Several plant laboratories capable of analyzing nutrient solutions are listed in Appendix 2. In addition, many universities are prepared to do such analyses.

The management of nutrient solutions through the use of electrical conductivity applies particularly to closed systems of NFT and subirrigation. However, it can be used to monitor open systems having large storage tanks of solution rather than those using proportioners.

3.8.5 Maintenance of the Solution Volume

The solution volume must be kept relatively constant in order to secure adequate plant growth. Plants take up much more water and at a much greater rate than the essential mineral elements. As water is removed from the nutrient solution, the volume of the solution naturally decreases. This effects an increase in the total solution concentration and in the concentration of the individual nutrient ions.

The average daily water-loss can range from 5 to 30 percent, depending upon the volume of the unit and the number and type of plants. Water to compensate for this loss can be added daily as long as the solution is used. Using NFT systems, workers can more accurately determine the rate of water uptake by plants. In England, Spensley and co-workers (1978) found that on a clear summer's day fully grown tomatoes consumed 1.33 liters (1/3 gallon) of water per plant. Winsor and associates (1980) determined that tomato plants lost through evapotranspiration 15 ml/plant/hour during the night, rising to a maximum of 134 ml/plant/hour at midday on a clear summer day. Adams (1980) calculated that cucumbers consumed water at approximately twice the rate for tomatoes due to their greater leaf area. Water uptake reached a maximum of 230 ml/plant/hour during maximum light intensity and temperatures of the afternoon. A rule-of-thumb estimate of water usage in a greenhouse is about 1 liter/sq ft/day for vine crops such as tomatoes and cucumbers. Experienced commercial growers can add water weekly or they can attach an automatic float-valve assembly to the inlet valve to the nutrient tank which will fill it daily. When water is added weekly, water in excess of the original volume of the solution is introduced. The solution is then allowed to concentrate as the plants remove water to below the original solution level. Usually the best procedure is to allow the solution volume to fluctuate equally on both sides of the original level. It follows that a solution-testing technique must be used in conjunction with this method of regulation of the solution volume.

References

Adams, P. 1980. Nutrient uptake by cucumbers from recirculating solutions. *Acta Hort.* 98:119–126.

Alt, D. 1980. Changes in the composition of the nutrient solution during plant growth—an important factor in soilless culture. *Proc. of the 5th Int. Congress on Soilless Culture,* Wageningen, May 1980, pp. 97–109.

Bauerle, W.L. and P. Flynn. 1990. Precision nutrient control—A window into the future. *Proc. of the 11th Ann. Conf. of the Hydroponic Society of America,* Vancouver, B.C. pp. 25–32.

Berry. W.L. 1989. Nutrient control and maintenance in solution culture. *Proc. of the 10th Ann. Conf. of the Hydroponic Society of America,* Tucson, AZ. pp. 1–6.

Butler, J.N. 1964. *Solubility and pH calculations.* Reading, MA: Addison-Wesley.

Muckle, M.E. 1990. *Hydroponic nutrients—easy ways to make your own.* rev. ed. Princeton, B.C.: Growers Press.

Resh, H.M. 1990. *Hydroponic home food gardens.* Santa Barbara, CA: Woodbridge Press.

Ryall, D. 1993. Commercially growing tomatoes in rockwool in British Columbia. *Proc. of the 14th Ann. Conf. of the Hydroponic Society of America,* Portland, OR. pp. 33–39.

Schaum, D., Beckmann, C. O., and J. L. Rosenberg. 1962. *Schaum's outline of theory and problems of college chemistry,* 4th ed. New York, NY: Schaum.

Schippers, P.A. 1991. Practical aspects to fertilization and irrigation systems. *Proc. of the 12th Ann. Conf. of the Hydroponic Society of America,* St. Charles. IL.

Schon, M. 1992. Tailoring nutrient solutions to meet the demands of your plants. *Proc. of the 13th Ann. Conf. of the Hydroponic Society of America,* Orlando, FL. pp. 1–7.

Schwarz, M. 1968. *Guide to commercial hydroponics.* Jerusalem: Israel Univ. Press.

Sienko, M.J. and R.A. Plane. 1961. *Chemistry,* 2nd ed. New York, NY: McGraw-Hill.

Spensley, K., G.W Winsor, and A.J. Cooper. 1978. Nutrient film techique crop culture in flowing nutrient solution. *Outlook on Agriculture.* 9:299–305.

Steiner, A.A. 1980. The selective capacity of plants for ions and its importance for the composition and treatment of the nutrient solution. *Proc. of the 5th Int. Congress on Soilless Culture,* Wageningen, May 1980, pp. 83–95.

Ulises Durany, Carol. 1982. *Hidroponia–Cultivo de plantas sin tierra,* 4th ed. Barcelona, Spain: Editorial Sintes, S.A. 106 pp.

Wallace, A. 1989. Regulation of micronutrients and use of chelating agents in solution cultures. *Proc. of the 10th Ann. Conf. of the Hydroponic Society of America,* Tucson, AZ. pp. 14–26.

Wallace, G.A. and A. Wallace. 1984. Maintenance of iron and other micronutrients in hydroponic nutrients solutions (Tomatoes, cucumbers). *Journal of Plant Nutrition* 7(1/5): 575–585.

Wilcox, G. 1991. Nutrient control in hydroponic systems. *Proc. of the 12th Ann. Conf. of the Hydroponic Society of America,* St. Charles, IL. pp. 50-53.

Winsor, G.W., P. Adams, and D. Massey. 1980. *New light on nutrition. Suppl. Grower.* 93(8):99, 103.

Withrow, R.B., and A.P. Withrow. 1948. *Nutriculture.* Lafayette, IN: Purdue Univ. Agr. Expt. Stn. Publ. S. C. 328.

Chapter 4

The Medium

A soilless medium, such as water, foam, gravel, rockwool, sand, sawdust, peat, perlite, pumice, peanut hulls, polyester matting, or vermiculite must provide oxygen, water, nutrients and support for plant roots just as soil does. The nutrient solution will provide water, nutrients and to some extent oxygen. How each of these soilless cultural methods satisfies plant needs will be discussed in detail in the chapters to follow.

4.1 Medium Characteristics

Moisture retention of a medium is determined by particle size, shape and porosity. Water is retained on the surface of the particles and within the pore space. The smaller the particles, the closer they will pack, the greater the surface area and pore space, and hence the greater the water retention. Irregular-shaped particles have a greater surface area and hence higher water retention than smooth, round particles. Porous materials can store water within the particles themselves, therefore water retention is high. While the medium must be capable of good water retention, it must also be capable of good drainage. Therefore, excessively fine materials must be avoided so as to prevent excessive water retention and lack of oxygen movement within the medium.

The choice of the medium will be determined by availability, cost, quality, and the type of hydroponic method to be employed. A gravel subirrigation system can use very coarse material, whereas a gravel trickle-irrigation system must use a finer material (see Chapter 7).

The medium must not contain any toxic materials. Sawdust, for example, often contains a high sodium chloride content due to logs having remained in salt water for a long period of time. The salt content must be tested and if any amount of sodium chloride is present, it will be necessary to leach it through with fresh water. Gravel and sand of calcareous (limestone) origin should be avoided. Such materials have a very high content of calcium carbonate ($CaCO_3$), which is released from the medium into the nutrient solution, resulting in high *p*H. This increased alkalinity ties up iron, causing iron deficiency in plants. Such materials can be pretreated with water-leaching, acid-leaching or soaking in a phosphate solution. This will buffer the release of carbonate ions. Nonetheless, this procedure is only a short-term solution and eventually nutritional problems will arise. This problem makes gravel and sand culture very difficult in some areas such as the Caribbean, where the materials are all of calcareous origin. The best gravel or sand is that of igneous (volcanic) origin.

The media must be of sufficient hardness in order to be durable for a long time. Soft aggregates which disintegrate easily should be avoided. They lose their structure and their particle size decreases, which results in compaction leading to poor root aeration. Again, aggregates of granitic origin are best, especially those high in quartz, calcite and feldspars. If a hydroponic system is to be set up out of doors, particles having sharp edges should be avoided, since wind can abrade the plant stem and crown against them, leading to injury and a port of entry for plant parasites. If a relatively sharp medium must be used, the top 2 inches (5 cm) should consist of smooth-edged medium so the area where most plant movement takes place will be protected from abrasion.

Igneous gravel and sand have little influence on the nutrient solution *p*H, whereas calcareous material will buffer the nutrient solution *p*H at about 7.5. Treatment with phosphate solutions can bring the *p*H down to 6.8 (see Chapter 7).

4.2 Water Characteristics

The water quality is of prime concern in hydroponic growing. Water with sodium chloride content of 50 ppm or greater is not suitable for optimum plant growth. As sodium chloride content is increased, plant growth is restricted; high levels result finally in death of the plant. Some plants are less sensitive to salt levels than others. For instance, herbs such as watercress and mint tolerate higher levels of sodium and chlorine than do tomatoes and cucumbers. Besides the sodium chloride content, the total dissolved solutes in the nutrient solution must be considered.

Hardness is a measure of the carbonate ion (HCO_3^-) content. As mentioned earlier, as hardness increases (*p*H increases), certain ions such as iron become unavailable. In particular, ground water which lies in calcareous and dolomitic limestone strata could contain high levels of calcium and magnesium carbonate which may be higher than or equal to the normal levels used in the nutrient solution.

Hard water contains salts of calcium and magnesium. Normally such waters are just as suitable as soft water for growing plants. Calcium and magnesium both are essential nutrient elements and ordinarily the amount present in hard waters is much less than that used in nutrient solutions. Most hard waters contain calcium and magnesium as carbonates or sulfates. While the sulfate ion is an essential nutrient, the carbonate is not. In low concentrations the carbonate is not injurious to plants. In fact, some carbonate and/or bicarbonate in the raw water is helpful in stabilizing the *p*H of the nutrient solution. The carbonate or bicarbonate found in most waters causes the *p*H to rise or remain high. With the presence of these ions, the *p*H resists dropping. This stabilizing effect is called "buffering capacity." Smith (1987) advises to take advantage of this buffering action on *p*H by maintaining the carbonate/bicarbonate level about 30–50 ppm, which is sufficient to prevent sudden fluctuations in *p*H. With very pure raw water, having little or no carbonate or bicarbonate, he suggests adding some potassium carbonate or bicarbonate to obtain this level to improve the buffering capacity.

Before any water is used, an analysis should be made for at least calcium, magnesium, iron, carbonate, sulfate and chloride. If a commercial hydroponic complex is being planned, water should be analyzed for all major and minor elements. Once the level of each ion has been determined, correspondingly less of each of these elements should be added to make up the nutrient solution. For example, the magnesium concentration of some well waters is so high that it is not necessary to add any to the nutrient solution.

The naturally occurring dissolved salts in the water accumulate with the additions of makeup water. Over a period of time this buildup will exceed the optimum levels for plant growth, and the nutrient solution will have to be changed to avoid injuring the plants.

The concentration of salts in a nutrient solution may be specified in terms of ppm, millimolar (mM), and milliequivalents per liter (me/l). Parts per million, as mentioned in Chapter 3, is based on a specified number of units by weight of salt for each million parts of solution. The millimolar unit of concentration involves the molecular weight of the substance. One mole of a substance is a weight in grams numerically equal to the molecular weight and commonly is called a gram molecular weight. A solution of one molar concentration is that which has one mole of the substance dissolved in one liter of solution. A millimolar concentration is 1/1,000 as concentrated as a molar concentration and amounts to one mole in 1,000 liters of solution.

Both solutions (KNO_3) and (KCl) would have the same number of potassium ions, and the chloride ions in one solution would be the same in number as the nitrate ions of the other. In the absorption of nutrients, it is the concentration of ions and not the weight of the element or ion that is of significance.

Where there are bivalent ions present in a compound it is better to use milliequivalents per liter. Milliequivalents per liter is similar to the millimolar unit, but involves gram equivalents instead of the mole or gram molecular weight. The gram equivalent is the gram molecular weight divided by the valency (number of charges on the ion). The gram equivalent of a salt like KCl, which consists of singly valent ions (K^+, Cl^-), is numerically the same as the mole, but where there are bivalent ions ($SO_4^=$) as in potassium sulfate (K_2SO_4), a gram equivalent would contain numerically only half as much as a mole. Thus two solutions of K_2SO_4 and KCl, having the same milliequivalent per liter concentration, would have in solution the same concentration of potassium ions, but half as many sulfate ions ($SO_4^=$) as chloride ions (Cl^-).

Milliequivalents per liter (me/l) is the most meaningful method of expressing the major constituents of water. This is a measure of the chemical equivalence of an ion.

The total concentration of a nutrient solution or water source occasionally is specified in terms of its potential osmotic pressure. It is a measure of the availability or activity of the water. The osmotic pressure difference between cells usually determines the direction in which water will diffuse. The osmotic pressure is proportional to the number of solute particles in solution and depends upon the number of ions per unit volume for inorganic materials. Osmotic pressure is usually given in atmosphere units, where 1 atmosphere (atm.) is 14.7 pounds per square inch.

In summary,

1 molar (M) $= \frac{\text{M.W.}}{1\ \text{l}}$

e.g.: KNO_3; M.W. = 101

$1\ \text{M} = \frac{101\ \text{g}}{1\ \text{l}}$

e.g.: KCl; M.W. = 74.6

$1\ \text{M} = \frac{74.6\ \text{g}}{1\ \text{l}}$

1 millimolar (mM) $= \frac{\text{M.W.}}{1{,}000\ \text{l}}$

e.g.: KNO_3

$1\ \text{mM} = \frac{101\ \text{g}}{1{,}000\ \text{l}}$ $10\ \text{mM} = \frac{1{,}010\ \text{g}}{1{,}000\ \text{l}}$

e.g.: KCl

$1\ \text{mM} = \frac{74.6\ \text{g}}{1{,}000\ \text{l}}$ $10\ \text{mM} = \frac{746\ \text{g}}{1{,}000\ \text{l}}$

where: M.W. = molecular weight of the compound.

$$\textbf{1 equivalent (Eq)} = \frac{M.W.}{Valence}$$

e.g.: K_2SO_4; M.W. = 174.3
Valence = 2

$$1\ Eq = \frac{174.3}{2} = 87.15$$

$$\textbf{1 milliequivalent (me)} = \frac{Eq}{1{,}000}$$

$$\text{e.g.: } K_2SO_4; \quad me = \frac{87.15}{1{,}000} = 0.08715$$

$$\textbf{1 milliequivalent/l (me/l)} = \frac{me}{1\ l} = \frac{Eq}{1{,}000\ l}$$

$$1\ me/l = \frac{ppm}{Eq}$$

e.g.: 100 ppm of $SO_4^{=}$; Eq of $SO_4^{=}$ is: $\frac{96}{2} = 48$

$$me/l = \frac{100}{48} = 2.08$$

The use of saline waters for hydroponic growing of crops has been investigated by several workers (Victor 1973; Schwarz 1968). Possibilities of using saline water of 3,000 ppm total salt were investigated by Schwarz (1968). Salt tolerance of varieties, stage of development, addition of nutrient absent in the raw water, and frequency of irrigation are some factors to be considered in using saline water. The concentration expressed by 0.4 atm. osmotic pressure was recommended as best for good development of tomato plants in tropical areas (Steiner 1968).

Saline waters are those containing sodium chloride. Highly saline water can be used for hydroponics, but a number of considerations are involved. The plants that can be grown are limited to salt-tolerant and moderately salt-tolerant species, such as carnation, tomato, cucumber and lettuce. Even among salt-tolerant species, one variety may be more tolerant than another. A grower should conduct his own varietal trials to determine the most salt-tolerant varieties.

Salt tolerance is also dependent on plant growth stage. No mature cucumber plants have yet been shown to adapt gradually to saline conditions; however, Schwarz (1968) points out that cucumbers started in nonsaline conditions may be irrigated with solutions of gradually increasing salinity until the desired level is reached. The younger the plant, the easier it adapts to saline conditions. Schwarz (1968) reported that tomatoes and cucumbers generally take about 20 percent longer to germinate under saline than under nonsaline conditions.

Depending upon the species and variety and on the salinity of nutrient solutions, yields may be lowered by 10 to 25 percent in saline conditions. Schwarz (1968) reported yield reductions of 10 to 15 percent in tomatoes and lettuce and 20 to 25 percent in cucumbers grown with water containing 3,000 ppm salts.

The total solute concentration (high osmotic pressure), by leading to a reduction in water uptake, is responsible for the inhibitory effect of saline solutions on plant growth. Schwarz (1968) found that extremely high osmotic pressures (over 10 atm.) for short periods are less damaging than long periods of moderately high pressures (4–5 atm.). Symptoms of salt toxicity are a general stunting of growth with smaller and darker green leaves, a marginal leaf burn and a blueing and bleaching of the plant tissues.

Salinity may inhibit uptake of certain ions. High sulfate concentrations promote the uptake of sodium (leading to sodium toxicity), decrease the uptake of calcium (leading to calcium deficiency, especially in lettuce), and interfere with potassium uptake. High calcium concentrations in nutrient solutions also affect potassium uptake. High total salt contents are thought to affect calcium uptake, leading to "blossom-end-rot" symptoms in tomatoes. Saline conditions reduce the availability of certain microelements, especially iron, so that additional iron must be added. In addition to chloride and sodium toxicity, boron toxicity is relatively common with some saline waters.

Schwarz (1968) reports that saline waters have some favorable effects on cucumbers and tomatoes in giving them a sweeter taste than those grown with freshwater solutions. Lettuce heads are generally more solid, and carnations are longer lasting. Plants grown in saline solutions apparently have a much higher tolerance to zinc and copper, so that zinc and copper levels previously considered toxic may be present without causing injury.

4.3 Irrigation

The frequency of irrigation cycles depends upon: the nature of the plant; plant stage of growth; weather conditions (greenhouses)—particularly light intensity, daylength, and temperatures; and type of medium.

Plants that are more succulent with an abundance of leaves require more frequent irrigation as they lose water rapidly through evapotranspiration from their leaves. The greater the leaf area, the more water they will consume. As plants mature, producing a large canopy of leaves and developing fruit, their water demand increases.

Under greenhouse conditions of high light intensity, generally accompanied by high temperatures, especially during summer months, the evapotranspiration rate of plants is greatly increased, and as a result the water uptake also increases significantly. As pointed out in Chapter 3 under the section "Maintenance of the Solution Volume," researchers have found that a mature cucumber plant may take up as much as 230 ml of water per hour during midday conditions.

The water retention of the medium is a factor in determining the frequency and duration of irrigation. Finer media such as peat, foam or rockwool will retain more moisture than coarser ones of sawdust, perlite, vermiculite, sand or gravel. A water culture system, such as the nutrient film technique (NFT), must flow continuously to provide adequate water. In this case, only the plant root mat contains a small amount of residual moisture should the flow of solution be interrupted. Coarse aggregates may need watering as often as once every hour during the day, while a fine medium such as peat could get by on one to two irrigations per day under similar conditions.

Both the frequency and duration of irrigation cycles are important. The frequency of cycles must be sufficient to prevent any water deficit by the plants between cycles, but must be long enough to provide adequate drainage of the medium so that proper oxygenation of plant roots occurs. Wilting of plants indicates a possible water deficit. However, root dieback caused by diseases, pests or lack of oxygen, can also cause wilting, so the health of the roots should always be examined when wilting occurs. Healthy roots will appear white, firm and fibrous. No browning of sections or tips of roots should be present.

The duration of any given irrigation cycle must be sufficient to provide adequate leaching of the medium. With some of the finer media, such as foam or rockwool, 20 to 30 percent runoff is needed to flush excessive nutrients through the substrate. If this is not done, salt levels will build up, causing slowing of growth or even a toxicity in the plants. More specific details are presented for each type of hydroponic system in the following chapters.

4.4 Pumping of Nutrient Solution into Beds

As mentioned earlier, the nutrient solution must provide water, nutrients, and oxygen to the plants. The frequency of irrigations will depend on the nature of the medium, the size of the crop and the weather conditions. To provide these plant needs most efficiently during each irrigation cycle, the solution must moisten the bed uniformly and drain completely and rapidly so that oxygen will be available to the plant roots.

Details of pumping frequency are discussed more thoroughly in the chapters on each soilless system. In all cases, free water must not remain in the medium. The voids (air spaces between the particles) should be filled with moist air, not water, in order to maintain oxygen concentration around the roots at high levels. In most systems, irrigation should be done mainly during the daylight hours, with only several or none at night (depending upon the medium).

Ideally, moisture levels in the medium can be maintained at optimum levels through a feedback system. Such a system has a moisture-sensing device, such as a tensiometer or electrode, placed in the medium. This apparatus is connected to an electrical circuit which activates a valve or pump which waters the plants when the moisture falls below a preset level. In this way, optimum moisture conditions are maintained within a fairly narrow range of variation.

In greenhouse culture, the temperature of the nutrient solution in contact with the roots should not fall below the high air temperature of the house. Immersion heaters can be placed in the sump to heat the nutrient solution, but care must be taken not to use heating elements such as lead which may react electrolytically with the nutrient solution to release toxic amounts of ions into the solution. Heat lamps could be used instead of immersion heaters. In general, the nutrient solution should be maintained between 60°–65° F (15.5°-18° C). Recent experiments have demonstrated that by heating the nutrient solution, air temperatures can be decreased to conserve heat in the greenhouses. During the winter season in the greenhouses, some crops, such as tomatoes, grow better if the temperature of the root medium is maintained a few degrees above the average night air temperature.

In no case should heating lines or electric heating cables be placed in the bed itself. Such a placement causes localized high temperatures around the heating elements, which injure the roots and may cause lead toxicity if lead cables are used.

4.5 Sterilization of Medium

When crops are grown for extended periods in any aggregate, soil-borne pathogenic microorganisms accumulate in the medium, and the chances of a disease occurring increase with each successive crop. It may be possible to grow several crops successively without sterilization between them; however, for best results the medium should be sterilized between each crop to prevent any possible disease carry-over. The most common methods of sterilization are steam and chemical.

If a greenhouse is being heated by a central hot water or steam boiler, sterilization by steam would be the most economical. A steam converter attachment would be installed on the boiler and steam pipes run to the greenhouse with outlet attachments at each bed. A steam line is run down the center of each bed and covered with canvas or some other heat-resistant material. Steam is then injected along the entire length of the bed at 180° F (82° C) for at least half an hour. This surface steaming is effective to a depth of 8 inches (20 cm) for sawdust beds, but only to 4 inches (10 cm) deep for a 3:1 sand-sawdust mixture. Where surface steaming is not effective, install a permanent tile or perforated rigid pipe in the bottom of the bed through which steam can be injected.

Several chemicals may also be used in place of steam for sterilization, keeping in mind that some of these chemicals, particularly chloropicrin and methyl bromide, are toxic to humans and should be applied only by persons trained in their use. In all cases, precautions prescribed by the manufacturer should be observed.

Formaldehyde is a good fungicide but is not reliable for killing nematodes or insects. A mixture of 1 gallon (3.785 liters) of commercial formalin (40 percent strength) with 50 gallons (189 liters) of water is applied to the medium at the rate of 2 to 4 quarts per square foot (20.4–40.7 liters per square meter). The treated area should be covered immediately with an airtight material for 24 hours or more. Following the treatment, about 2 weeks should be allowed for drying and airing before planting.

Chloropicrin is applied as a liquid by use of an injector, which should put 2 to 4 milliliters into holes 3 to 6 inches (8–15 cm) deep, spaced 9 to 12 inches (23–30 cm) apart, or it may be applied at the rate of 5 milliliters per cubic foot (5 ml per 0.0283 cubic meter) of medium. Chloropicrin changes to a gas which penetrates the medium. The gas should be confined by sprinkling the medium surface with water, and then covering it with an airtight material for 3 days. Seven to 10 days is required for thorough aeration of the medium before it can be planted. Chloropicrin is effective against nematodes, insects, some weed seeds, *Verticillium*, and most other resistant fungi. Chloropicrin fumes are very toxic to living plant tissue.

Methyl bromide will kill most nematodes, insects, weed seeds, and some fungi, but it will not kill *Verticillium*. It is injected at 1 to 4 pounds per 100 square feet (0.454–1.818 kg per 9.3 square meters) into an open vessel under a plastic cover placed over the medium to be treated. The cover must be kept sealed for 48 hours. Pressurized canisters having an injection tube for dispersing into the plastic-covered bed are available. Penetration is very good, extending to a depth of 12 inches (30.5 cm).

Materials are available which contain a mixture of methyl bromide and chloropicrin. These combinations are effective in controlling weeds, insects, nematodes and fungi. Aeration for 10 to 14 days is required following their application.

Vapam, a water-soluble fumigant, will kill weeds, most fungi and nematodes. It is applied as a spray to the surface of the medium through irrigation systems or with injection equipment at a rate of 1 quart (0.95 l) of Vapam in 2 to 3 gallons (7.6 to 11.4 l) of water sprinkled uniformly over 100 square feet (9.3 square meters) of area. After application, the Vapam is sealed with additional water. Two weeks after application, the area can be planted.

In gravel cultural systems, common bleach (calcium or sodium hypochlorite) or hydrochloric acid used for swimming pools can be used. A concentration of 10,000 ppm of available chlorine is made up in the sump tank and the beds are thoroughly moistened for one-half hour. The beds must then be thoroughly leached with fresh water to eliminate any chlorine before planting. Therefore, the entire sterilization process can be done in a very short time. With gravel culture, it is advisable to remove the gravel after 4 to 5 years' use and clean all decayed roots from the beds.

References

Chapman, H.D., and P.F. Pratt. 1961. *Methods of analysis for soils, plants, and waters.* Univ. of Calif. Div. of Agr. Sc.

Schwarz, M. 1968. *Guide to commercial hydroponics.* Jerusalem: Israel Univ. Press.

Smith, D.L. 1987. *Rockwool in horticulture.* London: Grower Books.

Steiner, A.A. 1968. Soilless culture. *Proc. of the 6th Coll. Int. Potash Inst.*, Florence, pp. 324–41.

Victor, R.S. 1973. Growing tomatoes using calcareous gravel and neutral gravel with high saline water in the Bahamas. *Proc. of the 3rd International Congr. on Soilless Culture.* Sassari, Italy, May 7–12, 1973, pp. 213–17.

Withrow, R.B., and A P. Withrow. 1948. *Nutriculture.* Lafayette, IN: Purdue Univ. Agr. Expt. Stn. Publ. S.C. 328.

Chapter 5

Water Culture

5.1 Introduction

Of all the soilless methods, water culture, by definition, is true hydroponics. Water culture includes aeroponics. In aeroponic systems, plant roots are suspended into a closed dark chamber in which jets of nutrient solution are periodically sprayed over them to maintain 100 percent relative humidity. In water culture, plant roots are suspended in a liquid medium (nutrient solution) while their crowns are supported in a thin layer of inert medium.The nutrient film technique (NFT) is a form of water culture in which plant roots are contained in a relatively small channel through which a thin "film" of solution passes.

For successful operation a number of plant requirements must be met:

1. Root aeration:

This may be achieved in one of two ways. First, forced aeration (by a pump or compressor) is used to bubble air into the nutrient solution through a perforated pipe or airstone placed at the bottom of the bed or in the nutrient tank. Second, the nutrient solution is circulated with a pump through the beds and back to a reservoir. A series of baffles placed at the end of the beds will aerate the water as it returns to the reservoir. A rate of about one to two complete changes per hour is required for a bed 100 feet (30.5 meters) long containing from 4 to 6 inches (10 to 15 cm) of nutrient solution. Best results can be achieved in a system in which the nutrient solution is pumped into the beds and allowed to flow past the plant roots continuously. In this way, freshly aerated solution will be in constant contact with the plant roots.

2. Root darkness:

Plants can function normally with their roots exposed to light during the daytime, provided they are always at 100 percent relative humidity. However, light will promote the growth of algae, which interferes with plant growth by competing for nutrients, reducing solution acidity, creating odors, competing for oxygen from the nutrient solution at night, and producing toxic products through its decomposition which could interfere with plant growth. To eliminate algae growth, construct beds of or cover containers with opaque materials.

3. Plant Support:

Plants may be supported by the use of a litter tray which sits above the nutrient solution as part of the bed in an early water-culture system (see Figure 5.1).

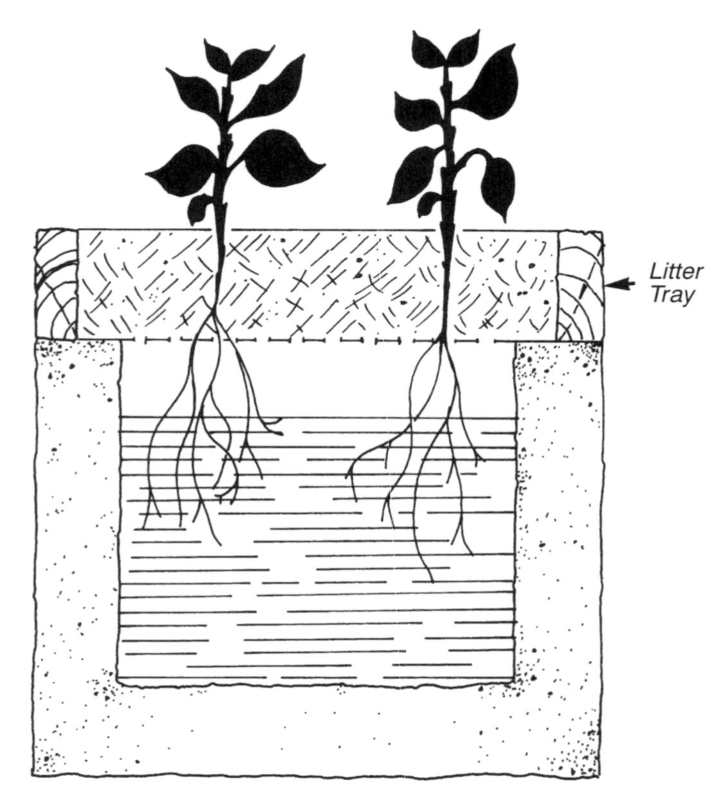

Figure 5.1. Cross section of a typical water-culture bed.

5.2 Early Commercial Methods

Water culture was one of the earliest methods of hydroponics used both in laboratory experiments and in commercial crop production. Concrete beds coated with bituminous paint were used for commercial growing. Bed widths varied from 12 to 42 inches (30.5 to 107 cm), lengths from 25 to 100 feet (7.6 to 30.5 m) and depth from 6 to 9 inches (15 to 23 cm).

A solution depth of 4 to 6 inches (10 to 15 cm) with an air space from 2 to 3 inches (5 to 7.6 cm) was generally used. On top of the bed was a litter tray for support of the plant, as illustrated in Figure 5.1.

The litter tray was a wooden framework which sat on top of the bed. The tray was usually 2 to 4 inches (5 to 10 cm) deep, with wire mesh across the bottom to support the porous material contained in the tray. Early trays used 1-inch (2.5-cm) chicken wire coated with asphalt paint to prevent zinc from being released from the galvanized wire into the nutrient solution. The type of litter material used depended upon the nature of the plant root system and whether direct seeding or transplanting was followed. Materials used were excelsior, straw, wood shavings, coarse sawdust, chaff, peat moss, sphagnum moss, dried hay or rice hulls. A finer upper layer with a coarser lower layer could be used to facilitate seed germination when direct sowing into the tray was practiced.

Today, plastic screen could be used instead of galvanized wire to overcome any zinc problems. Also styrofoam and other plastic particles could be used for the litter.

Initially, upon seeding or transplanting, the solution level was maintained at a higher level than normal within to 1 inch (1 to 2.5 cm) of the bottom of the tray but not close enough to wet the tray. As the roots elongated, the solution volume was reduced gradually until a 2- to 3-inch (5- to 7.6-cm) air space existed between the top of the solution and the bottom of the tray. An overflow pipe at the end of the bed allowed adjustment of the solution level.

5.3 Commercial Systems in Japan

At a time when Japan's total greenhouse area was 27,079 hectares, water-cultural systems were installed in four percent of the glasshouse area and one percent of the plastic house area. The plastic house area accounted for 25,795 hectares, or 95 percent. Therefore, water cultural systems amounted to 51 hectares under glass and 258 hectares under plastic, a total of slightly over 300 hectares (309), or more than 750 acres. About 2,000 greenhouses used water culture.

Typical water-culture units in such systems are a series of troughs which are 80 cm (31.5 inches) wide and about 3m (9.8 feet) long filled with nutrient solution to a depth of about 6 to 8 cm (2.4 to 3.2 inches) as shown in Figure 5.2. These troughs are made of rigid plastic and are available commercially.

Plants are inserted through holes in a styrofoam lid which caps the whole length (Figs. 5.3, 5.4). The nutrient solution is pumped around the troughs for ten minutes every hour, mainly for aeration. The troughs are always full of nutrient solution in which roots are suspended (Fig. 5.5).

Figure 5.2. Japanese water-culture system. Rigid plastic trough with styrofoam lid.

An aeration hose can be installed in the troughs to increase oxygenation. This hose has two 2-mm holes along its length every 4 cm (1.57 in.) (Fig. 5.6). The Ninomiya Agricultural Research Station doing work on such a water-culture system recommended an aeration cycle of one 15-minute period every two hours during the day and once at midnight.

Seedlings are started in perlite and transplanted into the styrofoam lids once they are large enough to support themselves, generally four to six weeks for cucumbers and tomatoes (Fig. 5.5).

The principal crops produced in water culture are tomato, Japanese honewort (*Cryptotaenia japonica*), cucumber, and salad crops.

5.4 The Ein Gedi System

The Ein Gedi System (EGS) of water culture was developed in Israel (Soffer and Levinger 1980) to overcome some of the main disadvantages of hydroponic systems and be applicable to regions having a limited

Figure 5.3. Commercial water-culture system in Japan using rigid plastic troughs and styrofoam lids.

Figure 5.4. Position of plants (melons) in styrofoam lid of Japanese water-culture system.

Figure 5.5. Roots suspended into nutrient solution of underlying plastic trough—the roots supported by plastic mesh container.

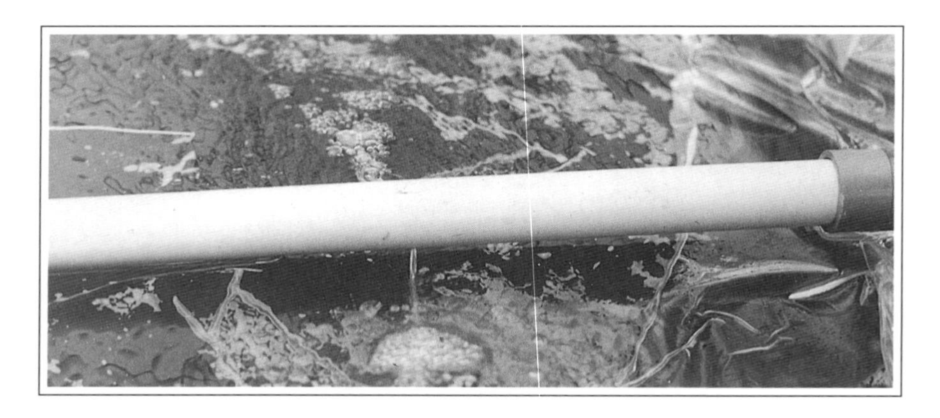

Figure 5.6. Pipe in trough for aeration of nutrient solution.

supply of water. The main disadvantages of hydroponic systems the EGS were to overcome were minimum buffering action of soilless media, small water-holding capacity of substrates such as gravel and sand, and insufficient oxygen available to plant roots, especially for full grown plants at increasing water temperatures.

The EGS channel is a self-contained unit so that it can easily be raised from the ground level to any level desired. The system having a cover made of polystyrene allows starting the crop from seeds, bulbs, plants, and rooting of cuttings. Sterilization is fast, simple and inexpensive with a minimum of down time between crops as the channels can be scrubbed clean with a disinfectant. The channel is constructed of durable reinforced double-layer polypropylene (Fig. 5.7).

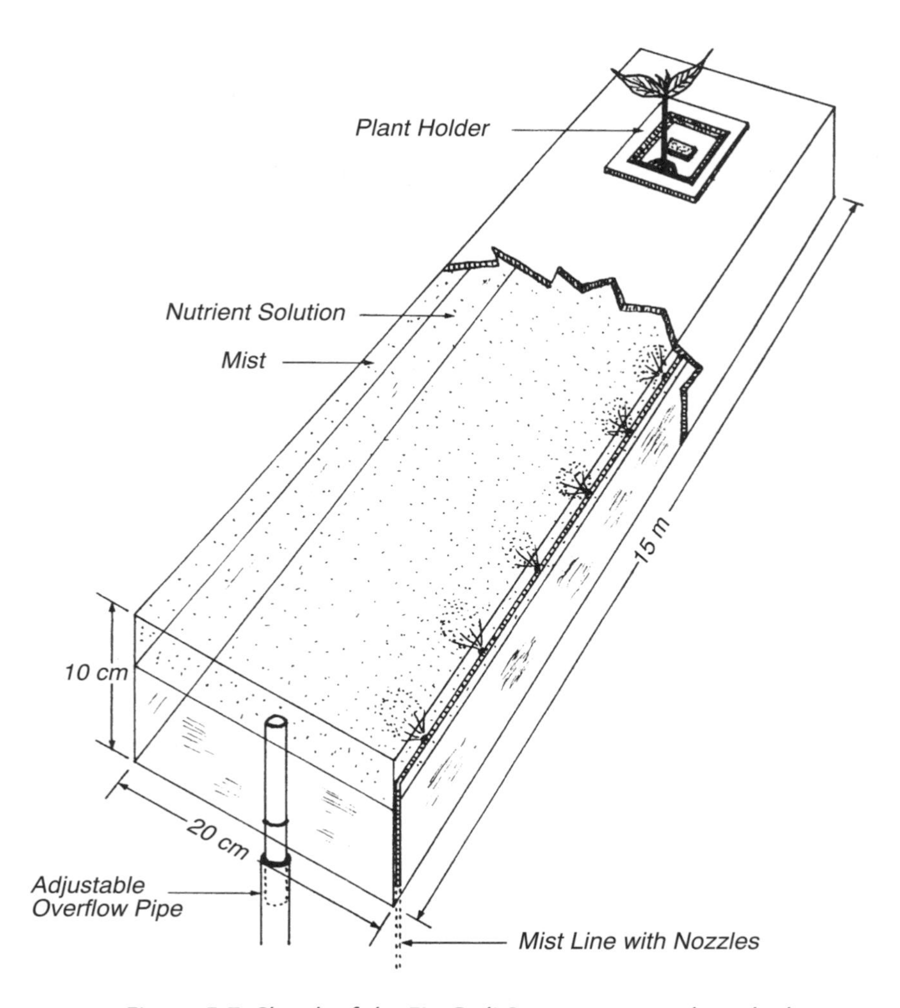

Figure 5.7. Sketch of the Ein Gedi System water-culture bed.

The EGS is a system of continually aerated flowing nutrient solution in which plant roots are immersed. A large volume of nutrient solution per plant reduces fluctuations in nutrient levels. Continued flow provides homogeneity of the solution composition and oxygen to the root surface. Aeration of the solution is achieved by spraying the solution throughout the entire growth trough. This produces a nutrient mist on top of a flowing solution. A 7-liter solution per tomato plant (2 plants/sq m or 10.76 sq ft), at a flow rate of not less than 7 liters (1.8 U.S. gallons)/plant/hour gives good buffering capacity of the root medium. This relatively large volume of solution continuously within the channel gives a safeguard against any mechanical or electrical breakdown.

In 1980, over 30 units of various sizes of this system were operating in Israel, the Netherlands, Norway, and Italy, for tomatoes, cucumbers, ornamental plants, and propagation. While some were of a semi-commercial scale, all were experimental to evaluate the economic feasibility of the system.

5.5 Raceway, Raft or Floating System

This system of water culture is a modification of the Japanese system (Section 5.3). Dr. Merle Jensen of the Environmental Research Laboratory (ERL) of the University of Arizona, Tucson, developed a prototype raceway lettuce production system during 1981–1982. He projected that such a system could produce 4.5 million heads of lettuce per year per hectare.

The system consists of relatively deep (15–20 cm or 6–8 inches) beds holding a large volume of nutrient solution. The solution in the beds is fairly static with a circulation of 2–3 liters per minute. The bed dimensions are about 60 cm wide by 20 cm deep by 30 m long (24 in. × 8 in. × 98 ft.) (Fig. 5.8). Such a bed has a volume of 3.6 cubic meters (127 cubic

Figure 5.8. Raft system of bibb lettuce. Lettuce from far right to left is 4, 3, 2, and 1 day after transplanting. Seedlings are 12–14 days old before transplanting. (Courtesy of Hoppmann Hydroponics, Waverly, Florida.)

feet) which is equivalent to 3,600 liters (about 950 U.S. gallons). Therefore, the flow through each bed at 2–3 liters per minute amounts to an exchange rate of one per 24-hour period.

The nutrient solution from the beds is recirculated through a nutrient tank of 4,000 to 5,000 liters (1,000 to 1,250 U.S. gallons). There the solution is aerated by an air pump, chilled with a refrigeration unit, and is then pumped back to the far ends of each bed. On its return to the beds, the nutrient solution is passed through an untraviolet sterilizer (Fig. 5.9). These sterilizers are manufactured by several companies (Appendix 2) for use in soft drink, beer, distillery, aquaculture, clothing dye, and cosmetic industries. They are effective against many bacteria, fungi, some viruses, and protozoa such as nematodes. However, their effectiveness against plant pathogenic organisms has not been fully documented.

Figure 5.9 Raft system of water culture. Chiller unit on left in front of acid- and stock-solution tanks with individual injectors. Circulating pumps in center and U.V. sterilizer on right. (Courtesy of Hoppmann Hydroponics, Waverly, Florida.)

Some work has been done to evaluate the effectiveness of a UV sterilizer against selected species of pathogenic and nonpathogenic fungi commonly associated with greenhouse crops (Mohyuddin 1985). The unit significantly reduced or eliminated the following fungi from an aqueous solution: *Botrytis cinerea, Cladosporium sp., Fusarium spp., Sclerotinia sclerotiorum, Verticillium albo-atrum*, and several others. Similar research is currently being carried out on fungi, bacteria and algae.

Pythium infection of roots causes stunting of plants (Fig. 5.10). Ultraviolet sterilization of nutrient solutions does not combat this disease organism. It can only be controlled by sterilization between crops of all beds, pipes, tanks, etc., with a 10-percent-bleach solution.

Figure 5.10. Pythium *infection of bibb lettuce. Healthy plant on right, infected plant on left. Note the difference in growth of the head and roots between the healthy and infected plants.*

The cost of these sterilization units varies with their capacity. The larger volume of water they can effectively treat, the larger the unit and higher the price.

One side effect of the use of UV sterilizers is their effect on a few of the micronutrients. Mohyuddin (1985) found that the boron and manganese contents in a nutrient solution were reduced by more than 20 percent over a period of 24 hours of sterilization. The most significant effect was on iron, which was precipitated as hydrous ferric oxide. Nearly 100 percent of the iron was affected. The iron precipitate coated lines and the quartz sleeve of the sterilizer, thereby reducing the UV transmission. Such precipitate could be removed with a filter. This, however, does not resolve the problem of the loss of iron from the nutrient solution during UV sterilization. Further work is needed to resolve the problem of iron precipitation by the UV sterilizer. Perhaps another form of iron chelate could be used that would not break down.

The *p*H and electrical conductivity (EC) are monitored with sensors in the return line. Automatic injection of nitric acid (HNO_3), sulfuric acid (H_2SO_4), phosphoric acid (H_3PO_4), or potassium hydroxide (KOH) are used to adjust the *p*H. Similarly, the EC is raised by injection of calcium nitrate and a mixture of the remaining nutrients from two separate stock concentrate tanks (Fig. 5.9) to keep it at 1.2–1.3 mMhos.

Jensen (personal communication) found that the use of a water chiller in the nutrient tank could be used to maintain the temperature of the nutrient solution between 70 and 75° F (21–23° C). This cooling of the solution would prevent bolting of lettuce in desert and tropical regions (Fig. 5.9). European bibb lettuce was grown commercially by Hoppmann Hydroponics, Ltd., in Florida, at temperatures exceeding 110° F (43° C) in the greenhouses by use of the floating system using a water chiller in the nutrient tank.

Water-chiller units cool, aerate and circulate the solution. They are available in ⅙-, ⅓- and 1-horsepower units. The 1-horsepower unit is capable of cooling 1,000 gallons of water in a temperature range from 35–70° F (2–21° C). For nutrient solutions, units with stainless-steel drive shaft, evaporator tube, and circulator blade should be used. These units are used in aquarium tanks for the raising of fish.

Jensen (1985) successfully grew a leaf lettuce (Waldemann's Green) and three varieties of European bibb (Ostinata, Salina, Summer Bibb). Recent communication with Jensen revealed that an operation in Norway is growing over one acre of head lettuce using the floating system.

Hoppmann Hydroponics in Florida produced bibb lettuce on a 30- to 34-day cycle in the beds after transplanting (Figs. 5.11–5.13). Lettuce can be sown using rockwool cubes, jiffy pellets, or directly in a peat mix medium in plastic seeding trays. While rockwool cubes placed in 240 compartment trays are easily sown by automatic sowing equipment, the cost of rockwool is greater than using a peat mix in 273 compartment trays, which can also be sown automatically using pelletized seed.

Figure 5.11. Lettuce 6 days after transplanting. The solution inlet pipe to the bed is in the foreground. Note the wire hook attached to the first board (raft) to allow pulling of the boards during harvest.

Figure 5.12. Lettuce 12 days after transplanting.

Figure 5.13. Lettuce 32 days after transplanting, ready for harvest. (Courtesy of Hoppmann Hydroponics, Waverly, Florida.)

Seedlings should be 12–14 days old before transplanting to the beds. The seedling trays can be bottom irrigated through the use of a capillary mat (Fig. 5.14). This prevents overhead watering, which can burn the seedlings under high solar conditions of tropical and desert regions. Seedlings should be irrigated with a dilute nutrient solution when the cotyledons unfold. The capillary mats can either be replaced or sterilized between seedling crops (12–14 days) using a 10-percent-bleach solution to eliminate algae, fungal spores, and insects such as fungus gnats. Seeding and transplanting is carried out daily so that continuous production is achieved. Late evening transplanting after sunset will assure successful "take." Plants will have time to acclimate before full-light conditions of the following day. This is especially significant if transplants are bare-rooted as is the case when the lettuce is sown in a peat medium. Bare-rooted plants are placed in paper supports which are then placed at the 1-inch (2.5-cm) diameter holes of the styrofoam "rafts" (Fig. 5.15).

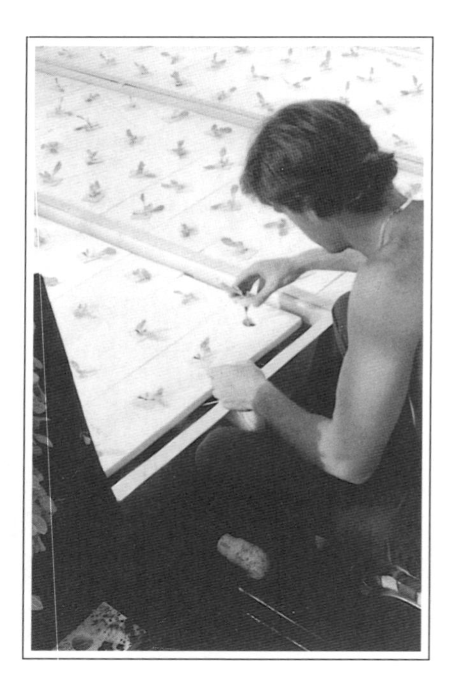

Figure 5.14. Lettuce seedlings, 10–12 days old, in the foreground, are almost ready for transplanting. Lettuce is seeded into a peat plug mix in #273 trays (273 compartments) and set on a capillary mat in beds for watering.
Figure 5.15. Lettuce seedlings being planted into 1-inch holes in "rafts." (Courtesy Hoppmann Hydroponics, Waverly, Florida.)

Figure 5.16. A "raft" supporting four lettuce plants. Note the vigorous, healthy root growth. (Courtesy of Hoppmann Hydroponics, Waverly, Florida.)

The "floats" or "rafts" are 1-inch × 6-inch × 24-inches (2.5-cm × 15-cm x 61-cm) boards of styrofoam (Fig. 5.16). The styrofoam may be ordered in specific dimensions from the manufacturer if large quantities are purchased. They will cut the boards to the required dimensions with holes for placement of the plants. A high-density "roofmate" type material (blue in color) used in house construction is most suitable.

The rafts insulate the underlying solution in the bed and are a moveable system of transplanting and harvesting. Maintenance of cool-solution temperatures (optimum 75° F or 23° C) is a prime factor in preventing bolting of lettuce in hot climates. This is achieved by the water chiller in the nutrient tank, the large volume of solution in the beds, and the insulation from solar radiation by the rafts.

The rafts simplify harvesting. A string placed under the rafts in the bed, attached to three or four wire hooks which in turn are attached in several places to the rafts along the entire bed length, can be pulled in by a boat winch at the harvesting end of the greenhouse (Figs. 5.11, 5.17). The rafts float the mature lettuce plants *in situ* within the beds. Easier flotation is possible by raising the solution level in the bed prior to harvesting, by plugging the return pipe of the nutrient-circulation system.

During transplanting, the boards are pushed along the bed from the harvesting end as the plants are placed in them (Fig. 5.15). Long lines of floats with growing lettuce are readily moved.

Figure 5.17. A boat winch used to reel in string which pulls boards along the bed. (Courtesy of Hoppmann Hydroponics, Waverly, Florida.)

Between crops, the boards must be cleaned and sterilized by hosing with water prior to dipping them in a 10-percent-bleach solution. Similarly, the beds must be drained and cleaned after each harvest (Fig. 5.18). A new nutrient solution is prepared in the bed after cleaning and is ready for transplanting the same day.

Sowing, transplanting, and harvesting must be coordinated to get a continuous daily cycle. The growing period in the beds may vary from 28 to 35 days for bibb lettuce, depending on sunlight and temperature conditions. In semi-tropical, tropical, and desert regions where sunlight is abundant and day-length averages between 14 to 16 hours, it is possible to obtain 10 to 12 crops annually, whereas in the temperate climates where sunlight is lower and day-length may be 8 hours or less during the winter months, it may only be possible to reach 7 to 8 crops annually. In temperate regions, winter crops may take 13 weeks (including 40 days from sowing seed to transplanting), whereas summer crops take 5 weeks (including 12 days from sowing to transplanting). By use of supplementary artificial lights, the sowing-transplanting period could be shortened during the winter to perhaps half the time.

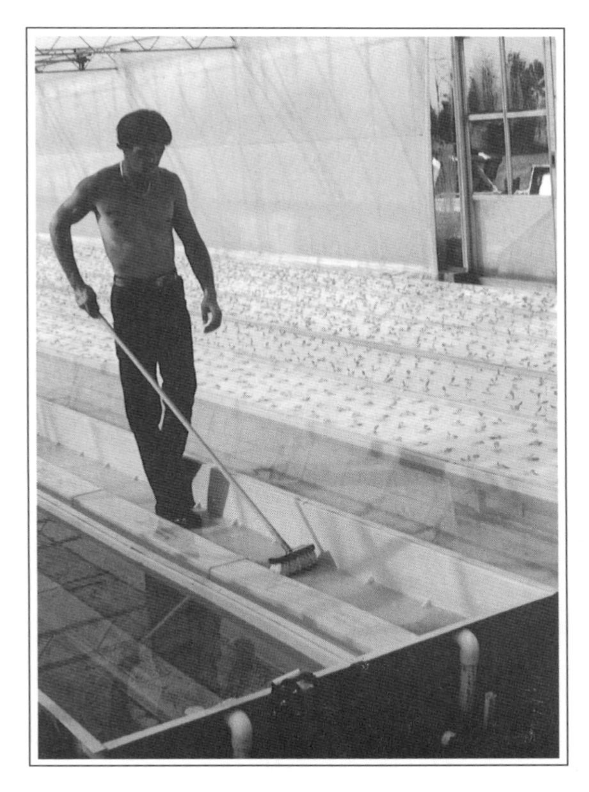

Figure 5.18. Beds are cleaned between crops with a 10-percent-bleach solution. (Courtesy of Hoppmann Hydroponics, Waverly, Florida.)

Individual heads are packaged in plastic bags. Some growers leave about one inch of roots on the plants when packaging, with the expectation of longer shelf-life than without roots. However, some consumers may not like the presence of roots. In Arizona (Collins and Jensen 1983) the roots-on package did not increase shelf-life, and it was found to be three times more expensive to prepare and pack. The roots-on packaging was not well accepted by wholesalers or retailers, and added to the cost of transportation due to increased volume and weight. Generally, 24 head are packaged to a case.

The raceway system maximizes the usage of greenhouse floor space for production of lettuce or other low profile crops. For example, a greenhouse of one acre (43,560 square feet) with dimensions of 108 feet × 403 feet (33 m × 123 m) allowing front and back aisles of 8 feet (2.4 m) and 2 feet (0.6 m) respectively, with 2-foot access aisles every 29 feet (11 sets of beds), has a useable production area of 37,000 square feet (3,439 sq m). This is 84 percent utilization of greenhouse floor area. Such a growing

Figure 5.19. Watercress growing in raft culture.

area could produce 112,100 head of lettuce per crop. That is, 2.6 head per square foot (28 head/sq m) of greenhouse area, or 3 head per square foot (32 head/sq m) of the available growing area.

The raft system has been used by Del Mar Farms near San Diego, California, to grow watercress. This farm integrated an aquaculture fish-farming operation with a hydroponic watercress system. They grew the watercress on floating styrofoam boards in beds 8 feet (2.4 meters) wide by 120 feet (37 meters) long (Fig. 5.19). Each bed was covered by a hoop house (Fig. 5.20). The polyethylene covering of the hoop houses was rolled down during cool nights to maintain optimum temperatures. The water from the fish was circulated through the watercress beds before returning it to the fish tanks. Some fertilizer was added to the watercress beds, but not sufficient to harm the fish. The solution was aerated by use of a "water wheel" in each bed as shown in Figure 5.20. The plants, therefore, acted as a filtering system utilizing the wastes from the fish.

The main advantage of the raft system is its ability to grow lettuce and cool-season crops in tropical climates by use of the chiller unit, cooling the nutrient solution. The high planting density (utilization of greenhouse floor area) is a second factor.

The principal disadvantage is a higher capital cost than conventional NFT systems.

Figure 5.20. Raft culture of watercress in a hoop house with a "water wheel" in bed (foreground) to aerate the nutrient solution.

5.6 Aeroponics

Aeroponics is the growing of plants in an opaque trough or supporting container in which their roots are suspended and bathed in a nutrient mist rather than a nutrient solution. This culture is widely used in laboratory studies in plant physiology, but not as commonly used as other methods on a commercial scale. Several Italian companies are, however, using aeroponics in the growing of numerous vegetable crops such as lettuce, cucumbers, melons and tomatoes.

Innovative work with aeroponic cultivation continues at the Environmental Research Laboratory (ERL) of the University of Arizona in Tucson. Working with Walt Disney Productions, Dr. Carl Hodges, director of the ERL and Dr. Merle Jensen, research horticulturist, developed concepts for presenting leading-edge agricultural technologies to the public in an entertaining way. The ERL helped create two attractions, the "Listen to The Land" boat cruise and the "Tomorrow's Harvest Tour," for The Land—a major facility at Epcot Center at Walt Disney World near Orlando, Florida.

Both Purdue University and the ERL are researching controlled environment life-support systems to be used in space stations. These programs sponsored by NASA are called Controlled Ecological Life Support Systems (CELSS). The University of Arizona's Environmental Research Laboratory is developing systems to support visitors to Mars. These experiments must be carried out in completely closed chambers. Closed systems of liquid hydroponics must be used so that surplus solution is recovered, replenished and recycled. Generally, this is the nutrient film technique (NFT) or modifications of it.

Other work at the ERL includes lettuce growing on styrofoam A-frames in a greenhouse. Roots fall toward an A-frame's center and are periodically misted with nutrients (Fig. 5.21). This aeroponic system increases the heads of lettuce that can be grown in a greenhouse space, and is similar to the "cascade" and moveable NFT systems that decrease unit production costs. Another way ERL scientists increase greenhouse output is to grow melons on A-frames, while lettuce in styrofoam boards is floating on water beneath the melons (Fig. 5.22).

Figure 5.21. Styrofoam A-frame with lettuce root system growing through styrofoam sheet.

Figure 5.22. Melons on A-frame with lettuce in styrofoam boards floating beneath.

(Photographs courtesy © Walt Disney Productions. World rights reserved.)

They also grew lettuce on the interior of a large drum rotating around an artificial light source. Lettuce was planted through holes in an inner drum that rotated at 50 rpm. As the inner drum rotated, the plant tops grew toward the center. The roots, sealed between the two drums, were misted regularly with nutrients (Fig. 5.23). The ERL has also experimented with a conveyor-belt plant system of hanging vertical pipes (see Chapter 6). The plants suspended in the pipes are periodically cycled past spray nozzles that mist nutrients into the pipe tops. At the same time, they pass above an open tank so that excess solution can be recycled (Figs. 6.22, 6.23).

Tomatoes are grown in A-frame mist cabinets (Fig. 5.24). While such cabinets would not be commercially feasible for growing tomatoes, cucumbers or melons, they demonstrate how aeroponics can grow plants with large, healthy root systems.

Figure 5.23. Bibb lettuce growing in rotation drum.

Figute 5.24. Healthy roots of tomatoes growing in aeroponic A-frame mist cabinet.

(Photographs courtesy © Walt Disney Productions. World rights reserved.)

With increasing fuel costs for heating greenhouses, new growing systems such as NFT and aeroponics will be rapidly adopted in the future to save energy and increase production volume.

5.7 Hydroponic Grass Units

The growing of grains with a nutrient solution within an enclosed environmentally controlled chamber or unit has become of commercial significance as a source of year-round fresh grass feed for animals.

Grains such as oats, barley, rye, wheat, sorghum or corn are pre-soaked for 24 hours prior to being placed in growing trays (about 0.5 square meters) for 6 days. The trays may be watered manually on shelves with excess nutrient solution draining to waste (Fig. 5.25), or the entire system of trays may be mounted on rotating drums which are automatically fed with a nutrient solution which is recycled. Light is provided artificially by use of cool-white fluorescent lighting. After 6 days of growth, the grain (grass) has grown to 4–5 inches (15–20 centimeters) and is ready for harvesting and feeding to the animals (Fig. 5.26).

Figure 5.25. Shelves of grass-growing trays. (Courtesy of La Serenisima, Buenos Aires.)

Figure 5.26. Six-day-old grass ready for feeding to livestock. (Courtesy of La Serenisima, Buenos Aires.)

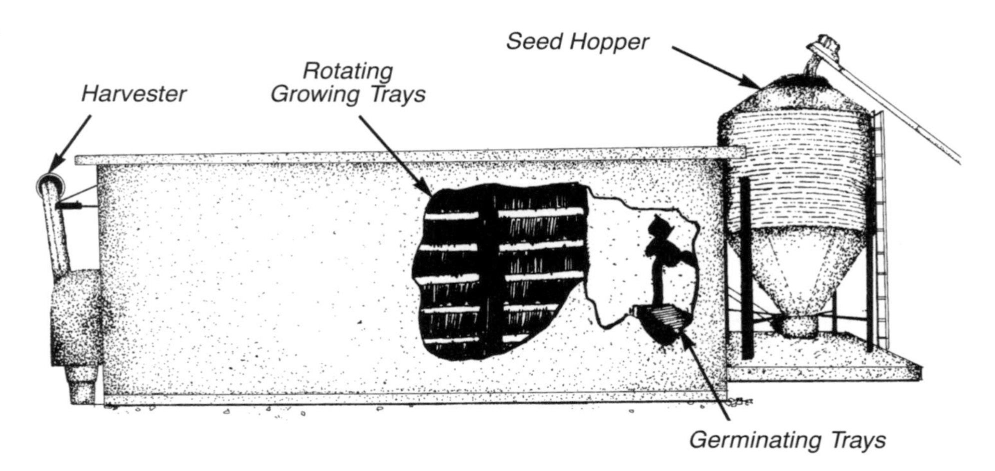

Figure 5.27. An automatic commercial grass-growing unit.

Various commercial grass-growing units are available in a number of sizes. A 20-foot long by 8-foot high by 10-foot wide (6.0 meters by 2.4 meters by 3.6 meters) unit is shown in Figure 5.27.

In this unit, a four-fold bank of six layered trays separated by about 30 centimeters (12 inches) is rotating under artificial lighting. Each layer has a set of 5 trays 0.9 meters by 0.45 meters (36 inches by 18 inches), giving a total area per layer of 2.0 square meters (22.5 square feet). Into each layer (5 trays) about 11.3 kilograms (25 pounds) of grain is placed daily. The temperature is maintained at 22° to 25° C and the relative humidity at 65–70 percent.

This unit of four banks of 30 trays is said to produce up to one-half ton (450 kilograms) per day of fresh green grass from 100 pounds (45 kilograms) of grain. The grass is fed to animals in its entirety—roots, seeds and green foliage.

The cropping schedule is set up so that one series of trays is harvested each day and at the same time one series is also seeded. In this way, continuous production 365 days a year is possible.

It is stated (Arano 1976) that each kilogram (2.2 pounds) of grass is equivalent nutritionally to 3 kilograms (6.6 pounds) of fresh alfalfa. Arano also states that 16 to 18 kilograms (35–40 pounds) of grass is sufficient as the daily food requirement for one cow in milk production.

He speculates that a standard grass-growing unit of 6 separate tiers each having 40 trays could feed 80 cows all year. In a test of milk production with a diet of grass versus one of normal feed (such as grain, hay and silage), the group of 60 cows on the grass diet increased their milk production by 10.07 percent over those on the normal diet. In addition, the group fed on grass produced a butterfat content of 14.26 percent higher than those fed the regular diet.

The grass-growing units have proven to be beneficial to other animals besides dairy cows. Race horses fed on grass performed better, and zoo animals which are accustomed to a grass diet in their normal habitat were more healthy in confinement when fed fresh grass year round.

Evidence is given (Arano 1976) that the hydroponic grass units produce animal feed at about one-half the cost of that produced conventionally. This is based upon the larger amounts of fuel needed in the production and transportation of traditional animal feeds. The grass-growing units enable animal producers to grow feed year round *in situ*. No storage of hay or silage is necessary since fresh grass is produced daily. This grass can be grown in a very small area compared to field-grown grasses and feeds. Costs of insecticides, fertilizers, machinery for cultivation and harvesting, and labor of field-grown feeds are estimated to be at least 10 times greater than that of hydroponically grown grass.

References

Arano, C.A. 1976. Raciones hidroponicas. Buenos Aires, Argentina: *La Serenisima* 29:13–19.

———. 1976. Cultivos hidroponicos. *La Serenisima* 31:4–19.

———. 1976. Forraje verde hidroponico (FVH). *La Serenisima* 35:19.

Jensen, Merle H. 1980. Tomorrow's agriculture today. *Am. Veg. Grower* 28(3):16–19, 62, 63.

Jensen, Merle H. and W.L. Collins. 1985. Hydroponic vegetable production. *Hort. Reviews* 7:483–557.

Mohyuddin, Mirza. 1985. Crop cultivars and disease control. *Hydroponics Worldwide: State of the Art in Soilless Crop Production,* Ed., Adam J. Savage. Honolulu, HI: Int. Center for Special Studies, pp 42–50.

Morgan, J.V. and A. Tan. 1982. Production of greenhouse lettuce at high densities in hydroponics. P.1620 (abstr.). *Proc. of the 21st Inter. Hort. Cong.*, Hamburg, Vol. 1.

Resh, H.M. 1988. Water Culture Systems. *Proc. of the 9th Ann. Conf. of the Hydroponic Society of America*, San Francisco, CA, April 22–24, 1988, pp. 47–51.

Soffer, H. and D. Levinger. 1980. The Ein Gedi System—Research and development of a hydroponic system. *Proc. of the 5th Int. Congress on Soilless Culture*, Wageningen, May 1980, pp. 241–252.

Vincenzoni, A. 1976. La colonna di coltura nuova tecnica aeroponica. *Proc. of the 4th Int. Congress on Soilless Culture,* Las Palmas, Oct. 25–Nov. 1, 1976.

Zobel, R.W., P.D. Tridici and J.G. Torrey. 1976. Method for growing plants aeroponically. *Plant Physiol.* 57:344–46.

Chapter 6

Nutrient Film Technique

6.1 Introduction

The Nutrient Film Technique (NFT) is a water-cultural technique in which plants are grown with their root systems contained in a plastic film (trough) through which nutrient solution is continuously circulated.

Work on NFT cropping was pioneered by Allen Cooper at the Glasshouse Crops Research Institute in Littlehampton, England, in 1965. The term *nutrient film technique* was coined at the Glasshouse Crops Research Institute to stress that the depth of liquid flowing past the roots of the plants should be very shallow in order to ensure that sufficient oxygen would be supplied to the plant roots. Other workers (Schippers 1977) call it *nutrient flow technique* since the nutrient solution is continuously circulated.

6.2 Early NFT System

In the earliest NFT system, a catchment trench was dug across the middle of the greenhouse floor and the ground sloped on either side toward this trench. The least acceptable slope was about 1 in 100. A steeper gradient, up to 1 in 25, was favored to reduce the effect of localized depressions. The trench was lined with an expanded-polystyrene sheet and polyethylene film. The "layflat" polyethylene troughs were placed on strips of hardboard 8 inches (20 cm) wide at the normal spacing of plant rows sloping toward the central catchment trench from each side (Fig. 6.1).

Fourteen-inch (35.5-cm) layflat was prepared by punching out holes along one edge at the within-row plant spacing, as shown in Figure 6.2-A. The line of holes was then relocated centrally along the layflat as shown in plan view (Fig. 6.2-B) and cross section (Fig. 6.2-C). The edges of the layflat were then turned upward and into the center and secured together with a small piece of PVC tape at 10-foot (3-meter) intervals (Fig. 6.2-D).

Small slits were cut in the upper surface of the upper folds to prevent possible ethylene buildup in the layflat which would cause premature

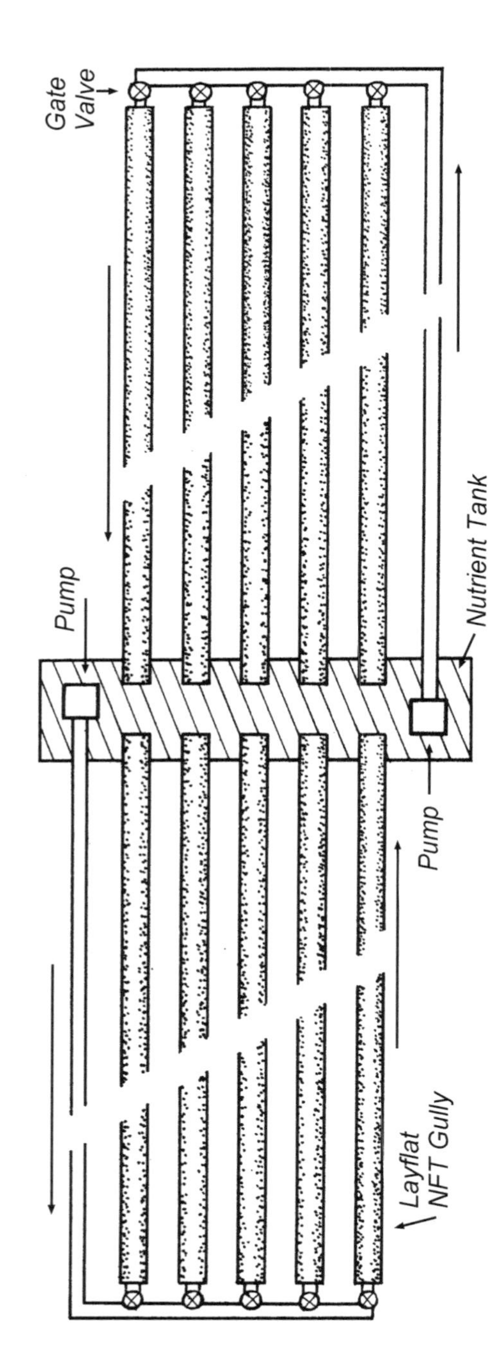

Figure 6.1. Plan of NFT system in a greenhouse. (Adapted from The Grower, *London.)*

root senescence. The lower end of each line of layflat hung down into the catchment trench, while the upper end of each line was turned upward and over and sealed with PVC tape to prevent the loss of nutrient solution.

Nutrient solution in the catchment trench was circulated to the upper ends of each layflat by two small submersible pumps and ABS plastic piping, as shown in Fig.6.1.

Gate valves installed at the inlets to each layflat line regulated the flow of nutrient solution evenly into each line. The roots of the plants were inserted into the layflat through the planting holes shown in Figure 6.2. The upper parts of the plants were supported in the normal way by tie strings suspended from support cables in the greenhouse.

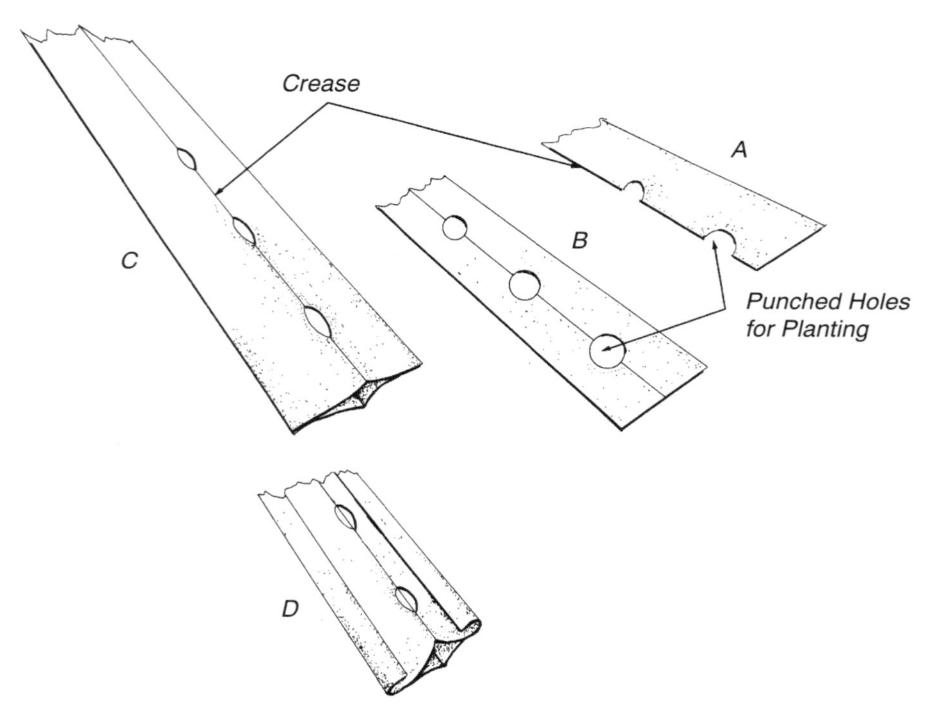

Figure 6.2. Details of preparation of "layflat" polyethylene into a NFT trough. (Adapted from The Grower, *London.)*

6.3 Later NFT Systems

Although good crops were obtained from the earlier NFT system technique, experience showed that ethylene buildup inside the layflat caused root damage which led to blossom-end-rot fruit in tomatoes and reduced yields. This buildup of ethylene necessitated a modification of the technique to improve ventilation.

The layflat was replaced by long narrow sheets of black polyethylene film laid on the ground, each corresponding to a row of plants. Peat pots, peat pellets (Jiffy-7s) or rockwool cubes were placed on the polyethylene at each plant position in the row. The edges of the polyethylene were then turned up and around the sides of these growing pots or cubes, and stapled between every other pot (or cube) to form a gully through which a thin stream of nutrient solution flowed, as shown in Figure 6.3.

Plants were seeded directly into the pots or cubes, and once they had grown to several true leaves, they were placed, with their containers, directly onto the polyethylene at which time the gully was formed as described above.

The use of pots or cubes serves a number of purposes. First, they sup-

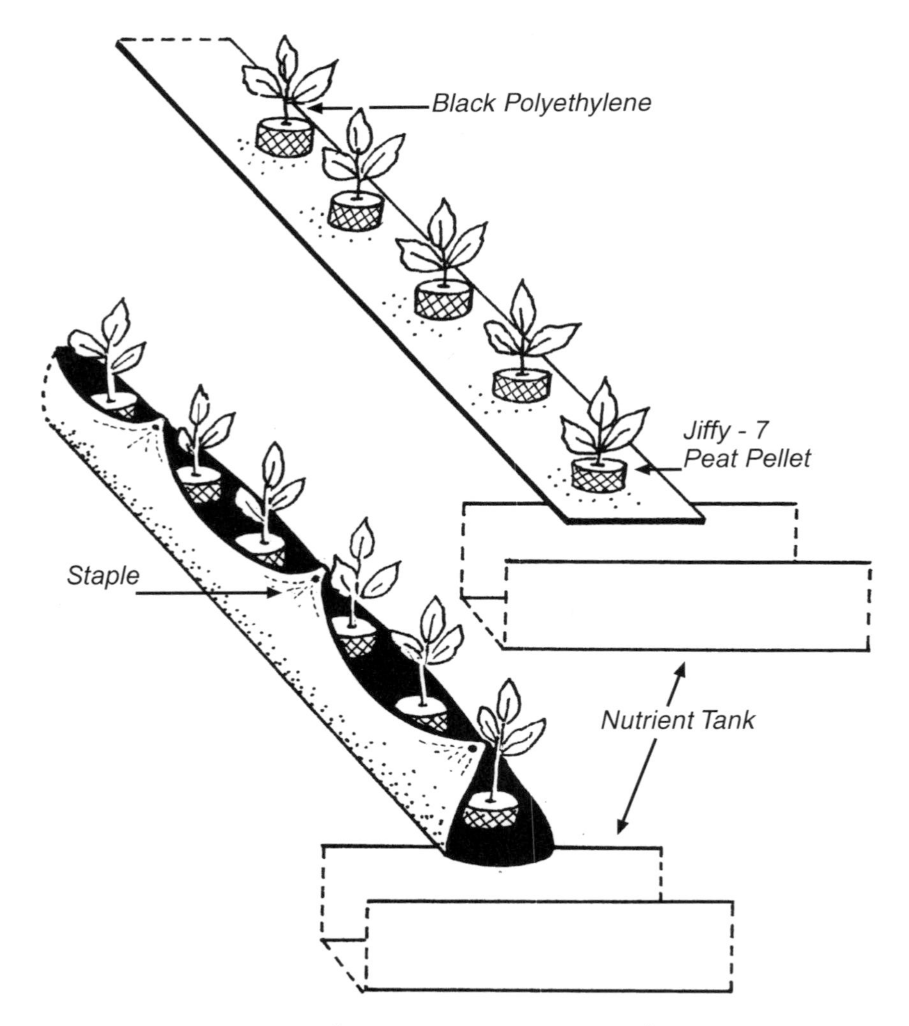

Figure 6.3. NFT gully using growing pots or cubes to support plants and the gully. (Adapted from The Grower*, London.)*

port the polyethylene film so that when the edges of the film are stapled together, a gully is formed. Second, support of the film by the pots or cubes keeps the film apart and therefore allows good ventilation of air within the gully, thus reducing ethylene buildup. Third, they support the young seedling or cutting in the early stages of growth.

The technique can be extended to pot-plant production, enabling such plants to be automatically watered and fed in a very simple, inexpensive system.

In a greenhouse, a series of gullies are laid out on a slope to a catchment tank as shown in Figure 6.4. As in the early NFT system, a submersible pump is placed in the bottom of the tank and ABS or PVC piping is used to conduct the nutrient solution being constantly pumped to the upper ends of the gullies. By suspending the flow pipe several inches (5 cm) above the gullies and drilling small holes above each gully, the nutrient solution enters each gully. This facilitates checking for blocked outlets and aerates the solution as it falls through the air. The top end of each gully is sealed by rolling up tightly a few inches (5 cm) of the polyethylene and packing some soil under the roll.

The lower end of the gully is made into a chute which projects out over

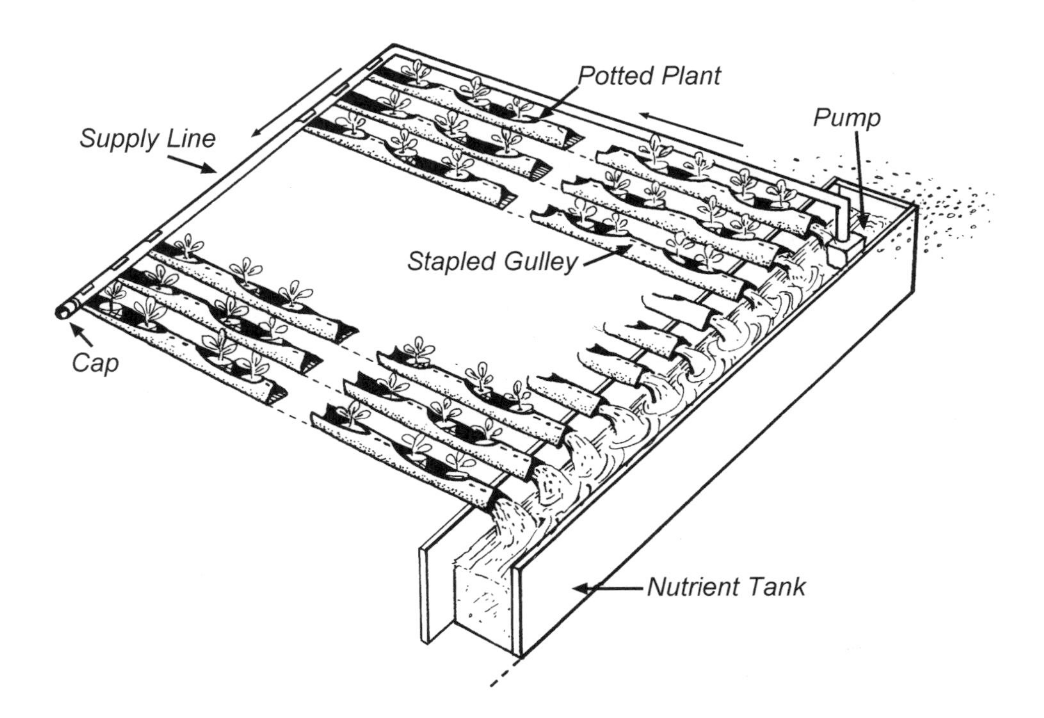

Figure 6.4. Layout of a series of NFT gullies and nutrient solution tank. (Adapted from The Grower, *London.)*

the trench by stapling the top edges of the polyethylene together every few inches (5 cm) for the last 18 inches (46 cm). In this way, rapid discharge aerates the solution as it falls into the trench. A constant volume of water is maintained in the trench by a plastic float-valve.

Roots quickly grow through the pots (or cubes) and spread out in the shallow stream of nutrient solution flowing down the gullies. They merge to form a single, thick, continuous mat in the bottom of the gullies which provides support for large plants.

A further modification of this NFT system is by use of a supporting wire to hold the gully. The ground must be prepared with a slope of 1 in 25 and be evenly smooth so that there are no localized depressions. At both ends of each plant row, a wooden peg is driven into the ground until it is projecting approximately 4 inches (10 cm). Between these two pegs, two strands of fine wire are stretched along the length of the row. A 6-inch (15-cm) length of steel rod is inserted between two strands of wire in the center of the row to serve as a turnbuckle to adjust the wire until it is taut.

In long rows, intermediate pegs can be driven into the ground to provide extra support. An 18-inch (46-cm) wide strip of black polyethylene film is laid on the ground along the length of the row. The edges of the polyethylene are attached to the wire with clips to form the gully down which the shallow stream of nutrient solution flows, as illustrated in Figure 6.5.

To supply nutrient solution to the gullies, a plastic pipe is supported just above the higher inlet-ends of the gullies at right angles. In this pipe, a 1/16-inch (1.6-mm) hole is drilled above each gully in such a position that the nutrient solution is discharged into each gully (Fig. 6.6).

The outlet ends of the gullies hang down into a catchment gutter which conducts the solution back to the nutrient reservoir or directly into the reservoir, depending on the size of the operation. The catchment gutters and reservoir must be covered with an opaque material such as black polyethylene in order to exclude light and prevent the growth of algae. The outlet ends of the gullies are inserted through slits in the polyethylene cover.

By use of support wires and clips, the gully walls can be kept closed at the top to prevent light from adversely affecting plant roots. At the same time, there is adequate space above the plant roots to provide ventilation for the escape of ethylene, and for gas exchange.

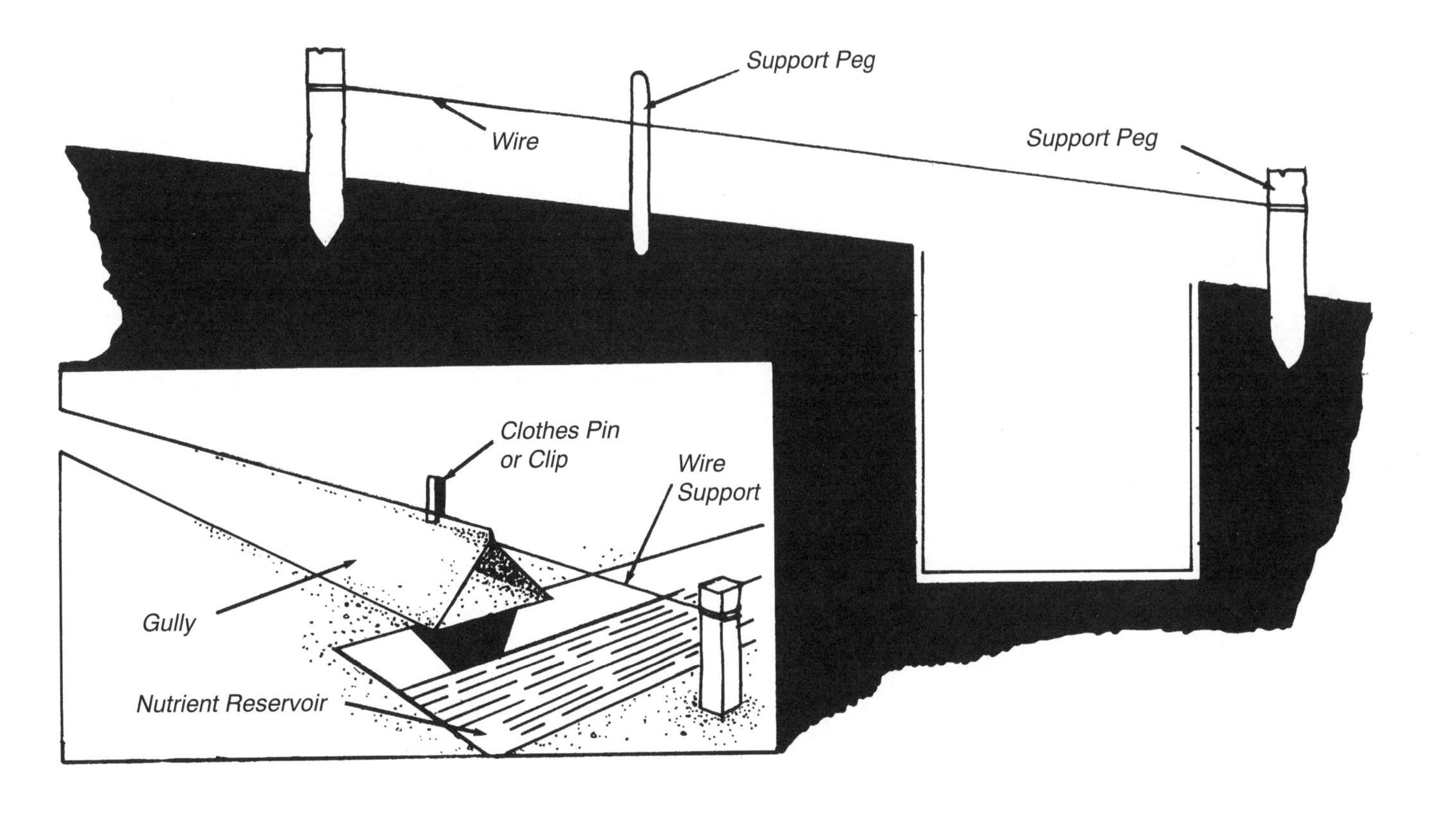

Figure 6.5. Wire-supported NFT gully. (Adapted from The Grower, *London.)*

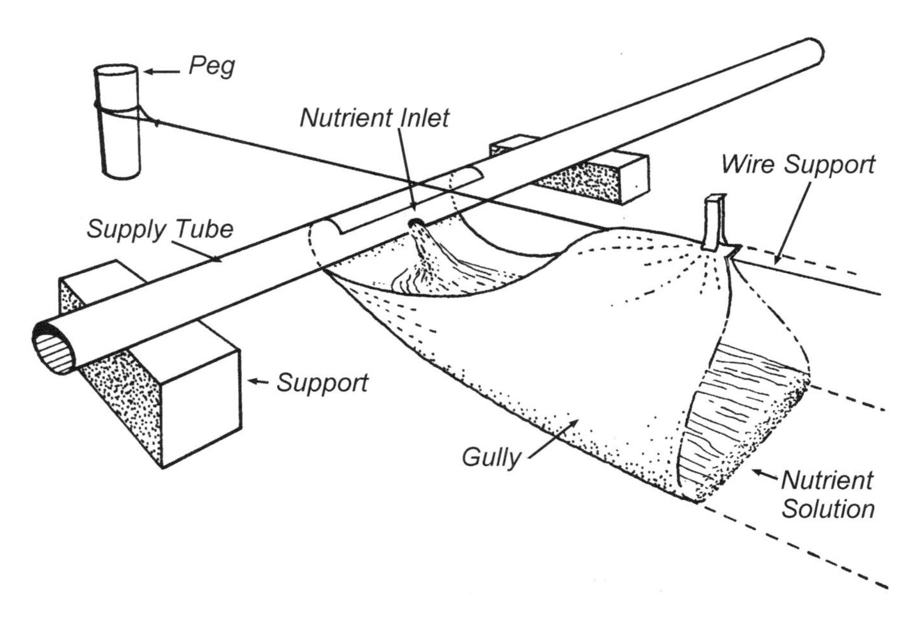

Figure 6.6. Inlet supply of nutrient solution to each gully. (Courtesy of the Glasshouse Crops Research Institute, Littlehampton, England.)

6.4 Other NFT Modifications

Other workers (Resh 1976, Adamson *) have carried out experiments, independently, in studies attempting to solve the problem of inadequate root oxygenation. The greatest problem facing the NFT systems has been root dieback caused by inadequate oxygen in the nutrient solution immediately around the roots. This root dieback results in water stress within the plant, which causes wilting and blossom-end rot (a sinking-in of the fruit at the blossom end with a resultant leathery depression) of fruit in crops such as tomatoes. The problem is associated with large root mats developing in the bottom of the NFT canal (gully), which impedes the flow of water, especially at the interface between the gully and plant roots. This is termed "puddling." This stagnation of water allows plant roots to use up all the available oxygen in that area, and with slow exchange of water a shortage of oxygen develops.

The solution to this poor root aeration was believed to be in providing a supporting medium in the gully through which roots could grow laterally without concentrating on the root-gully interface. A number of fibrous mats were tested with reasonable success (Resh 1976). However, the most suitable type of matting is difficult to obtain commercially. The matting material must satisfy a number of criteria: It must be readily avail-

* R.M. Adamson 1976; personal communication.

able in large quantities and inexpensive; it must be dense and fairly rigid to give support to the plant roots and prevent their rapid growth downward; it must retain sufficient moisture to prevent root desiccation, yet not hold excessive gravitational water or puddling will occur. Several matting materials were tested, but none adequately satisfied the above criteria. Some of the materials were too porous and roots still grew along the bottom of the matting at the gully-mat interface. All of them were too expensive to justify their use commercially since they must be disposed of at the end of each crop. Also the cost of labor for installation was prohibitive. The aims of the NFT systems are low cost and simplicity.

While the use of soaker hoses on the top of the matting for even water distribution makes the NFT system too expensive and complicated for commercial use, spaghetti feeders to each plant is a practical approach.

A NFT canal with a false bottom or collecting gully underneath may be the answer to the problem. In this way water would move vertically past the plant root instead of laterally, assuring good root aeration (Fig. 6.7).

Such a system has been developed on an experimental scale in Denmark.* The Danish system is constructed of rigid ABS plastic having a molded nutrient feed line, CO_2 enrichment line, growing conveyor belt, collecting trough and solution heating tubes, as shown in Fig. 6.8.

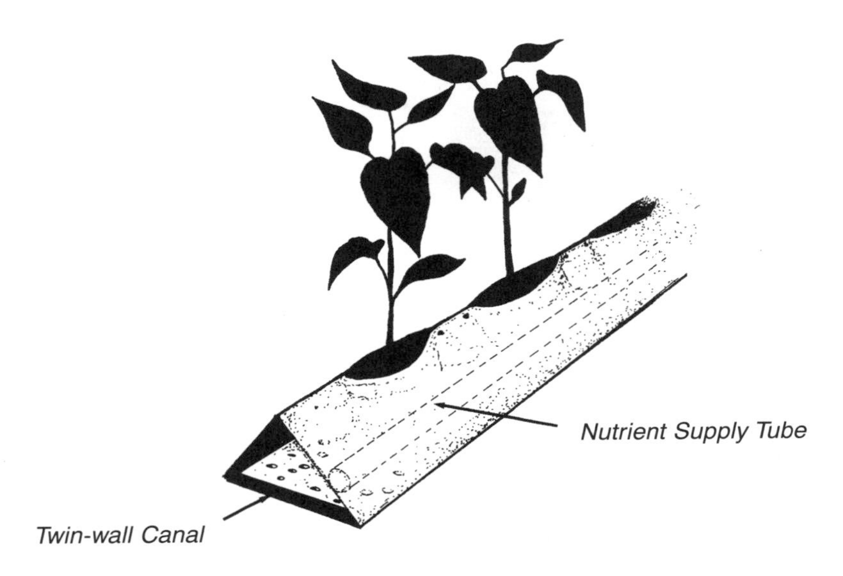

Figure 6.7. A twin-wall NFT canal system.

* Bent Vestergaard 1976: personal communication.

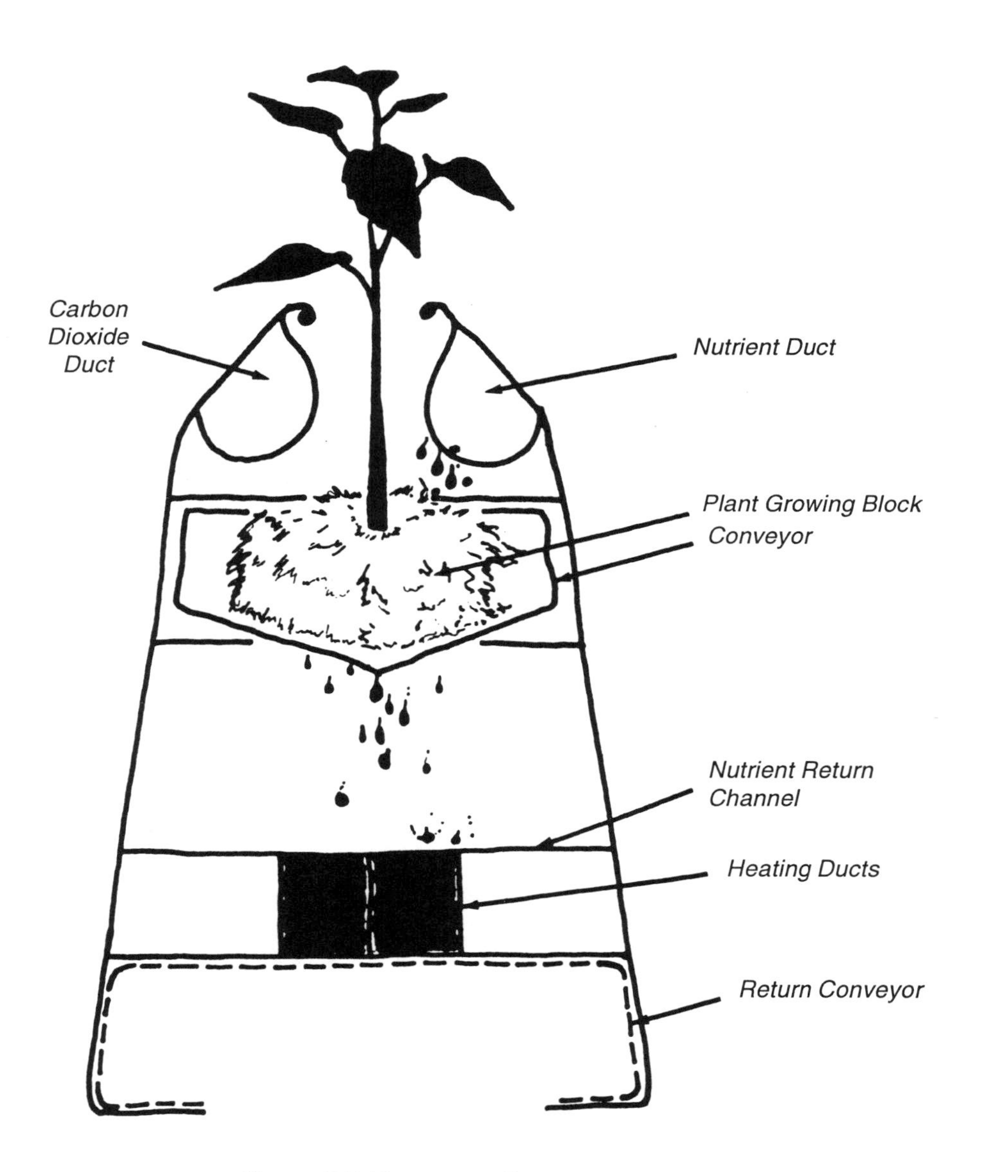

Figure 6.8. Vestergaard's conveyor NFT system. (Scale 3×)

6.5 Commercial NFT Systems

During the late 1970s, NFT cropping was being tested, or was in commercial use, in over 68 countries (Cooper 1976). Today, various modifications of the NFT system are operating throughout the world. The NFT systems are producing greenhouse vegetables and some flower crops such as roses. In 1991, 100 acres (40 hectares) of tomatoes in the United Kingdom were grown with NFT.

One of the leaders in the manufacture of commercial NFT systems in the late 1970s was Soil-less Cultivation Systems Ltd. of Aldershot, England. They worked together with Allen Cooper in the design and development of their "Hydrocanal." Their Hydrocanal system is marketed as a complete package with all the growing requirements. The other components are the "Nutrient Feed System," comprised of pumps, valves, filters, tanks, a range of flexible ducting, monitoring and dosage equipment, and prepackaged nutrients. All components are designed to form an integrated whole, balanced precisely to give optimum results and minimum operational difficulties.

The Hydrocanal is extruded polyethylene, which when unreeled forms a self-supporting, continuous growing canal. It is available in reel lengths of 25, 50 and 100 meters (82, 164 and 328 feet) and widths of 7½ centimeters (3 inches), 15 centimeters (6 inches) and 22½ centimeters (9 inches). The life span of the Hydrocanal has been projected to 5 to 6 years under climatic conditions similar to those of England, while in warmer countries 3 to 4 years are expected. The Hydrocanal is supplied prepunched with small holes in its outer edges to facilitate fastening studs at 10-cm or 4-in. spacings. These fastening studs incorporate hook rings for attaching wire strings to plants. The Hydrocanal fastener studs ensure a continuous opening along the canal, thus enabling good air exchange to plant roots. Secondly, they protect young seedlings against abrasion from the top edge of the canal and they are used for attachment of plant-support strings.

The site should be prepared similarly to that described in earlier NFT systems above. A long and narrow rectangular land area is ideally suited to the Hydrocanal system. Ideally there should be a two-way fall on the land with the greatest fall along the length of the rectangle and a slight fall across the width. In this way, one corner of the rectangle will be lower than the other three. A nutrient-tank cistern is located in the lowest corner. The size of the tank is determined by the area of the rectangle on which the hydrocanal system is placed—the larger the area, the greater the volume of the cistern.

The slope along the length of the rectangle is graded so that there are no lengthwise localized depressions. The steeper the slope, the less care is needed in grading. However, with a minimum gradient of 1 in 100 there must be virtually no localized depressions.

The Hydrocanals are laid on the ground along the length of the rectangle. The cistern (catchment tank) is filled with nutrient solution which is pumped through flexible plastic pipe up to the top edge of the rectangle. Solution is discharged from this flow pipe into the gullies through small flexible discharge pipes (Fig. 6.9). The solution flows down the gullies by gravity. At the ends of the gullies the solution is discharged into a large-diameter flexible-plastic catchment pipe (Fig. 6.10), down which it flows by gravity back to the catchment tank. A layout of an NFT system is shown in Figure 6.11.

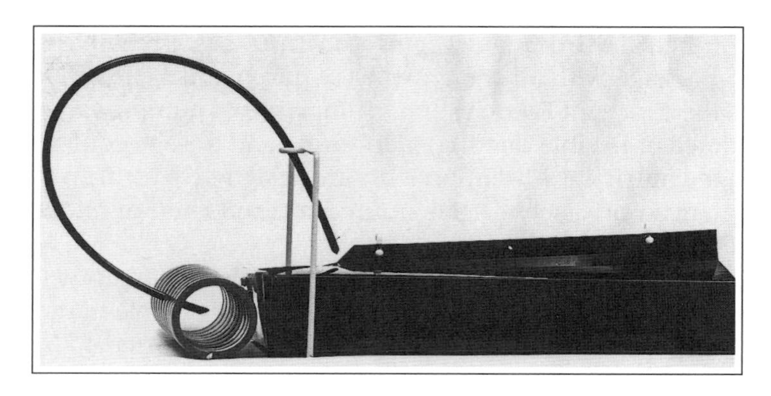

Figure 6.9. Supply end of the Soil-less Cultivation System (SCS)Hydrocanal. (Courtesy of Soil-less Cultivation Systems Ltd., Aldershot, England.)

Figure 6.10. Catchment end of the SCS Hydrocanal. (Courtesy of Soil-less Cultivation Systems Ltd., Aldershot, England.)

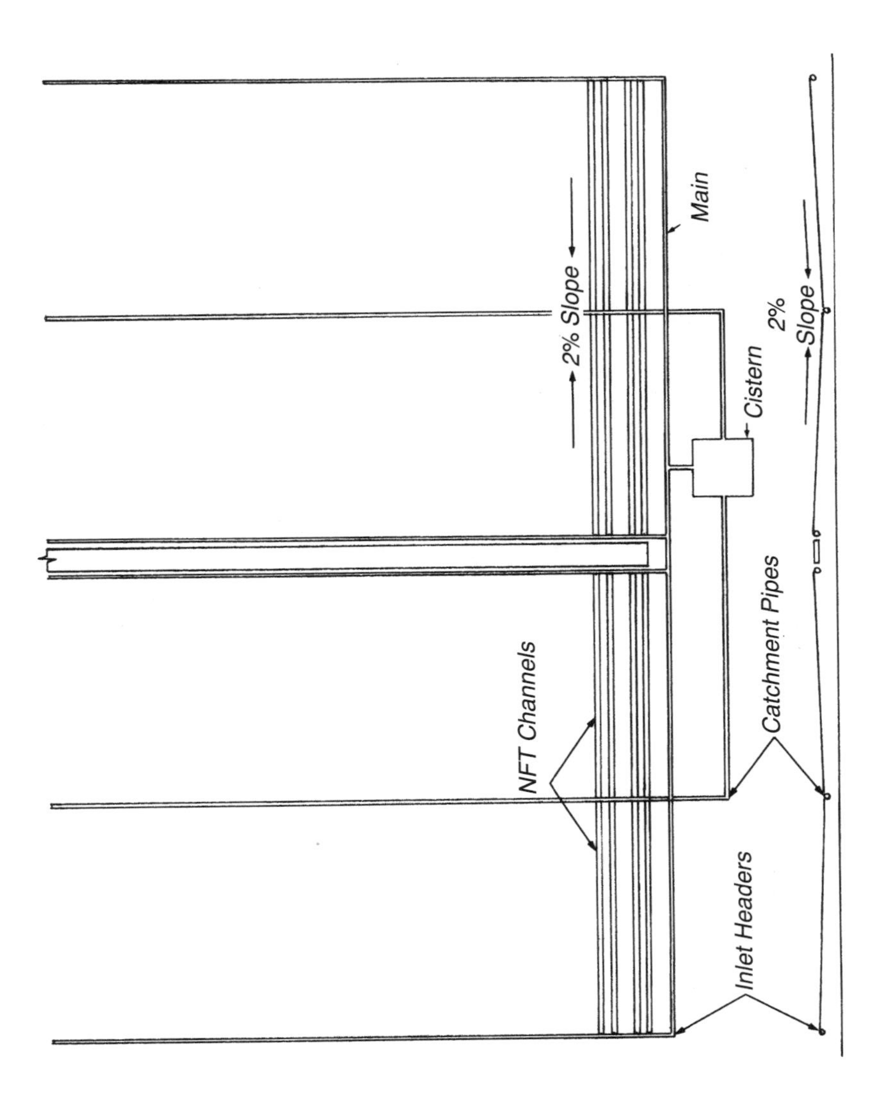

Figure 6.11. Layout of an NFT system.

The film of nutrient solution must not rise above the upper surface of the plant root mat. This ensures that all the roots are moist but that the upper surface of the root mat is in the air. Thus, no matter how long the gully is, there should be no shortage of oxygen supply to the roots even at the end of the row (Fig. 6.12).

A capillary matting is placed along the bottom of the Hydrocanal, as shown in Figure 6.13. It provides even distribution of the nutrient solution, encourages oxygenation by complete surface coverage of roots, secures roots as they grow out of the growing containers and ensures water retention for some time if a pump fails (Fig. 6.14).

Figure 6.12. The Hydrocanal inlet end with a dense, healthy root mat receiving adequate oxygen. (Courtesy of Glasshouse Crops Research Institute, Littlehampton, England.)

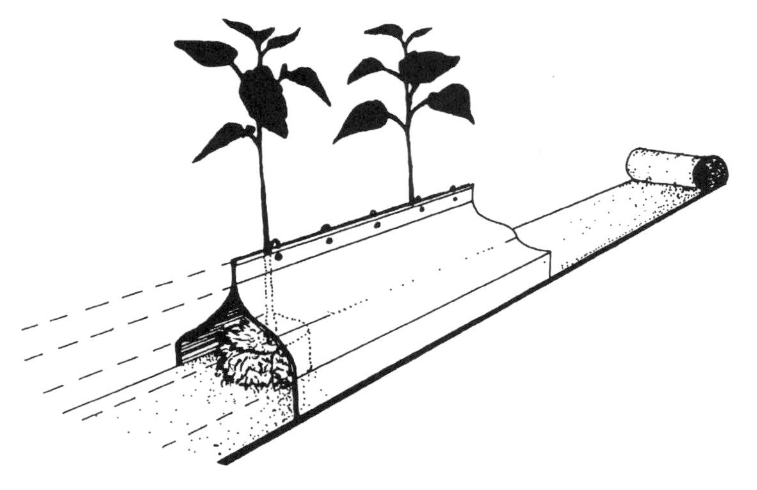

Figure 6.13. The Hydrocanal capillary-matting system.

Figure 6.14. Hydrocanal capillary matting with plant roots growing from Jiffy-7 peat pellet into matting. (Courtesy of Glasshouse Crops Research Institute, Littlehampton, England.)

Various systems have been designed for ground level and elevated positions, with or without gradient support devices according to natural level, as shown in Fig. 6.15. Sheet galvanized metal ducting supports the nutrient-fluid canal, and gradients can be adjusted to provide the 1 in 100 minimum operating slope.

Soil-less Cultivation Systems' Hydrocanal systems have been introduced to growers in many parts of the United Kingdom. Installations range from small-scale experimental systems to relatively large commercial schemes. In the United Kingdom and Channel Islands in 1978 there were 4 hectares (about 10 acres) using this technique, mainly in trial units but there was one 0.3 hectare (about ¾ acre) unit and four 0.2 hectares (about ½ acre), as shown in Figures 6.16 to 6.19. A diversity of crops, foliage plants and shrubs may be grown with the Hydrocanal system (Figs. 6.20 and 6.21).

The prices of Hydrocanal and Nutrient Feed Systems depend upon the type of Hydrocanal used (long-life or disposable), and the size of the system. In general, the costs—especially for small areas—are relatively expensive.

Conventional NFT channels are now generally formed by use of white-on-black polyethylene of 6- to 10-mil (0.15–0.25 mm) thickness. The thicker polyethylene gives a smoother, more even base for more uniform spreading of solution. The film is placed with the white side facing outwards to reflect light and reduce solar insolation. The sides are raised and supported by stapling together between the plants to form a gully, as described for the Hydrocanal system. The width of the channel should be from 15 to 30 centimeters (6 to 12 inches), depending on the crop grown. Generally, the narrower channels are suitable for smaller plants such as lettuce, while the wider channels are suitable for tomatoes and cucumbers, to avoid thick root mats developing in the channel, which may impede the flow of nutrient solution. The nutrient solution should flow at a rate of 1–2 liters (¼–½ gallon) per minute. Generally, with a slope of 1 in 50, the length of the channels should not exceed 20–25 meters (65–80 feet) or nutrient gradients may become evident, and solution oxygenation reduced. While it is possible to introduce the solution into the channels at several points along their length with longer channels, risk of oxygen deficit and temperature buildup in the solution exists.

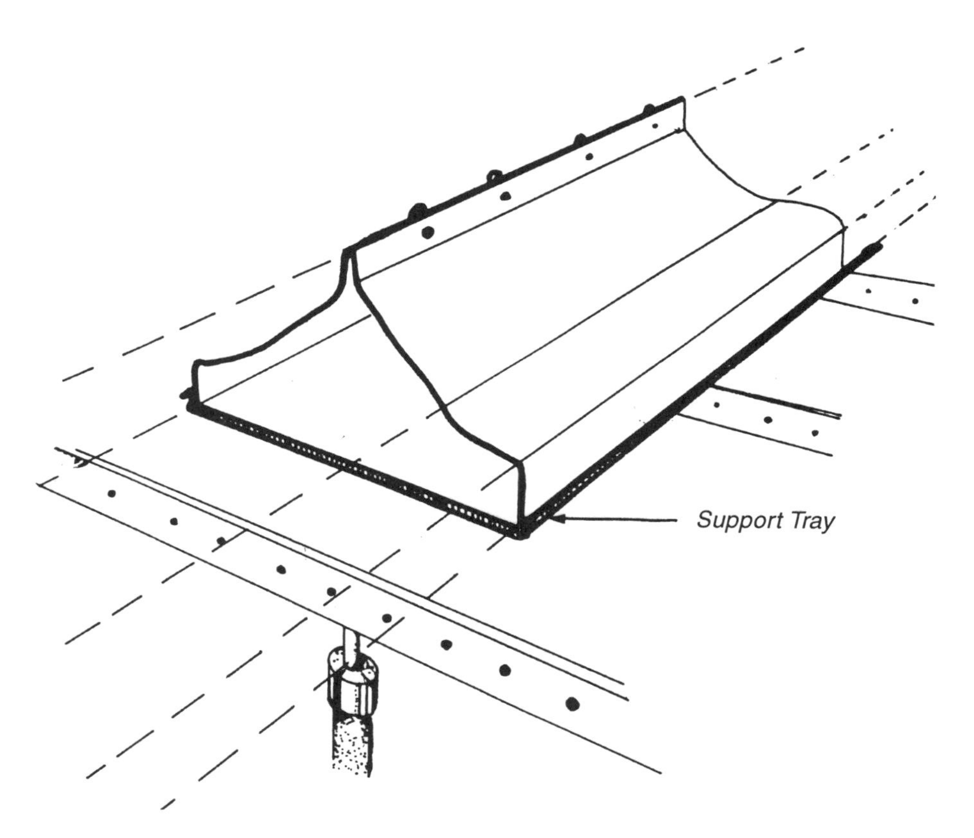

Figure 6.15. The Hydrocanal elevated ground system which offers an autogradient support device.

Figure 6.16 and Figure 6.17. Hydrocanal Basic Ground System 100. (Courtesy of Soil-less Cultivation Systems Ltd., Aldershot, England.)

Figure 6.18. Catchment end of the Hydrocanal with tomato plants.
Figure 6.19. Supply end of the Hydrocanal with tomato plants.
(Courtesy of Glasshouse Crops Research Institute, Littlehampton, England.)

Figure 6.20. A wide range of vegetable and ornamental crops may be grown in the NFT system.

Figure 6.21. Unrooted cuttings of shrubs and foliage plants have been rooted and grown successfully using the NFT.

(Courtesy of Glasshouse Crops Research Institute, Littlehampton, England.)

To prevent plants from drying out during transplanting, a narrow strip of capillary matting 5 centimeters (2 in) wide should be placed across the width of the channel directly under the transplant. This ensures that the flow of nutrient solution, as it flows down the channel, does not bypass the plant.

Growers have become increasingly aware of the need to maintain relatively constant temperatures at the base of the plants, especially tomatoes, cucumbers and peppers. The use of 1-inch thick styrofoam boards underneath the black-on-white polyethylene NFT channel is becoming a standard practice to prevent heat loss from the bottom of the channel, as shown in Figure 6.22.

Figure 6.22. Use of styrofoam boards under NFT channels. Note the heating duct bringing hot air to the base of the tomato plants. (Courtesy of R&P Hydrofarm, McMinnville, Oregon.)

Tomatoes, which are lowered during their continued growth, need their stems supported above the growing channel. Several methods are being used to keep the stems about 12 inches (30 cm) above the channels. One is by use of heavy-gauge wire and another is 1-inch PVC piping supports (Fig. 6.23). Cross members are located near each plant so their centers are the same as those of the plant spacing.

The PVC frame also supports heating ducts (Figs. 6.22 and 6.23) and with the use of wire hooks, the upper seam of the NFT channels (Fig. 6.24). Two NFT channels are placed next to each other with the stem support

Figure 6.23. Tomato stems are supported above the NFT channels by a 1-inch diameter PVC piping frame. These cross members also support the heating tubes. (Courtesy of R&P Hydrofarm, McMinnville, Oregon.)

Figure 6.24. Wire hooks attached to pipe frame hold the upper seam of the NFT channel. (Courtesy of R&P Hydrofarm, McMinnville, Oregon.)

frame spanning both channels. Nutrient solution enters the NFT channel on the higher end via a 1-inch PVC line and exits into a 4-inch PVC catchment pipe which conducts it back to a 2,000-U.S. gallon (7,570-liter) tank (Figs. 6.22 and 6.25).

Figure 6.25. Outlet ends of NFT channels exit into 4-inch diameter PVC catchment pipe. Inlet line (on right) runs between NFT channels to upper inlet end. (Courtesy of R&P Hydrofarm, McMinnville, Oregon.)

Otsuki Greenhouses Ltd. of Surrey, B.C. operate almost 2.5 acres (1 hectare) producing tomatoes and peppers. The tomatoes are grown in a modified NFT system using partial slabs of rockwool (Fig. 6.26). Using several methods of NFT since 1979, the rockwool blocks offered most consistent growth with some safeguard against desiccation should the recirculation of nutrient solution fail due to mechanical breakdown. The 82-ft (25-m) NFT channels have a 2% slope. Seeds are sown into 1-inch (2.5

cm) rockwool cubes, and after 2 weeks are transplanted to 3-inch (7.5-cm) rockwool blocks. Three to four weeks later the seedlings are placed on top of a section of rockwool slab located inside the NFT channel (Fig. 6.27). The rockwool slab measures approximately 4 inches × 4 inches × 8 inches (10 cm × 10 cm × 20 cm).

Similar to a rockwool culture system, as is discussed in Chapter 10, a drip-irrigation system applies the nutrient solution to the top of each rockwool block (Fig. 6.26). The solution drains vertically down through the rockwool block and slab into the NFT channel (Fig. 6.27) where it is recirculated to a nutrient tank via a catchment pipe.This provides better root aeration and positions the crown of the plant above the nutrient solution.

Successful growers using the NFT system begin with the basic design. With experience in growing specific crops under their climatic conditions, they modify the system to achieve consistent production.

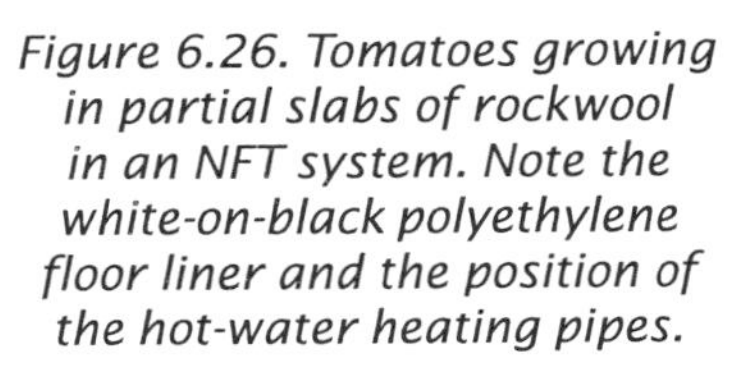

Figure 6.26. Tomatoes growing in partial slabs of rockwool in an NFT system. Note the white-on-black polyethylene floor liner and the position of the hot-water heating pipes.

Figure 6.27. Open NFT channel of white-on-black polyethylene showing small rockwool slabs and blocks in which the tomatoes are rooted. Note the drip-irrigation system with one emitter on top of each rockwool block.

(Courtesy of Otsuki Greenhouses Ltd., Surrey, B.C., Canada.)

6.6 Nutrient Flow Technique: Vertical Pipes, A-frame or Cascade Systems

Since 1978, rapid progress has been made with the commercial development of nutrient film technique (NFT). A text by Dr. Allen Cooper, *The ABC of NFT*, provides much technical information and speculation on the potential future uses of NFT. Independent of Dr. Cooper's work has been research by Dr. P. A. Schippers on what he terms the "Nutrient Flow Technique" at the Long Island Horticultural Research Laboratory of Cornell University at Riverhead, New York.

With the increasing cost of heating greenhouses in North America, Dr. Schippers has looked at modifications to the NFT system in an effort to save greenhouse space. By increasing the number of plants that can be grown in a given area of greenhouse, the cost per plant can be reduced. Experimental work was performed with lettuce in vertical pipes, down through which the nutrient solution dripped, moistening and feeding the plants. The principle is similar to that of the sack-culture system discussed in Chapter 11, but without a medium. Twenty-five to thirty plants were grown in each pipe, five feet (1.5 m) in length. A similar system is shown in Figures 6.28 and 6.29 of the Environmental Research Laboratory in Tucson, Arizona.

A nutrient solution is sprayed into the vertical pipes traveling on a conveyor system. As the pipes carrying the plants pass the point where nutrients are sprayed onto their roots within the pipe, they travel above a nutrient collector reservoir, thus the immediate drainage flows back into the system.

Schippers (1977) speculated that while this system was technically feasible, commercial application did not seem imminent at the time, especially in lettuce production, since differences in light intensity caused differences in growth between top and bottom plants. But he suggested that other crops which are harvested over a longer period, such as strawberries, peas, beans, etc. might offer better prospects.

He pursued different ways to increase the use of vertical space in the greenhouse by use of the NFT, particularly for low-growing crops such as lettuce. He built what he termed the "cascade" system. Growing troughs of 3-inch (7.6-cm) diameter PVC plastic pipe were cut open and suspended one above the other up to eight high. The nutrient solution entered the high end of the slightly sloping top pipe, exited at the low end of that pipe into the high end of the next one and so on to the reservoir from where it was pumped. The system was successful with lettuce, radish, peas and other crops. This system, shown in Figure 6.30, is used for fairly small plants.

Figure 6.28. Lettuce in vertical pipe showing nutrient-mist nozzle at top.

Figure 6.29. Vertical pipes pass above a collector reservoir for recycling excess nutrient solution.

(Photos courtesy of Walt Disney Productions. World rights reserved.)

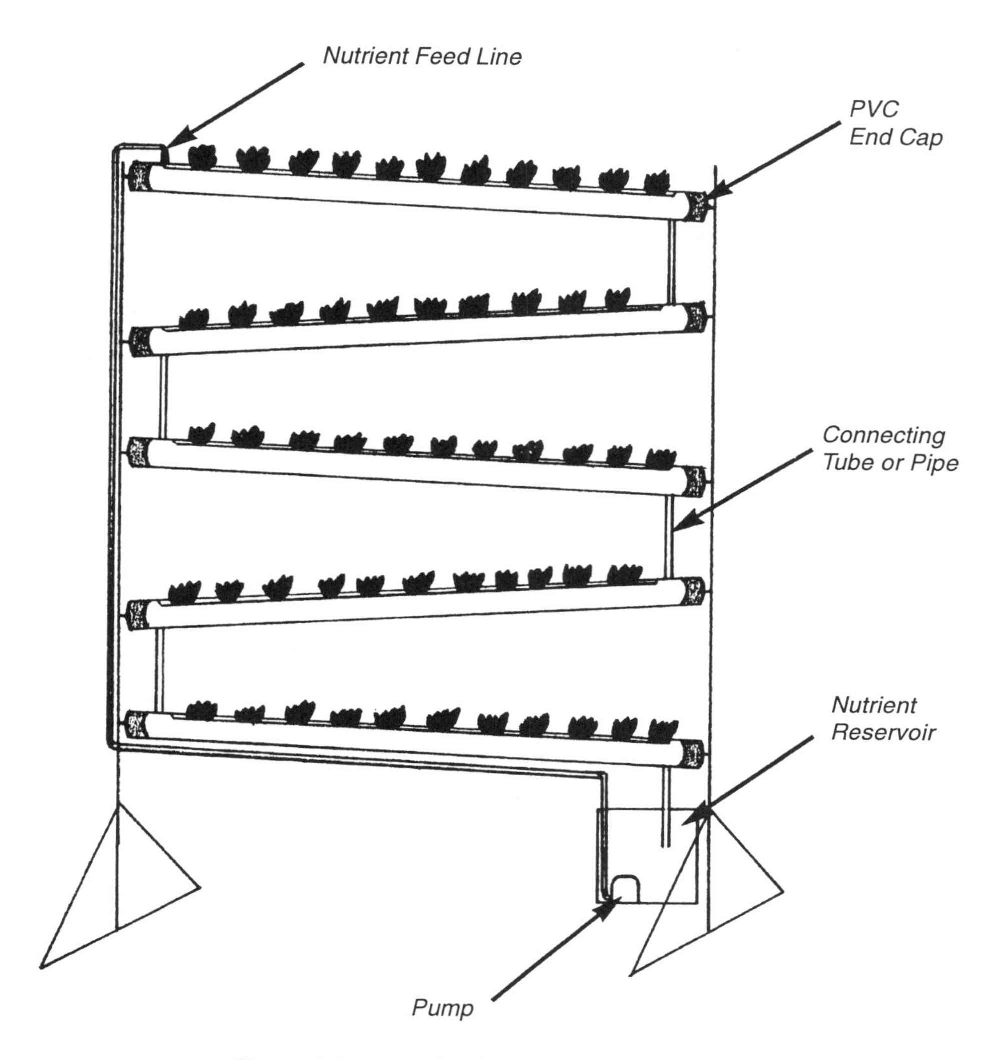

Figure 6.30. Details of a "cascade" NFT system.

This system can be made more efficient in the utilization of greenhouse space by mounting the growing channels on A-frames. The A-frames must be oriented north-south so that shade from one side will not be cast upon the other. The system is suitable only for low-profile plants such as lettuce, strawberries, spinach, and some herbs.

Several factors must be considered in the design of the A-frame system. The base of the frame must be wide enough to eliminate any mutual shading of one tier over the next lower one. The tiers of growing channels must be separated from each other with adequate distance to allow for the mature height of the crop grown. That is, the plants of a lower tier must not grow into those of the above tier. Finally, since this is basically an NFT system, all principles of oxygenation, nutrition, and optimum solution temperature apply. The total channel length for any combination of tiers should not exceed 30 meters (100 feet) in order to provide sufficient oxygenation. A minimum slope of 2 percent is necessary to provide adequate solution flow. If the A-frame unit is 30 meters (100 feet) in length, solution must be pumped into an inlet end of each channel and returned from the outlet end into a central nutrient tank.

If the A-frame is less than 100 feet, a series of channels occupying several tiers may be connected to obtain 100 feet in length. The growing channels should be spirally arranged on the A-frame. They are parallel to each other with opposite slopes from one side to the other of the A-frame similar to an expanded coiled spring (Fig. 6.31). Each subunit of approximately 100 feet in length will have an inlet and outlet to the nutrient tank. An inlet line from the main header to the pump connects to each subunit. Similarly, outlet lines from each subunit connect to a common return to the nutrient tank (Fig. 6.32). The unit illustrated in Figure 6.31 was 4 meters (13 feet) long. It used 2½-inch (6.35-cm) diameter PVC pipe as its growing channels, mounted on a 1½-inch × ⅛-inch angle-iron frame.

Plants may be seeded in rockwool cubes (1-inch × 1-inch × 1½-inch—200 per 1020 flat) or in mesh pots filled with a Peat-Lite medium. They are transplanted directly into the growing channels with the rockwool or pots suspended into the troughs so that the nutrient solution is touching the base of the pot or cube and will rise into the medium through capillary action (Fig. 6.33).

The slope of the growing troughs should be about 1 in 30 so that a rapid flow of solution will maintain high oxygenation. These units may be used in hot desert or tropical regions if a chiller refrigeration unit is placed in the nutrient tank to cool the solution temperature to 21–23° C (70–75° F). Solution temperatures may also be reduced by insulating the growing channels. If the channels are constructed of PVC pipes, styrofoam insulation with a reflective aluminum foil cover could be wrapped around the channels to reflect solar radiation.

The unit illustrated in Figure 6.31 was designed to grow lettuce at 20-centimeter (8-in) centers within the growing channels. Nine growing beds were placed on each side of the A-frame, giving a total of 18 beds. Six beds were joined into a subunit with its own inlet and outlet. Each bed of 4 meters (13 feet) contained 20 plants, totalling 360 lettuce in an area of 4 meters by 1.8 meters or 7.2 square meters (77 sq ft). This is equivalent to 50 head per square meter (4.6 lettuce/sq ft). This compares to 32 lettuce per square meter (3 head/sq ft) using the floating or raft water culture system and 18 plants per square meter (1.67 plants/sq ft) using the double-row NFT system.

Figure 6.31. "Cascade" NFT system mounted on A-frame supporting structure. PVC plastic pipes of 2½-inch diameter, arranged spirally and parallel to each other, with opposing slopes of one side from the other.

Figure 6.32. Supply lines from pump on right, and return lines to tank on left, attached to each subunit of three coils of pipe.

Figure 6.33. A-frame cascade system with lettuce seedlings growing in mesh pots supported by the pot rims, in PVC channels.

These figures are based on available growing area without allowance for aisles. Considering a greenhouse of one acre, for example, 132 feet by 330 feet (40 m by 100 m) in dimensions, the usable growing area would be: length 132 feet – 12 feet walkways = 120 feet (36.6 m); number of rows of A-frames is: 330 feet ÷ (6 feet frames + 2.5 feet aisles) = 38. Actual growing area is 120 feet × 6 feet × 38 = 27,360 square feet or 63 percent. Total number of plants per acre is:

$$360 \times \frac{27{,}360}{77} = 128{,}000 \text{ plants,}$$

which is 4.6 plants per square foot (49 plants per square meter) of growing area, or 3 plants/square foot (32 plants/square meter) of greenhouse area. This compares to 2.6 head/square foot (28 plants/square meter) of greenhouse area using the raft system.

While the productivity is greater, the capital cost of such a cascade system and the additional time required to clean each channel between crops does not make it economically feasible when almost the same plant density can be achieved using the floating (raft) system.

6.7 Moveable NFT

Schippers (1979) developed a system for lettuce using moveable gutters. Normally the lettuce is transplanted into fixed beds in the greenhouse, with spacing sufficiently wide to allow room for growth. This means that at the early growth stages after transplanting, much of the area is wasted and will only later fill in as the plants reach maturity. The nutrient flow (film) technique offers an alternative in the fact that the beds need not be in a fixed position. By adapting the distances between the gutters to the space requirements of the plants at various stages of growth, the plant population can be increased by about 50 percent. In this system a group of plants is moved, at intervals of one to two weeks, from one side of the greenhouse to the other at increased spacings and the abandoned section is occupied by the next lot, which is seeded a week or ten days later. The final end section is harvested, making room always for the section immediately adjacent to it. This system, of course, can only work for crops which can be sown and harvested at regular time intervals. A sketch of this moveable system is given in Figure 6.34.

The channels can be made from plastic gutters or purchased from manufacturers of PVC growing channels as described in Section 6.8, or made of aluminum downspouts cut lengthwise along the narrow sides. Several can be riveted together to increase the length. They can be lined with black polyethylene or painted with bituminous paint. The channels can be filled with perlite or left empty and covered with black polyethyl-

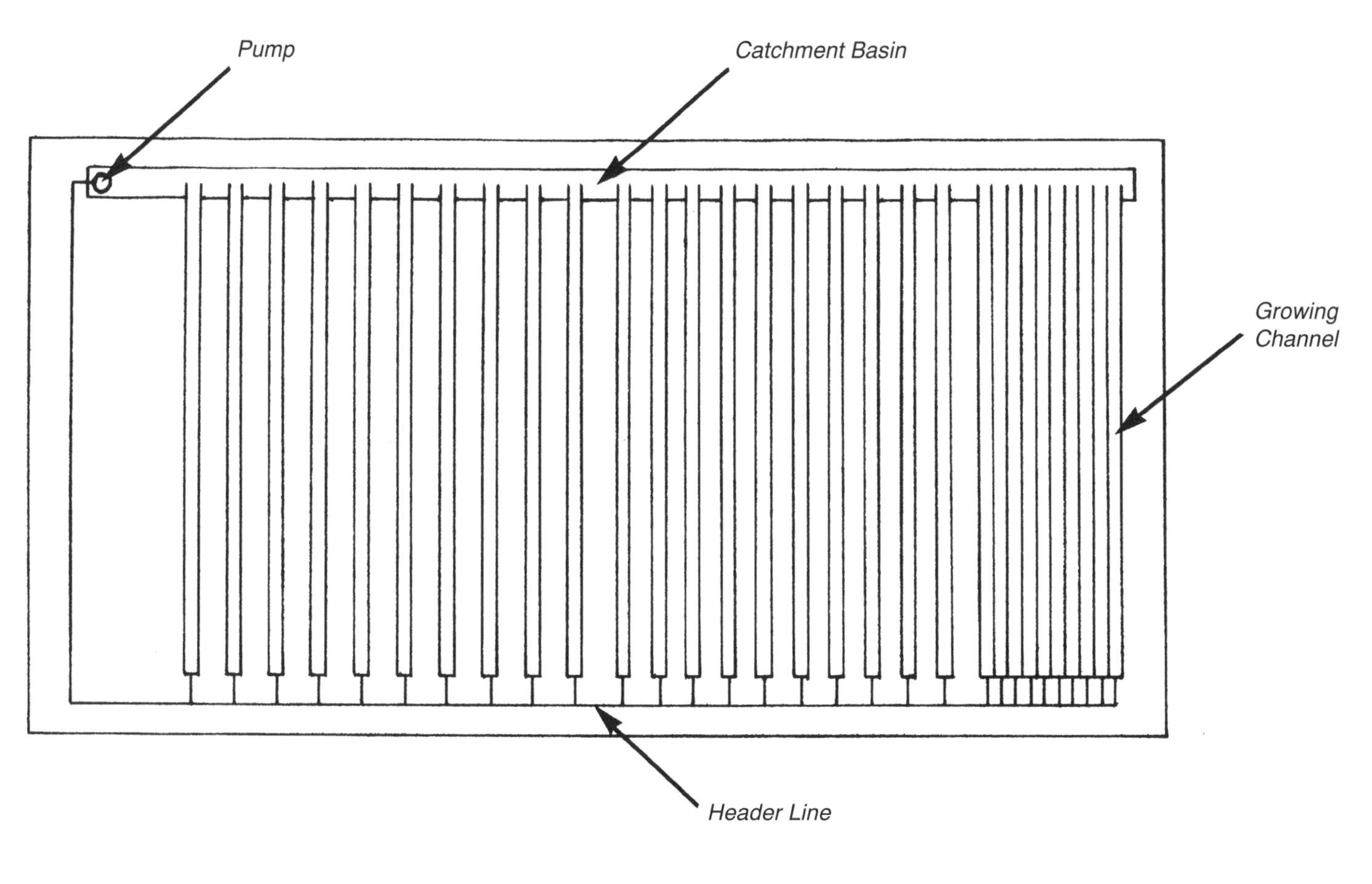

Figure 6.34. Plan view sketch of moveable NFT system.

ene punctured at the location of each plant. The catchment basin or nutrient reservoir may consist of a trench lined with vinyl or black polyethylene or it may be simply a 3-inch (7.6-cm) PVC pipe slit in half and used as a return channel to a nutrient tank at the end. The nutrient reservoir should be of sufficient volume to hold a liter of nutrient solution per plant for lettuce or other plants of comparable size.

It is generally better to have an automatic float-valve attached to a water inlet line to keep the reservoir level topped-up rather than depend on the nutrient solution lasting for a number of days. In this way when plants become mature under high light days; their high water demand will not cause the level of the nutrient solution to fall drastically with changes in its concentration.

A pump placed in the nutrient reservoir circulates the solution via a manifold system constructed of a 1-inch (2.54-cm) PVC header line with small ¼-inch (0.6-cm) diameter feeder lines to each channel. The growing channels must be sloped at least three percent. The low ends of the growing channels protrude slightly over the edge of the catchment channel so that excess solution drains back for recycling.

The width of the growing channels for tomatoes and cucumbers need not exceed 6 inches (15 cm) while that for lettuce need be only 3 inches (7.5 cm). For the growing of lettuce on a small scale, it is better to use two inches (5 cm) of perlite in the channels of a flat bed or cascade system. It gives better results and adds safety against any possible drying out due to pump failure or plugging of inlet tubes. For commercial growing it is more attractive to use beds without a substrate, since this would facilitate the removal of the old crop and sterilization of the beds before replanting.

Perlite is necessary for the growing of root crops. Fairly deep channels are needed with a thick layer of perlite. Eight-inch (20-cm) deep channels for potatoes, 6-inch (15-cm) for large carrots and 4-inch (10-cm) for smaller carrots.

A relatively simple method of constructing channels for small-scale purposes was suggested by Schippers (1979). He constructed channels by nailing at the proper distances, 1-inch by 2-inch (2.5 cm by 5 cm) wood strips with their narrow sides on a sheet of plywood. Channels are formed by pushing a sheet of black polyethylene down between the strips.

He made a further suggestion to use a flat-bed system for small plants which are seeded directly in place, such as radish, peas, beans, spinach, chives, etc. It consisted of a sheet of plywood or Masonite with a rim on three sides, lined with plastic and filled with several inches of perlite. The perlite is prevented from falling into the catchment trench at the low end by use of a nylon screen. An inlet tube should be used for each foot width of the bed.

A moveable growing system (Trof-Track System) designed for pot-plant and NFT production is manufactured by a British company, Trough Track Systems Ltd. It is based upon the moveable canal system developed by Schippers (1979). It consists of extruded aluminum troughs supported by wheeled carriers, which in turn, run on steel tracks. Parallel steel tracks run the entire length of the greenhouse and are supported on steel stanchions (Figs. 6.35, 6.36).

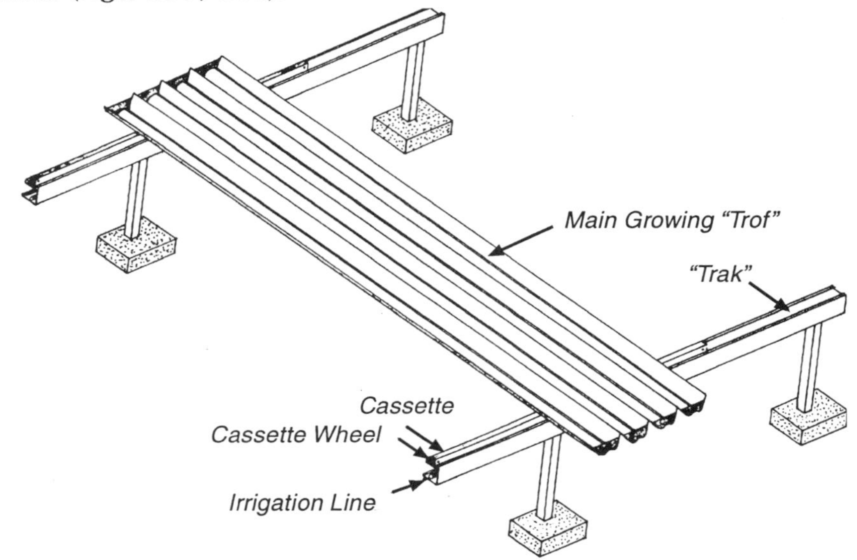

Figure 6.35. The Trof-Trak Mobile Growing System.

Troughs are 5.5 m (18 ft) long and wide enough to accommodate 14-cm (5½-inch) wide pots. The troughs have a "W" with a flattened middle cross-sectional form on which the pots rest. The side gullies thus formed give strength and act as water channels. Each trough is fed a nutrient solution through a single drip-irrigation line. The irrigation line runs the entire length of the greenhouse inside the track channel (Fig. 6.36).

As plants increase in size from the seedling stage, the carriers are moved along the greenhouse, and the distance between troughs increased to allow proper spacing of the plants as they reach maturity. Spacing may be done a number of times throughout the entire crop schedule. Such a system is especially suitable to low growing crops such as lettuce. With increased spacing of the troughs, aeration of the crop is improved. Mobility of the troughs allows free movement of workers among the plants. Such a moveable system is particularly useful in reducing energy costs of greenhouses during the early winter months. Plants can be spaced close together in a seedling area of the greenhouse with supplementary artifi-

Figure 6.36. Trof-Trak aluminum troughs supported by wheeled carriers which run on steel tracks supported by stanchions. Irrigation lines run from mains carried inside the support track. (Courtesy of TTS – Baggaley Ltd., Gloucestershire, England.)

cial lighting. The remainder of the greenhouse can be shut down. As the plants grow and require additional space, the troughs merely need to be spaced further apart, and transplanting can be eliminated. The manufacturer claims that the system can increase production for a given area by 35 per cent.

6.8 Gutter and Pipe NFT Channel Systems

A gutter NFT system may be constructed of conventional plastic eavestrough gutters used for homes (Fig. 6.37), or a system may be purchased from several manufacturers of a rigid PVC extruded growing channel, such as that of Rehau Plastics, Inc. (Appendix 2). This NFT channel is particularly suited for the growing of European bibb lettuce and herbs. The channels are available in any length but should not exceed 50 feet (15 m) as insufficient oxygenation, blockage of channels by roots, and nutrient gradients may occur. Channels of 50 feet length would be awkward to handle during harvesting and cleaning. The maximum practical length would be about 15 feet (4.6 m). Two sets of channels could be sloped back-to-back to a central catchment channel. In temperate climates, the channels could be longer if they were located on the greenhouse floor.

Figure 6.37. Use of plastic eaves troughs for homes as gutters for an NFT system to grow lettuce.

Figure 6.38. Rehau NFT channels—trickle-irrigation inlet lines from black-poly header pipe. European bibb lettuce, "Ostinata," about 45 days from seeding, ready for harvesting. Note the use of the "yellow sticky trap" for insect control. (Courtesy of Gourmet Hydroponics, Inc., Lake Wales, FL.)

Two cross-sectional dimensions are made: a larger channel 2⁹⁄₁₆ inches in width by 2 inches deep (6.5 cm × 5 cm), and a shallower one 2⁹⁄₁₆ inches by 1¼ inches deep (6.5 cm × 3 cm). The choice of channel depends upon the propagation block used. For example, the deeper channel is used with 2-inch diameter mesh pots filled with peat medium, rockwool cubes or Oasis blocks, while the shallower channel is suitable for "seed cells" made of ultra low density acetate-fiber mass, wrapped with cellophane. The channel has a plastic cover 2⁷⁄₁₆ inches wide (6.2 cm) with a ⅜-inch (1-cm) leg that sits on a ridge molded into the side of the channel. The smaller channel has the advantage of being 30 percent cheaper in cost than the larger one. Round or square holes may be custom punched in the cover. Seven-inch (18-cm) centers are standard.

Nutrient solution is pumped from a header from the nutrient tank. Emitters of 6 U.S. gallons/hour (22.7 liters/hr) are connected between the header pipe and trickle line (Fig. 6.38). Several filters, 100 mesh followed by a 200 mesh, should be installed downstream from the pump to prevent clogging of the emitters.

Figure 6.39. Basil in 12-ft Rehau channels at 7.5 inches (19 cm) within and 12 inches (30.5 cm) between rows. Note the trickle-irrigation inlet lines in the foreground.

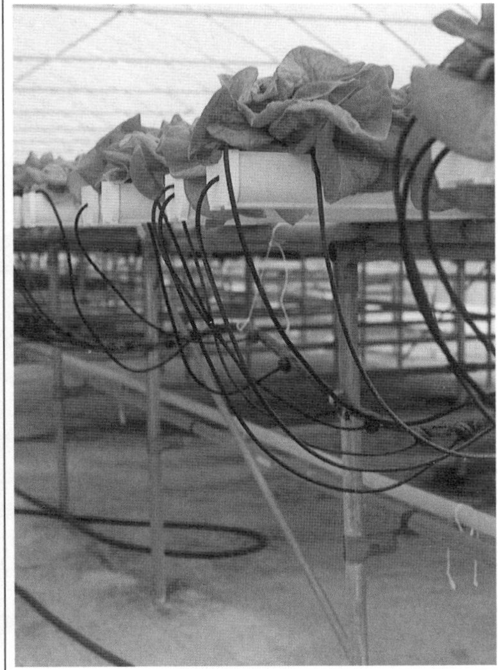

Figure 6.40. Trickle-feeding system to each Rehau channel, supported by a galvanized-iron-pipe bench.

(Courtesy of Gourmet Hydroponics, Inc., Lake Wales, Florida.)

A commercial grower, Gourmet Hydroponics, Inc., Lake Wales, Florida, uses 12-foot (3.6-m) channels to grow bibb lettuce and herbs (Figs. 6.38, 6.39). The channels are supported on a 3-foot (0.9-m) high galvanized-steel-pipe bench (Fig. 6.40). The lettuce is propagated in rockwool cubes (Fig. 6.41) and placed directly into the channels after approximately 14 days from germination.

Figure 6.41. Rockwool cubes used to start lettuce seedlings.

The lettuce roots grow along the channel (Figs. 6.42 and 6.43). The bottom of the channel is ribbed to give uniform distribution of nutrient flow. The channel has a 2 percent slope from inlet to outlet end. A catchment pipe at the outlet end returns the solution to a cistern.

Plants are spaced at 7.5-inch centers (19-cm) within the rows giving 20 head per 12-foot row, and 8 inches (20 cm) between rows. This system would utilize approximately 80 percent of the greenhouse floor area. In a one-acre greenhouse, allowing for walkways, two sets of 15-foot beds could be oriented perpendicularly to the gutters and posts of the greenhouse. The greenhouse would contain 88,740 plants using 7.5-inch by 8- inch spacing. Plant density would be 2 plants/square foot (21 plants/sq m) of greenhouse floor area, or 2.55 plants/square foot (27 plants/sq m) of growing area. This density is slightly less than the raft system but somewhat greater than the double-row NFT system.

Figure 6.42. Lettuce plant growing in rockwool cube. Plant is about three weeks old from seedling (one week after transplanting). Note the cover of the Rehau channel.

Figure 6.43. Cover removed from the Rehau channel show-ing how the lettuce roots grow along the bottom of the channel forming a root mat.

(Courtesy of Gourmet Hydroponics, Inc., Lake Wales, Florida.)

During harvesting the channels are removed with the plants intact (Fig. 6.44). This also allows the grower to select the more advanced plants to harvest earlier than smaller heads of the same age, should this occur during unfavorable weather conditions. After selectively harvesting, the channels may be replaced in their original position to allow the remaining lettuce to continue growing until they reach harvestable weight. Plants are cut off at the crown with a sharp knife (Fig. 6.45).

Any dead or yellowing base leaves are removed before moving them to the packing area. The channels are transported to a central cleaning-sterilizing vat where they are rinsed with clean water followed by soaking in a 10-percent bleach solution for at least one hour. The trays should be rinsed with water upon removing from sterilization and allowed to dry.

Lettuce is packed in polyethylene bags and placed in cardboard boxes for shipping. It must be refrigerated at 35° F (1.7° C). The shelf life for the bibb lettuce is 7–10 days.

Figure 6.44. Removal of entire NFT channel from growing table to allow easy harvesting. (Courtesy of Gourmet Hydroponics, Inc., Lake Wales, FL.)

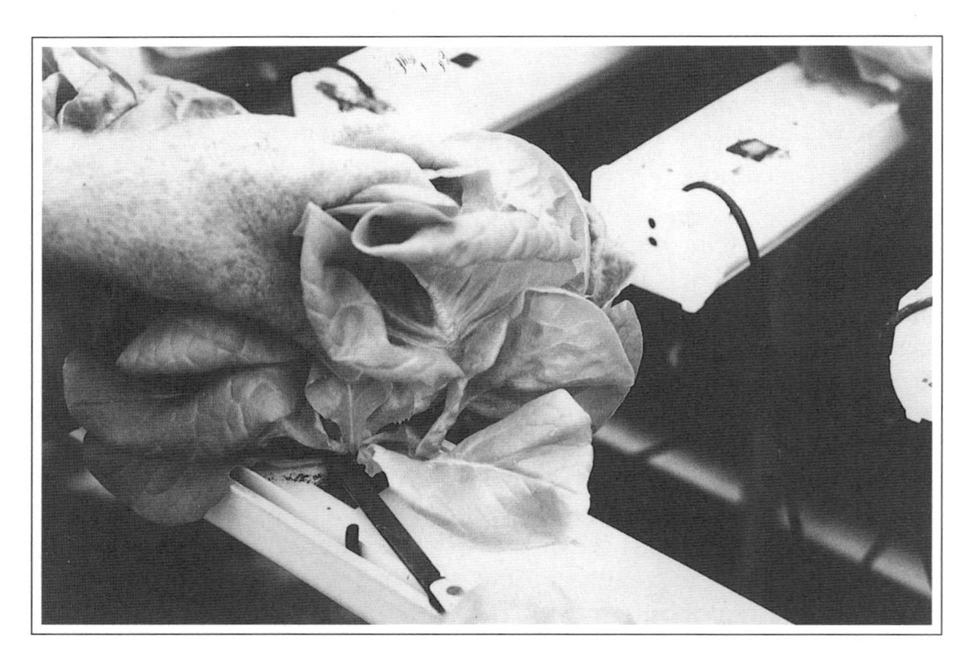

Figure 6.45. Cut base of plant at surface of NFT channel during harvesting. (Courtesy of Gourmet Hydroponics, Inc., Lake Wales, FL.)

PVC pipe systems are very similar in principle to the Rehau trough system of NFT. A series of 2-inch (5-cm) diameter PVC pipes have 1½-inch (3.8-cm) diameter holes drilled into the top face at centers equivalent to the appropriate spacing for lettuce, within each channel—generally, 6½ to 7 inches (16.5 to 18 cm). See Figures 6.46 and 6.47. Note that in Figure

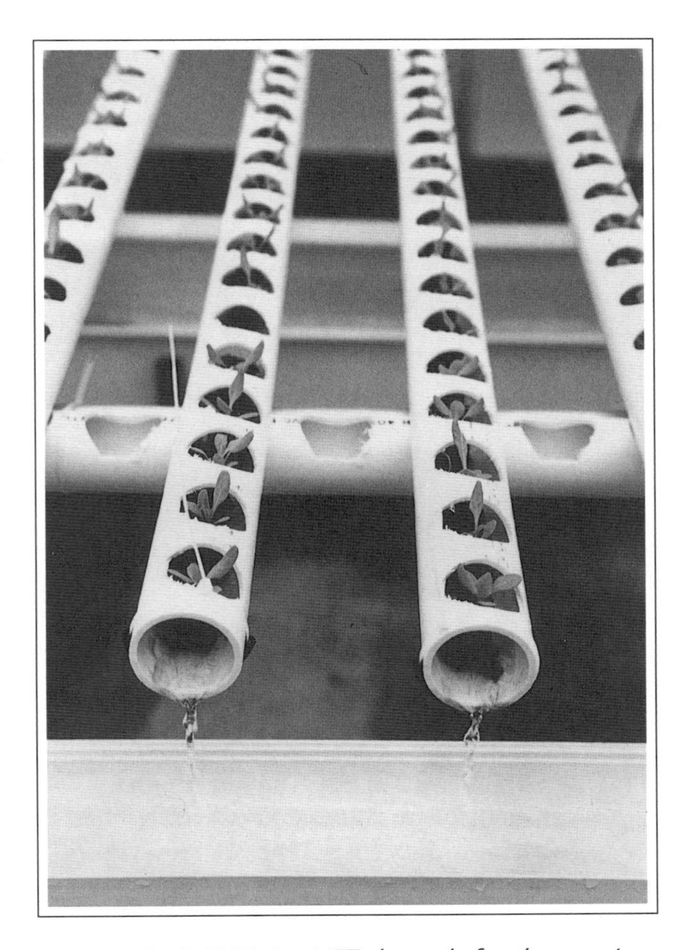

Figure 6.46. PVC pipe NFT channels for the growing of lettuce. Note the collection return channel in the foreground.

6.47, while there are 15 channels per bench, only 7 or 8 of them are planted. This was due to the pipes and holes being located too close together. Holes were made at 5½-inch (14-cm) centers within the pipes and pipes spaced 6 inches (15 cm) apart. Each pipe should be spaced 7 inches (18 cm) apart on the supporting bed frame with a total of 15 channels (pipes) so that they may be reached from the aisles on either side. The channels

Figure 6.47. A series of 2-inch diameter PVC pipes growing "Ostinata" bibb lettuce. A trickle-irrigation inlet system is used. Note that only every other channel was planted due to inadequate spacing between channels. (Courtesy of Garden Patch Produce, Sarasota, Florida.)

have a 2 percent slope and are 26 feet (7.9 m) long. Feeder tubes run to the inlet end of each pipe from a header and returning nutrient solution is collected in a catchment pipe which conducts the solution to a central cistern of 500 U.S. gallons (1,890 liters).

The *p*H and EC of the cistern are monitored and adjusted by injection of acid and stock solutions.

Lettuce seedlings are grown about 5 weeks (under Florida light conditions) before being transplanted to the growing channels. They are grown initially for 2½–3 weeks in Speedling-type trays with a vegetable-peat-plug mix, then bare-rooted and placed into other larger trays having 1-inch diameter cells (Fig. 6.48) for another 1½–2 weeks. These trays are set in shallow beds which can be flooded from the bottom to a level of 1 inch in depth for a period of 15 minutes and then drained for a comparable time. In this way, new roots develop, the seedlings grow larger and transplanting into the growing channels can be done bare-rooted without any supporting medium or device. The transplants are about 30–35 days old (Fig. 6.49).

The lettuce is grown to maturity within 14 days from transplanting. Two crops per month are possible from any given bed. In a greenhouse floor area of 6,800 square feet (632 sq m) Garden Patch Produce of Sarasota, Florida, grows over 11,000 head per month. That is, 5,500 head per two-week cropping period within the growing channels, which is equivalent to

Figure 6.48. Lettuce seedlings growing in trays fed by a sub-irrigation flooding system. Seedlings are about three to four weeks old. (Courtesy of Garden Patch Produce, Sarasota, FL.)

Figure 6.49. A lettuce transplant with large 1-inch diameter root ball ready for planting into the growing channels. The seedling is about three to four weeks old. (Courtesy of Garden Patch Produce, Sarasota, FL.)

0.8 head/square foot/ cycle. Since the cycle time in the beds is half that of the other NFT systems and raft system, on a similar monthly basis this production would be 1.6 head/square foot (17 head/sq m). This compares to 2.6 head/square foot (28 head/sq m) and 1.67 plants/square foot (18 plants/sq m) of the raft and double-row NFT systems respectively. The lesser plant density of this NFT system is due to the growing space lost in aisles and the area required to grow the seedlings.

A second growing system used by Garden Patch Produce is a multiple-gutter system made from molded fiberglass (Fig. 6.50). This is a series of double-channel gutters similar to the double-row NFT system described below. Channels are 85 feet (26 m) long with a 2 percent slope to the collection channel. Each pair of gutters has a vinyl cover. Plants are spaced at alternating 7-inch × 7-inch (18-cm × 18-cm) centers within each double gutter. Pairs of gutters are 9 inches (23 cm) apart.

Figure 6.50. Multiple-gutter NFT system made of molded fiberglass. (Courtesy of Garden Patch Produce, Sarasota, Florida.)

Seedlings for the system are started in Speedling trays with a peat mix as for the PVC pipe system. After 21–23 days they are bare-rooted and transplanted to a styrofoam floating system (Fig. 6.51). Seedlings are placed in about 1½-inch (4-cm) diameter pots cut from plastic compartmentalized trays. These pots, with a hole in the bottom, are set into the 2-inch (5-cm) thick styrofoam sheets floating on 1 inch of nutrient solution contained in galvanized-steel beds about 2 ft wide and 2½ inches deep (60 cm × 6 cm). The beds are coated with an epoxy resin to prevent reaction of nutrients with the metal.

After 1½–2 weeks the plants are taken from the transplant beds and placed into the double-gutter growing beds. The harvesting time and productivity is very similar to that of the pipe system.

Figure 6.51. Styrofoam floating seedling beds.
(Courtesy of Garden Patch Produce, Sarasota, FL.)

6.9 Double-Row NFT

At least one company in Holland (Reko) (Appendix 2) manufactures an NFT channel and cover to contain two rows of lettuce per channel. The semi-rigid black plastic trough may be used for rockwool culture or NFT culture of other crops such as tomatoes and/or cucumbers.

A white cover is impermeable to light and gives good light reflection (Figs. 6.52, 6.53) to reduce heat buildup in the channel.

The white covering can be rolled onto a drum of a planting machine. Two planters sit on the machine while transplanting. The machine travels down the house unrolling four covers as it moves along and places them on the NFT channels as the planters set the plants in the holes. Holes are cut alternatively in the cover at 10 inches (25 cm) between plants within the row and 9 inches (22.5 cm) between rows when used to grow European bibb lettuce. The NFT channels are spaced 8 inches (20 cm) apart to give a row spacing of 9 inches (22.5 cm). No aisles are made, so that in effect, the entire greenhouse growing area is utilized with lettuce rows 9 inches (22.5 cm) apart (Figs. 6.54, 6.55). These channels can easily be sterilized between crops with a 10-percent bleach solution.

Lettuce can be seeded into compressed peat blocks using an automated block compressor and seeder (Fig. 6.56). While most Dutch growers prefer

Figure 6.52. Double-row NFT system with bibb lettuce. The white cover is raised to show the root mat and black NFT channel. The root mat encompasses a paper-towel strip set in the channel bottom during transplanting, to obtain uniform solution distribution.

Figure 6.53. White cover of double-row NFT raised to show the size of the hole for inserting the plants. Young trans-plants are about three weeks old growing in peat blocks.

to use a compressed peat block to germinate seedlings, an alternative method is to sow the seeds into rockwool cubes. A propagation area of approximately 1,100 square feet (100 sq m) is sufficient to service 10,760 square feet (1,000 sq m) of production area. The propagation area should be provided with supplementary artificial lights and a misting irrigation system. The lettuce seedlings are transplanted to the NFT channels 12 to 40 days after sowing, depending upon the season (summer and winter respectively).

The NFT channels should be a maximum length of 50 feet (15 m) sloping towards a catchment return line. Two sets of channels slope to each catchment trench from each side (Fig. 6.54). The catchment trenches return the solution to a cistern where the solution is automatically adjusted with injectors for *p*H and EC. A pump circulates the nutrient solution continuously 24 hours per day. Uniform distribution of nutrient solution along

Figure 6.54. Black plastic NFT channels are spaced 8 inches (20 cm) apart. The inlet ends of the trickle-irrigation system appear in the foreground. NFT channels, 50 feet (15 m) long, slope toward a central catchment trench.

Figure 6.55. Maturing lettuce covering most of the greenhouse floor area.

Figure 6.56. Lettuce is seeded into compressed peat blocks. Lettuce at 2 to 3 days after germination.

the NFT channels is achieved by use of a strip of paper towel on the bottom of the channel. Nutrient solution enters the higher end of each channel and is recirculated via the catchment trench, similar to other conventional NFT systems.

A one-acre greenhouse of dimensions 132 feet by 330 feet (40.2 m by 100.6 m) having a 10-foot (3-m) central walkway and 2-foot (0.6-m) end aisles, has a usable length of 120 feet (36.6 m) which can contain two 60-foot (18.3-m) NFT channels for every 9 inches (22.5 cm) of greenhouse width.

Access every 27 feet (8.2 m) with 2-foot (0.6-m) wide aisles for pest control is equivalent to 36 rows of plants per 29 feet (8.8m) of greenhouse width, or a total of 410 rows. Rows of 120 feet (36.6 m) in length contain, at 10-inch (25-cm) centers, 144 plants. Therefore, the total plants per crop in a greenhouse of one acre is: 144 × 410 = 59,040 plants. The usable growing area is: 27 feet × 120 feet × 11 sections = 35,640 square feet, or 82 percent. Productivity is 1.66 plants/square foot (18 head/sq m) of growing area, or 1.36 plants/square foot (14.5 head/sq m) of greenhouse floor area.

Of the three systems (cascade, raft, double NFT), this system is the least expensive to set up.

The main advantage of this system is the lower capital cost to achieve a NFT system of relatively high-density planting.

6.10 Concrete NFT Channels

While capital costs of constructing concrete NFT channels are higher, maintenance is reduced. The NFT channels are formed in the concrete slab of a greenhouse floor on a 1 in 50 slope. The concrete is coated with epoxy resin to isolate it from the nutrient solution. Channels 4 inches (10 cm) wide by 1 inch (2.5 cm) deep are formed at 11-inch (28-cm) centers, giving a 7-inch (18-cm) aisle between each channel (Fig. 6.57). With a 10-inch (25-cm) spacing between rows, the same plant density is achieved as for the double-row NFT system. An operation of 10 acres (4 hectares) built in England produces about 8 million lettuce annually.

Channels are sloped from the inlet end toward a catchment trench at the far end and the solution is monitored, adjusted, and recirculated as for conventional NFT systems (Fig. 6.58). A white polyethylene cover through which holes are punched for the placement of lettuce transplants into the NFT channels is spread across the concrete floor (Figs. 6.57, 6.58). Lettuce is sown into compressed peat-soil blocks by use of machines, as is done for the double-row NFT system. Propagation procedures are also the same as for double-row NFT. The surfaces of the blocks are covered with perlite after sowing to maintain moisture and discourage algae growth by keeping the surface dry (Fig. 6.59).

Figure 6.57. A lettuce operation in England using concrete NFT channels. Four-inch-wide concrete channels are covered with white polyethylene, through which plants are placed.

Figure 6.58. Concrete NFT channels at the solution inlet end.

Figure 6.59. Lettuce seedlings, about two to three days after germination, started in compressed peat-soil blocks covered with perlite to maintain moisture and keep the surface dry.

Irrigation is provided with an overhead misting system (Fig. 6.60). As seedlings grow, they are spaced further apart in a checkerboard configuration to allow further growth before transplanting to the channels (Fig. 6.61).

Figure 6.60. Overhead irrigation-fertilization system for watering the seedlings.

Figure 6.61. Lettuce seedlings growing in peat-soil blocks spaced out for further growth.

The white polyethylene cover is used for several crops before replacing it and cleaning the channels. Planting and harvesting can be fully mechanized.

In addition to the higher capital cost, another disadvantage of this system is that it is designed specifically for lettuce or other low-profile crops that are planted at similar densities. Thus, its adaptability to growing other crops is limited.

6.11 Agri-Systems NFT

A highly efficient system of NFT was developed in the 1980s by Agri-Systems in Somis, California. Seedlings are grown in 154-celled "creamocup" trays. Holes are slit in the bottom with a special multi-bladed table saw. The trays are filled with coarse vermiculite and seeded with European bibb lettuce using a flat filler machine and automatic seeder. The seedlings are grown either in a small seedling greenhouse or in a controlled environment growth room with artificial lighting, automatic watering and temperature control (Fig. 6.62). Within three weeks, the seedlings are ready to transplant into the NFT channels in a greenhouse. The small cells containing the seedlings in the plastic trays are punched out of the trays, and the small lettuce plant with its cell is planted directly into a moveable tape within the growing channel (Figs. 6.63, 6.64). The slits in the bottom of the cells allow roots to grow through into the nutrient solution of the growing channel.

A planting-harvesting machine feeds a coiled heavy plastic tape into grooves of the growing channel. Holes have been punched in the tape at the correct spacing for the lettuce. The operator of the planting machine simply drops the cells containing seedlings into the holes of the tape as it feeds into the NFT channel (Fig. 6.64).

Agri-Systems tested lettuce production with two to five tiers of channels. They found that the five-tier system did not allow sufficient natural sunlight to enter the lower levels to support plant growth adequately in order to achieve a marketable product. The two-tier system was the most productive in utilizing all available light; however, still a large percentage of the lettuce in the lower level was not of top quality due to lack of light (Figs. 6.65, 6.66).

Planting was scheduled so that rows of various stages of maturity were mixed to allow maximum light penetration to the lower tier (Fig. 6.67). If plants of a similar age were placed in the channels adjacent to each other, insufficient light would pass to the lower tier as the lettuce matured to full size.

The company claimed that with their system, 8 million heads of lettuce could be grown annually on a 2½-acre (1-hectare) operation in comparison

Figure 6.62. A controlled-environment growth room. Lettuce seedlings in vermiculite in plastic trays with an ebb-and-flow hydroponic system. (Courtesy of F.W. Armstrong Greenhouses, Inc., Oak View, CA.)

Figure 6.63. Lettuce seedling growing in plastic cell. (Courtesy of Whittaker Corporation's Agri-Systems Division, Somis, CA.)

with 75,000 produced in the same area by conventional open-field farming. This is equivalent to 9 lettuce per square foot of greenhouse area per crop, or 72 head per annum if 8 crops are produced per year. If only 6.5 million are marketable from the total production of 7.5 to 8 million and an estimated 19 people are required to operate a 2.5-acre lettuce module, the estimated labor cost is just over $0.02 per head compared to $0.07 to $0.10 per head for field-grown lettuce.

It is doubtful that these objectives were reached, as the greenhouse operation was taken back by the original owners and the tier system of growing channels was abandoned in favor of a single level. Agri-Systems in the early 1990s helped establish two other greenhouse operations of about ½ acre each in the Somis area. Agri-Systems provided much of the NFT growing system components and the transfer of technology to these growers.

Figure 6.64. Transplanting lettuce seedlings into moving tape cover of NFT channels. (Courtesy of F.W. Armstrong Greenhouses, Inc., Oak View, CA.)

Figure 6.65. Agri-Systems lettuce production in four tiers of NFT troughs. (Courtesy of Whittaker Corporation's Agri-Systems Division, Somis, CA.)

Figure 6.67. Two tiers of NFT troughs with varying stages of plant maturity from seedlings to mature lettuce. Note the coiled moveable tapes and the recycling closed irrigation system.

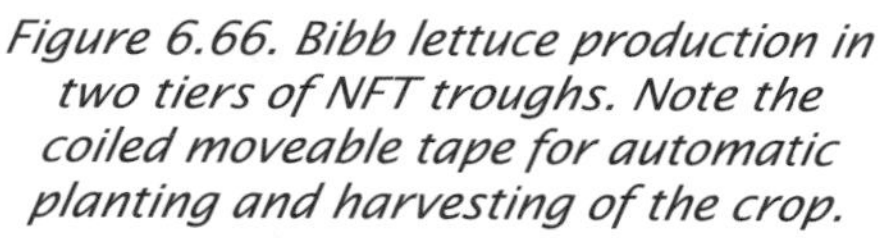

Figure 6.66. Bibb lettuce production in two tiers of NFT troughs. Note the coiled moveable tape for automatic planting and harvesting of the crop.

(Courtesy of Whittaker Corporation's Agri-Systems Division, Somis, CA.)

One of these greenhouses, F.W. Armstrong Greenhouses, operates about 20,000 square feet (1,860 sq m) of lettuce production. The NFT channels are raised about 3 feet (1 m) above the floor of the greenhouse with an aluminum supporting frame (Fig. 6.68). From the cooling pad end, fans blow cool air underneath the beds.

The fans bring air through the cooling pad and blow the cooled air down convection tubes located on the floor going across the house and under the lettuce channels. This positive-pressure cooling system is contained under the NFT channels by use of perimeter polyethylene curtains (Fig. 6.69). The air rises up through the greenhouse to exit by overhead vents in the sawtooth structure.

The NFT channels, constructed of aluminum, are approximately 3 inches (7.6 cm) wide by 2 inches (5 cm) deep, with heat sink ridges built into their bottoms to assist in cooling (Figs. 6.70, 6.71). Plants are supported in the flexible plastic tape that sits between several ridges on the inside top edge of the channel.

Figure 6.68. NFT channels raised about 3 feet by a metal frame. Note the convection tubes underneath the table and the cooling pad behind the deflector panels at the upper right. (Courtesy of F.W. Armstrong Greenhouses, Inc., Oak View, CA.)

Figure 6.69. Perimeter polyethylene curtain around beds contains cooled air under the beds. (Courtesy of F.W. Armstrong Greenhouses, Inc., Oak View, CA.)

Figure 6.70. Aluminum NFT channels exit to a catchment pipe which returns nutrient solution to a cistern. (Courtesy of F.W. Armstrong Greenhouses, Inc., Oak View, CA.)

Figure 6.71. Inlet ends of NFT channels. Note the heat-sink ridges at the bottom of the channels, inlet header with a feeder tube to each channel and coil of flexible-tape channel cover supported by ridges in top of channel. (Courtesy of F.W. Armstrong Greenhouses, Inc., Oak View, CA.)

The NFT channels were 150 feet (46 m) long, but later reduced to 73 feet (22 m) and 70 feet (21 m) with a 7-foot (2-m) aisle in the middle of the greenhouse. Each of the two sections contained 228 channels. In this way, two separate systems operate each with its own nutrient-solution tank. If a problem arises in one section, it is easier to isolate and remedy it without interrupting the production of the other section. Nutrient solution is pumped from a 500-gallon (1,890 l) cistern at the cooling pad end of the greenhouse to the inlet end of each channel via a 1½-inch (3.8-cm) diameter PVC header and a ¼-inch (0.6-cm) drip line (Fig. 6.71). The NFT channels have a 2% slope toward the catchment end where the nutrient solution is collected and returned to the nutrient tank (Fig. 6.70). Plants are spaced at 6 inches by 6 inches (15 cm by 15 cm) to make a total of 70,000 plants in the 20,000 square feet of greenhouse.

The cropping schedule for lettuce is from 28 to 38 days, depending upon weather conditions, especially sunlight hours, and day-length, with the shorter period between harvests occurring during the late spring-early summer months. Therefore, about 10 crops are expected annually. Armstrong Greenhouses harvests on the average 2,200 head daily. At 6-inch (15-cm) centers, there are approximately 140 plants per 70 ft (21 m) row. About 16 rows are harvested daily. This is based upon a 32-day cropping period.

A harvesting machine is used to pull the moveable tape from the NFT channel and roll it up as the operators cut the heads of lettuce from the tape (Figs. 6.72).

Figure 6.72. Use of harvesting-transplanting machine to pull rows of lettuce from the channels. Operators cut the lettuce off at their base as the tape passes to be rolled up below. (Courtesy of F.W. Armstrong Greenhouses, Inc., Oak View, CA.)

The lettuce is stored in a mobile refrigeration unit which then transports it to the packing area (Fig. 6.73). The lettuce is packaged individually in a heat-sealed polyethylene bag and shipped as 12 heads per box (Fig. 6.74).

Figure 6.73. Mobile refrigeration unit with storage crates.

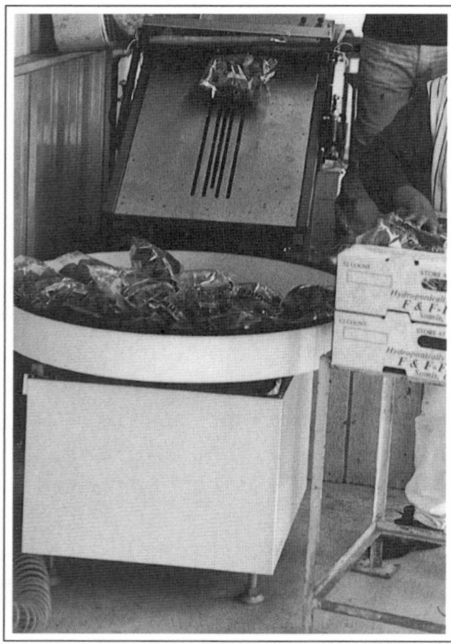

Figure 6.74. Packing of lettuce in heat-sealed plastic bags. Sorting table at bottom of heat-sealer machine.

(Courtesy of F.W. Armstrong Greenhouses, Inc., Oak View, CA.)

The moveable tapes are sterilized in a vat of 10 percent bleach solution. After rinsing and drying, they are placed one at a time on the transplanting machine and the tapes are threaded back into the NFT channels while placing the seedlings into them as described earlier (Fig. 6.64). The sequence of transplanting is to do every third row of channels until reaching the last row. Then go back and plant every second row of every three, and finally the last row of every set of three. In this way, smaller plants are next to the older more mature plants (Fig. 6.75). This gives better light to the crop and during harvest when the tape is pulled, the plants do not brush into one another.

Figure 6.75. Rows of lettuce showing sequence of planting dates. (Courtesy of F.W. Armstrong Greenhouses, Inc., Oak View, CA.)

6.12 Tube or Pipe Culture

Tube culture is a modification of the NFT and water culture. The principles are the same. The nutrient solution is pumped through 4-inch polyvinyl chloride (PVC) drain pipes covered with black polyethylene film. The PVC pipe is cut in half and black polyethylene placed over the top of it to prevent light from entering. Holes are cut in the top of the polyethylene through which seedlings grown in cubes are set into the nutrient solution flowing along the bottom of the PVC pipes. During crop changeover the polyethylene cover and plants are removed, the tube washed with a bleach-sterilizing agent and a new polyethylene cover replaced ready for new seedlings.

The "cascade" system of Schippers described earlier is a form of tube culture. Further developments for both commercial and hobby uses have been achieved by Homeland Industries, Inc. of Brooklyn, New York. They term their system "Moduleponic." It consists of two plastic pipes on a supporting stand with an air supply to the lower pipe containing the nutrient solution, as shown in Figure 6.76. The nutrient reservoir can be constructed of 3-inch (7.6-cm) PVC pipe while the growing bed should be of 4- to 5-inch (10-13 cm) diameter PVC pipe. While the system is not truly a water-culture system, since the upper pipe contains a gravel medium, it does closely resemble the tube-culture system.

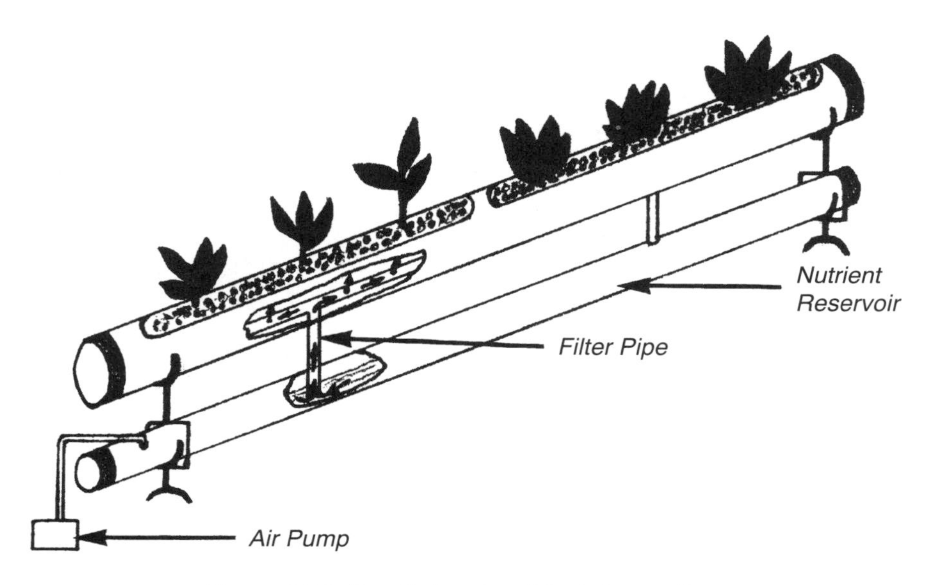

Figure 6.76. Detail sketch of "Moduleponic" system.

Air is pumped through an air supply pipe into the bottom nutrient reservoir pipe. A filter pipe joins the nutrient reservoir with the growing bed at intervals along the unit. This filter pipe is perforated with small holes. As air is pumped into the sealed nutrient reservoir, it is forced up through the filter pipe carrying nutrient solution with it. The nutrients are then distributed throughout the gravel bed by escaping through the perforated filter pipe. When the air pressure stops, the excess nutrient solution drains back down to the nutrient reservoir. The pump is activated by a time clock three to four times a day, depending upon the plant stage of growth and weather conditions.

A modification of tube culture has been used in Kenya in the production of grapes (Barrow 1980). A gravel substrate was used in a 10-inch (25-cm) PVC pipe. A segment was cut from the top of the pipe, and this segment was placed in the bottom of the pipe to facilitate filling and draining of the nutrient solution. Slots were cut at regular intervals along this segment.

Sixty-foot (18.3-m) sections were supported by concrete blocks every ten feet (3 m) to provide a slight slope towards the discharge end. For grapes, the spacing between each pair of pipes was 12 feet (3.66 m). A discharge pipe connected all the planting tubes to an underground catchment tank from which the nutrient solution was then pumped back to a storage tank. A main inlet pipe was connected to the inlet end of each bed from the storage tank. Rooted cuttings of grapes were planted in the tubes at a spacing of approximately two feet (61 cm). Plants were supported by an overhead trellis.

In 1989 about one acre of greenhouses, Massey Greenhouses, located in Pruit, Florida, was using a tube-culture system to grow tomatoes. They used 4-inch (10-cm) diameter PVC pipe for the growing channels. Nutrient solution was pumped from a central cistern through a 1½-inch (4-cm) PVC header pipe to the upper ends of the growing channels where it entered each channel via a ¼-inch drip-tube line (Fig. 6.77). Two sections of growing channels, each about 70 feet (21 meters) long, sloped toward a 4-inch (10-cm) diameter PVC central catchment pipe (Fig. 6.78). The catchment pipe returned the solution to the nutrient tank cistern. Tomatoes were seeded in growing cubes and later transplanted directly into the growing channels through 2-inch (5-cm) diameter holes cut in the top of the channels located at the correct spacing (Fig. 6.77).

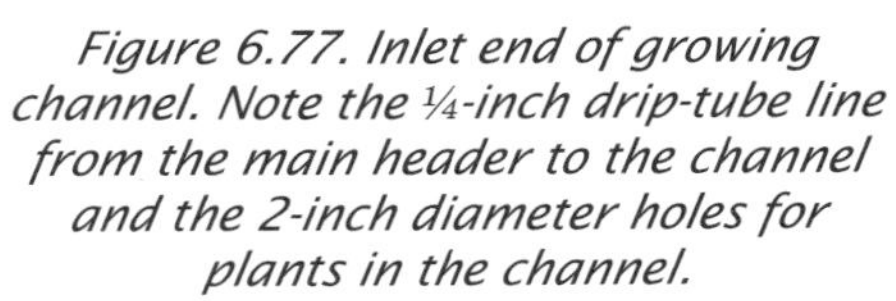

Figure 6.77. Inlet end of growing channel. Note the ¼-inch drip-tube line from the main header to the channel and the 2-inch diameter holes for plants in the channel.

Figure 6.78. Central catchment pipe conducting nutrient solution from the growing channels back to the cistern. Smaller pipes are inlet nutrient solution lines from the cistern to the inlet ends of the channels.

6.13 Ebb-and-Flow Systems

The ebb-and-flow technique is basically a subirrigation method. Nutrient solution is pumped into a shallow bed to a depth of about 1 inch (2–3 cm) for about 20 minutes and then allowed to drain back to the nutrient tank once the pumps are shut off. Ebb-and-flow benching systems are available from commercial manufacturers. These bench systems are particularly suited for the growing of seedling transplants and ornamental potted plants (Fig. 6.79). The bottom of the bench has small cross channels perpendicular to deeper channels (Figs. 6.80, 6.81). This allows uniform filling and complete drainage during the irrigation cycles. The deeper channels lead to the inlet-outlet pipes (Fig. 6.81). Tables are irrigated from both ends. The tables are 50 feet (15 m) in length supported by metal framing with concrete footings to maintain them perfectly level.

Figure 6.79. Ebb-and-flow benches used for production of ornamental transplants.

Figure 6.80. Fill-drain channels of ebb-and-flow bed. Flats of ornamental ferns are in the bed.

Another type of ebb-and-flow system is used by Bevo Farms Ltd. of Langley, B.C. to propagate transplants for commercial greenhouse and field growers. The entire floor of the 108,000 square foot (10,000 square meter) greenhouse is of concrete which forms the basis of the ebb-and-

Figure 6.81. Inlet-outlet of the bed.

flow recirculating system (Fig. 6.82). One layer of concrete bricks cemented together on edge form the sides of each bed approximately 20 feet by 100 feet (6 m by 30.5 m). A slight slope from the edge to the center enables flooding and draining to occur from the middle of each bed by way of an open flume partially covered with bricks. Solenoid valves on the main PVC headers control the irrigation cycles. The nutrient system is fully automated with a computerized controller. The greenhouse is heated with a natural-gas-fired boiler hot-water system with heating pipes buried in the concrete floor. This produces optimum temperatures for the various crops through bottom heating combined with an overhead system.

The greenhouse vegetable seedlings are started in rockwool cubes which are later transferred to rockwool blocks. Cucumbers are transferred within 4–5 days, tomatoes in 10–12 days and peppers in 12–14 days. The EC

Figure 6.82. Concrete floor ebb-and-flow system for growing transplants. Note the brick sides of the beds and the slight slope to the center fill-drain channel (spaced bricks on top). (Courtesy of Bevo Farms Ltd., Langley, B.C., Canada).

levels are maintained at 2.3–3.5 mMho for tomatoes, 2.5–3.5 mMho for peppers and 1.8–2.2 mMho for cucumbers, with the *p*H between 5.6 and 6.0. Cucumbers are grown for 4 weeks, tomatoes for 6 weeks and peppers for 8 weeks prior to delivery to the greenhouse growers. These crops are grown from October 1 through late December. Using plug trays, where no spacing is required, over 1 million plants can be produced every two weeks (Fig. 6.83). Field crops of lettuce, broccoli, cauliflower and celery, are propagated at the rate of 1.5 million plants weekly in the early spring. With peppers, tomatoes and cucumbers growing in rockwool blocks, which require approximately 8 inches by 8 inches (20 cm by 20 cm) spacing, 180,000 plants are produced each 4 weeks (Fig. 6.84).

Rose cuttings, cuttings for hanging baskets and ornamentals are also propagated.

6.14 The Ariel NFT

Dr. Allen Cooper, working with Nutrient Film Technology, Ltd. in England, developed a divided channel NFT system. According to Cooper (1985), the Ariel NFT system was developed to overcome a number of limitations inherent in conventional NFT systems:

Figure 6.83. Seedlings in plug trays in the ebb-and-flow hydroponic system. (Courtesy of Bevo Farms Ltd., Langley, B.C., Canada).

Figure 6.84. Greenhouse pepper seedlings in rockwool blocks in ebb-and-flow system. Note the presence of irrigation solution and the drain-fill channel at lower right. (Courtesy of Bevo Farms Ltd., Langley, B.C.)

1. The precise control of solution levels in a conventional NFT system to about one centimeter of depth is difficult to maintain. If the root mat is submerged completely in the nutrient solution, the plant in effect suffers oxygen deficit. The reduction of flow to maintain a one-centimeter depth of solution is difficult, especially as the root mat thickens and restricts the flow of solution. The root mat must be only half submerged in the solution to allow diffusion of gases into and out of the top half of the roots. The reasoning for this is that the diffusion rate of gases in air is 10,000 times more rapid than it is in water. If this level adjustment of the solution to keep half of the root mat exposed to the air is not maintained, the submerged roots will rapidly die, the plant wilts, and fungi such as *Pythium* and *Phytophthera* invade the roots, causing further death. This precise regulation of solution level which introduces a high probability of error can be overcome with the Ariel NFT system, Cooper believes.

2. Conventional NFT systems require complex monitoring and injection equipment. Continual maintenance and calibration of *p*H and EC equipment is necessary to prevent breakdown and loss of accuracy of measurements.

3. The need for continuous recirculation of the nutrient solution in conventional NFT demands an uninterrupted power supply. This is achieved with an automatic start-up standby generator and alarm systems to advise technicians of failures. This is usually part of a computer-operated greenhouse control package.

4. Skilled management is an integral part of successful conventional NFT crop production. Alteration of nutrient formulae during different crop stages, seasons and weather conditions are decisions made by management.

5. In conventional NFT systems, some solid rooting medium is needed for the propagation of the plants. Usually rockwool or compressed peat-soil blocks, peat pellets, or pots filled with a medium of sand, peat, perlite, or a mixture of these components is used. Such use of a propagation medium is often costly. The Ariel NFT system minimizes the use of propagation media and concentrates on bare-rooted seedlings. It should be pointed out, however, that the use of rockwool cubes for propagation of seedlings will reduce transplant "shock" and provide a small reservoir of nutrient solution, preventing desiccation.

The Ariel NFT system grows bare-rooted plants so that any changes in the root environment (adjustment of nutrient solution) will produce a rapid response in the plant. When a solid medium such as rockwool or peat is used, response time is slowed due to the interaction with the rooting medium. The significance of this may be questionable. An inexpensive propagation system was developed for the Ariel system by use of "germination pockets" made from an absorbent matting material (Cooper 1985). The "pockets" with a fold at the top into which the seed is placed are set astride ridges in trays. The trays are placed in germination chambers having controlled optimum temperature for germination. The work of seeding would be tedious if it could not be automated.

Once the seeds germinate, the pockets are removed from the germination chamber and placed at close spacing astride ridges in a propagation bed which has a ridge-and-furrow configuration. Nutrient solution is recirculated down the furrows of the propagation bed. The cost of propagation materials is less than in conventional NFT systems, and labor is less, since large numbers of plants can be transported in trays when they are removed from the propagation beds to the growing channels. But, the labor of sowing seeds may be greater than with conventional NFT. The propagation area with a high density planting can be placed in a small area of the greenhouse having supplementary artificial lighting during the winter months. The space requirements, however, will be a function of the plants grown and to what age they will be grown before transplanting to the final growing beds in the greenhouse. This method of propagation produces a plant with a divided root system necessary for the subsequent growing channels.

The growing channel consists of a divided, metal-channel base of aluminum or galvanized steel (or fired clay), depending upon cost and availability. The base comprises a central ridge with two side channels (Fig. 6.85). This cross-sectional configuration adds strength to the base so that it may be placed directly on the ground of a roughly prepared slope without any depressions occuring. The bases are overlapped down the slope, similar to roofing tiles.

Black-on-white polyethylene is laid in the base channels. Using a wide strip of polyethylene enables it to be rolled above each channel to form two air-filled rolls retained by the walls of the base on either side of the central ridge (Fig. 6.85). These rolls provide insulation against solar radiation to prevent overheating of the roots and solution, and at the same time reduce heat loss during the winter months, thus stabilizing the solution temperature.

Capillary matting is placed on the ridge, with its sides reaching to the bottom of each side channel. The plants are then placed astride the ridge with half of the divided root system on either side of the ridge. The polyethylene cover is then clipped together between the plants. These clips serve also as attachment points for support strings of plants normally trained vertically.

Mr. Fathi Farimani of Delta, B.C., experimenting with the Ariel NFT system, has made some unique discoveries. Growing channels were built of

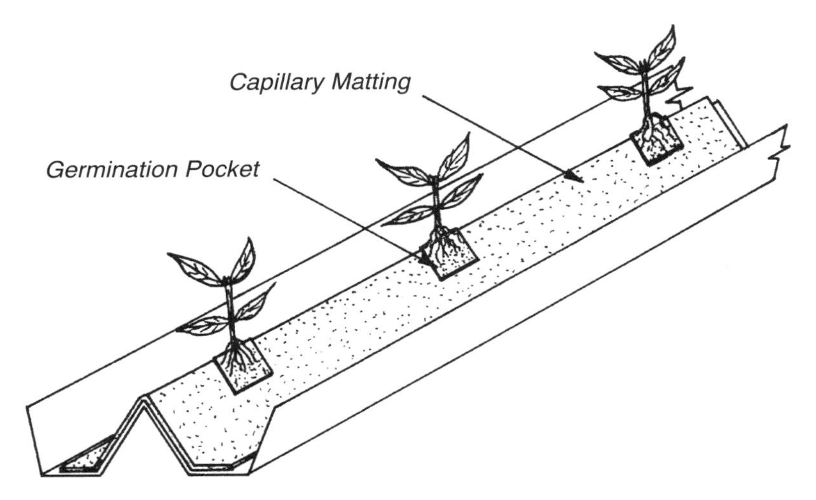

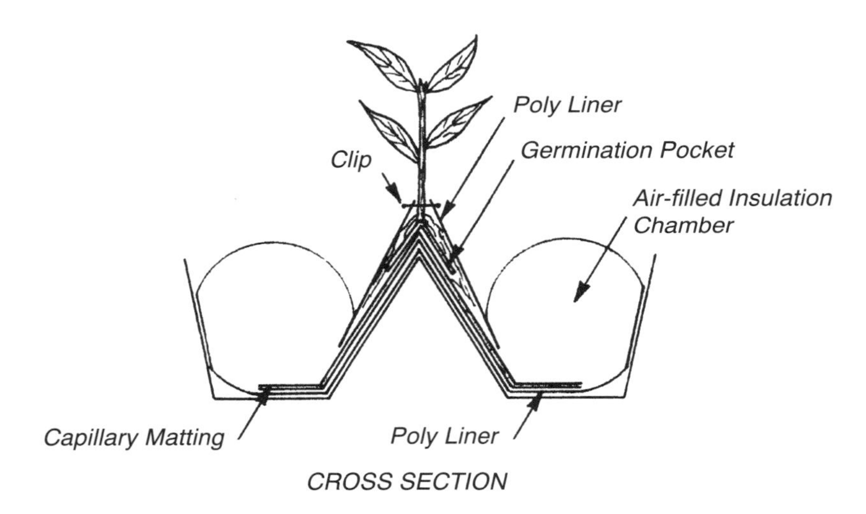

Figure 6.85. Profile section of Ariel NFT system.

plywood, lined with black-on-white polyethylene. They were approximately 12 inches wide by 10 inches deep (30 cm by 25 cm) and 30 feet (9 m) long (Fig. 6.86). Tomato seedlings were started in strips of capillary matting (Fig. 6.87) or in rockwool cubes which had a "V" shape cut from their bottom so that they would stay on the ridge of the growing channel. The trials with the capillary matting were more successful than the rockwool cubes. It was difficult to get capillary movement of the nutrient solution through the rockwool cubes unless the level of nutrient on one side of the channel was raised sufficiently to contact the rockwool cube.

Figure 6.86. Ariel NFT system with tomatoes. (Courtesy of Mr. Fathi Farimani, Delta, B.C., Canada.)

Figure 6.87. Tomato seedlings started in strips of capillary matting. (Courtesy of Mr. Fathi Farimani, Delta, B.C., Canada.)

Plants grown on capillary matting could be placed on the ridge of the NFT channel splitting the roots into both channels (Fig. 6.88). Their subsequent growth was satisfactory.

Figure 6.88. Tomato plants started in capillary matting have a split root system by placing them on the ridge of the NFT channels. (Courtesy of Mr. Fathi Farimani, Delta, B.C., Canada.)

Nutrient solution entered one end of the bed on both sides of the channel (Fig. 6.89). One side of the channel received a low-level nutrient solution while the other side received a high-level. But, they were not irrigated at the same time. While one side was irrigated, the other side had no solution as it was for air exchange. With the cover on the bed, a 100% relative humidity level was maintained in the channel without solution. As a result, roots were able to grow. The high-nutrient solution was circulated for 15 minutes, while the low-nutrient flowed for 3½ hours during each cycle. Nutrient tanks were located under the floor of the greenhouse. The pumping cycles were controlled by time clocks.

In the Ariel NFT system, with plants of a divided root system, the gaseous and aqueous phases are separated. The need to closely regulate the nutrient solution flow-rate in conventional NFT is eliminated. In the Ariel system the nutrient solution flows down only one side of the channel at a time. The time period for alteration of flow down the channels depends upon plant species and weather conditions. This alteration of flow in the

Figure 6.89. Inlets to Ariel NFT channel. Note the split roots of the plant and that one channel is dry while the other has received nutrient solution. (Courtesy of Mr. Fathi Farimani, Delta, B.C., Can.)

channels separates the gaseous and aqueous components of the root environment, and as a result, at all times one side is fully exposed to the air. The channel receiving the solution can use high-flow rates to completely submerge the roots without causing oxygen deficit to the plant as the other half of the root system has no solution flowing in it. Claims of a 10 percent increase in yields are attributed to this increased root oxygenation. Cooper (1985) has demonstrated that each half of the divided root system is capable of supporting all the plant needs.

Complicated monitoring and injection equipment has been replaced with three plastic bins, one for dilute acid (*p*H control) and two for nutrient stock solutions. Each bin is connected to the catchment pipe with a narrow-bore plastic tube from the side wall near the bottom of the bin. Every morning a solution sample is taken and the *p*H and EC are determined. The manager then determines the quantities of *p*H adjuster and stock solutions required for the subsequent 24-hour period. These quantities are placed in the bins each morning and drip slowly via the tubes into the catchment pipe.

Cooper (1985) states that the Ariel system is adaptable to third-world countries due to the simplification of the NFT system. He indicates that the need for pumps and electrical power can be eliminated by circulating the solution, using animal-powered water wheels, etc. Edwards (1985) speculates that technology transfer to third-world countries is possible whereby several skilled technicians working from a service center can provide the technical inputs to many growers using the Ariel system. Monitoring of nutritional needs and simple adjustment procedures would be provided to the growers.

6.15 Outdoor NFT Watercress

An herb grower, California Watercress, Inc., in Fillmore, CA, had been farming 60 acres (24 hectares) of watercress in conventional field beds, but by 1989–1990, due to drought conditions, had to reduce production to close to half. By 1990, the water table had dropped so low that the irrigation water would percolate below the plant roots before reaching the latter part of the growing beds. This lack of water prompted the company to look for alternative methods of growing which would more efficiently utilize the existing water. The author developed for the company a NFT hydroponic system using beds 9 feet (2.75 meters) wide by 500–600 feet (152–183 meters) long. This compared to the conventional field beds which measured 50–60 feet (15–18 meters) wide by in excess of 1,000 feet (305 meters) in length. The field beds were partitioned into 20 ft (6 m) widths using berms.

The hydroponic project was located adjacent to an existing well and source of electricity. The first section of the 3-acre (1.2-hectare) site was leveled with a scraper-leveler, but the subsequent expansion and future area was more accurately leveled with the assistance of laser equipment. The beds had 1 ft (30 cm) of slope lengthwise, but were level across the beds to avoid any localized depressions. Twenty-eight beds were constructed on an area of 3 acres. Four of the beds were 12 feet (3.65 m) in width, center to center of the berms, while the remaining were 9 feet (2.75 m). A 10-ft (3-m) wide road access was allowed after each section of beds.

Berms of approximately 1-ft (30-cm) wide by 6 inches (15 cm) in height were formed using a tractor and two disks on a three-point hitch. A similar berm was made at each end of the beds. The berms not only separated individual beds, but also provided access to the beds for plant maintenance. An underground irrigation system was located about 12 inches (30 cm) below the surface before covering the beds with a 6-mil black polyethylene. Rolls of polyethylene, 100 ft (30.5 m) by 20 ft (6 m), were used to cover two beds and their corresponding berms with sufficient overlap on the outside berms to seal the adjoining pair of beds (Fig. 6.90). The polyethylene on the berms was covered with 2-ft (61-cm) wide nursery weed mat and secured with 6-inch (15-cm) and 9-inch (23-cm) landscape staples (Fig. 6.91).

Figure 6.90. Laying of black polyethylene liner over the beds. (Courtesy of California Watercress, Inc., Fillmore, CA.)

With the raw water containing 150 to 200 ppm of sodium chloride, as well as having high levels of calcium and magnesium carbonate, a recirculating hydroponic system could not be used initially when production was badly needed. Later, closed systems were to be tested on a small-scale. As the growing of watercress required a constant flow of water, the continuous use of a nutrient solution in an open hydroponic system would not have been cost effective. A plan was designed to inject fertilizers periodically into the continuously flowing raw-water irrigation system. The system consisted of two main lines, one a 3-inch PVC raw-water line and the other a 2-inch PVC nutrient-solution line. The flow of water or nutrient

Figure 6.91. Author attaching weed mat on top of berm with landscape staples. (Courtesy of California Watercress, Inc., Fillmore, CA.)

solution to the beds was regulated by two 2-inch solenoid valves on these mains connected to 2-inch PVC submains going across underneath all of the beds at 100 ft (30.5 m) intervals down the bed length (Fig. 6.92). A normally-open solenoid valve on the raw-water line allowed continuous flow of raw water to the beds between fertilizing cycles. On a feeding cycle, a controller would send an electrical current to the valves of one station, closing the normally-open valve and opening the normally- closed valve. This would operate for the preset time interval, generally 5 minutes per cycle, sequentially for each of the five stations. Nutrient solution entered only one station at a time during any given cycle, while the other four stations irrigated with raw water.

Within the first berm of each section, at the edge of the road, a ¾-inch gate valve and riser from the 2-inch submain were located to direct the raw water or nutrient solution into each bed at 50-ft (15-m) intervals via ¾-inch black polyethylene lines (Fig. 6.93). As the hard water plugged drip emitters, ¼-inch drip-line tees and elbows inserted every 2 feet along the ¾-inch black poly lines, regulated the flow of water and solution (Fig. 6.94).

Due to insufficient depth of the well, its partial collapse, and inadequate pressure from its pump, a 5,000-gallon (18,925-l) storage tank with a booster pump to raise the pressure to 60 psi had to be installed. Water entered the tank through a 6-inch PVC main from the well pump and ex-

Figure 6.92. Three-inch diameter main raw-water line on left and 2-inch main nutrient-solution line on right. Two-inch solenoid valves on the 2-inch submains going across the beds.

Figure 6.93. Three-quarter-inch diameter header pipe with gate valve and ¾-inch black poly lines with emitters running across a section of 6 beds. Note the taping of the plastic liner joints.

(Courtesy of California Watercress, Inc., Fillmore, CA.)

ited by a 4-inch PVC main attached to the booster pump. Near the fertilizer injector system, a reduced tee on the end of the 4-inch main split it into two 3-inch mains. One main was positioned along the side of the first bed to provide raw water and the other main looped through the injector system to provide a 2-inch main, after its reduction downstream from the blending tank, to supply nutrient solution to the beds. It was placed beside the 3-inch main raw-water line as described earlier (Fig. 6.92).

The injector system consisted of an Anderson injector with 5 heads, a blending tank, stock solution tanks and a filter (Fig. 6.95) as described in detail in Chapter 3. Concentrated stock solutions of 200 times normal strength were the basis of the nutrient solution.

Two heads were used for stock solution A, two for stock B, and the fifth smaller head for acid to adjust the *p*H. When one of the normally- closed solenoid valves was opened by the time-clock controller, water would move through the injector system loop driving the injector. The resultant nutrient solution formed would then flow into the section of the hydroponic

Figure 6.94. Drip elbow used as an emitter on the ¾-inch black-poly feeder line. Watercress rooted cuttings have recently been placed in the bed at the right. (Courtesy of California Watercress, Inc., Fillmore, CA.)

Figure 6.95. Anderson injector (5 heads) with blending tank and stock solution tanks in back-ground. (Courtesy of California Watercress, Inc., Fillmore, CA.)

beds through the open valve, submain and drip lines. The 1,500-gallon (5,678-liter) stock tanks provided solution for one week.

A modified lettuce nutrient formulation was used for growing the watercress. As the raw water was high in calcium and magnesium carbonate, as well as boron, a small amount of calcium—but no magnesium or boron—was added. Stock A consisted of: potassium nitrate, calcium nitrate, ammonium nitrate, nitric acid and iron chelate. Stock B included: potassium nitrate, potassium sulfate, monopotassium phosphate, phosphoric acid and the micronutrients. The *p*H of the stock solutions was maintained below 6.0 as the raw water was very basic. A 10%-sulfuric-acid stock solution adjusted the final nutrient solution by entering the injector loop downstream from the entrances of stocks A and B.

The watercress was grown in a modified NFT system using a capillary-mat medium. As the black polyethylene bed liner did not provide anchoring for the plant roots and a thin film of water would not spread across the beds, some medium was needed. The beds were covered with a ⅛-inch thick, 100% polyester matting through which roots could grow and water could spread laterally (Fig. 6.96).

Because of the 1-ft (30.5-cm) slope in the beds lengthwise, water accu-

Figure 6.96. Placing capillary matting into the growing beds. (Courtesy of California Watercress, Inc., Fillmore, CA.)

mulated in the latter half of the beds. If was found, therefore, that the capillary matting was only required in the upper half of the beds and the lower half could be covered with a nursery weed matting which provided anchorage for the plants and spread the water across the beds. Once the plants became established, their roots would retain a lot of water, and in fact at the lower ends of the beds a 1-inch to 2-inch (2.5 cm–5 cm) depth of water was held. Between cropping, the mats were removed, cleaned and replaced after drying. They can be sterilized with a 10% bleach solution.

Watercress was seeded directly onto the capillary matting with a "whirlybird" hand seeder. Depending upon temperatures, germination would take place from 5 to 8 days. Initially, during the first phase of the project, beds were seeded followed by thinning and transplanting to new beds being constructed. It took about 6 weeks from seeding to tranplant stage of 3 inches (7–8 cm) in height. Transplants were pulled out of the matting, placed in crates and then spread onto the moistened matting of the new beds (Fig. 6.97).

It took 3 to 4 weeks from transplanting until first harvest. Thereafter, harvesting was done every 28 to 30 days so that continuous production of one bed per day could be achieved. During the winter months it would take up to 45 days between harvesting. While watercress can also be vegetatively propagated by use of stem sections after harvesting the top 7

Figure 6.97. Transplanting watercress seedlings in recently cleaned bed. (Courtesy of California Watercress, Inc., Fillmore, CA.)

to 8 inches (18–20 cm) of growth, more uniform growth was obtained by seedling transplants. Once the 28 beds were completed, the first three beds were used only for the growing of seedlings. Fifty feet (15 m) of seedling bed produced sufficient transplants for a 600-ft (183-m) grow bed. Due to the small size of watercress seed (less than 0.5 mm in diameter), the flow of water in the capillary matting carried the seed downstream to accumulate behind the drip lines and in areas having less depth of water. To overcome this problem, the seedling beds were filled with 1 inch (2.5 cm) of pea gravel, and overhead sprinklers were installed on the berms to unifomly moisten the medium during germination (Fig. 6.98). Transplants were easily removed from the gravel with roots intact.

Figure 6.98. Use of pea gravel in seedling beds. Note the 5,000-gallon water storage tank on the left. (Courtesy of California Watercress, Inc., Fillmore, CA.)

Plants were grown under two cropping cycles: a summer cycle from May through October and a winter cycle from November through April. With slower growth during the winter cycle, 4 to 5 harvests per crop were cut before replacement of the plants, whereas a single harvest cropping schedule was in effect in the summer. During the summer cycle, continuous production resulted from daily harvesting, seeding, transplanting and cleaning. Combined with the 25 production beds were three seedling beds growing the transplants. A 50-ft (15-m) section seeded each day produced transplants within 4 weeks.

Normal production averaged one dozen bundles per lineal foot of bed in the 9-ft (2.75-m) wide beds. During the summer cycle, daily production ranged from 500 to 600 dozens per bed. Watercress is cut by hand (Fig. 6.99), bundled with "twist ties" (Fig. 6.100) and transported to a cooling, washing and packing facility. In comparison to field-grown watercress, the hydroponic watercress was taller, larger-leaved, more tender and milder in flavor (Fig. 6.101). The hydroponic product, due to its succulence, must be handled more carefully to avoid bruising. Individual packaging, instead of bulk shipping, would help in avoiding damage during shipping.

Figure 6.99. Harvesting of watercress by hand. (Courtesy of California Watercress, Inc., Fillmore, CA.)

Figure 6.100. Bundling of watercress. (Courtesy of California Watercress, Inc., Fillmore, CA.)

Figure 6.101. Field-grown watercress on left compared to hydroponic-grown on the right. (Courtesy of California Watercress, Inc., Fillmore, CA.)

The most troublesome pests in this crop are aphids and fungus gnats. These were controlled using Pyrenone and Safer's Soap (M-pede) which could be applied up to one day prior to harvest. Floating duckweed was a problem if introduced by workers not washing their boots, or reusing crates that had some weeds adhering to them. This was resolved by one crew working only in the hydroponic beds, and by sterilizing the crates in a 10% bleach dip prior to reusing.

With careful management a vigorous crop would be ready to harvest within 30 days (Fig. 6.102).

Figure 6.102. Watercress 30 days after previous cut being harvested. (Courtesy of California Watercress, Inc., Fillmore, CA.)

6.16 NFT Basil and Mint

Basil and mint have also been grown with the NFT capillary-mat system using recirculation of the nutrient solution. The facility was located in a greenhouse 30 ft by 156 ft (9 m by 48 m). The greenhouse having 6 ft (1.8 m) sidewalls, was covered with double polyethylene. The polyethylene on the sides was raised during warm weather for ventilation.

The hydroponic components included a 1,800-U.S. gallon (6,800-liter) concrete nutrient-tank, drip irrigation, growing beds, and catchment-return pipes. Forty beds, constructed of lumber, measured 6 ft by 12 ft (1.8 m by 3.65 m) on one side of a 3 ft (0.9 m) central passageway and 6 ft by 14 ft (1.8 m by 4.27 m) on the other side (Fig. 6.103). The beds were 36 inches (91 cm) high for ease of working with the plants. Nutrient solution was pumped from the nutrient tank via a 1½-inch PVC main buried beneath the floor and teed at each pair of beds to a ¾-inch riser up to the beds (Fig. 6.104). Black polyethylene ½-inch diameter lines ran from the riser along the center of each 3-foot-wide growing channel formed by a 2 inch by 2 inch wooden partition on the top of the plywood bed. A gate valve on each riser and ¼-inch tees every 12 inches (30.5 cm) along the black poly line (Fig. 6.103) regulated the flow of solution.

Figure 6.103. Basil growing in NFT capillary-mat system. Note the raised beds, ¾-inch diameter risers and ½-inch diameter black poly feeder lines with elbow and tee emitters. (Courtesy of California Watercress, Inc., Fillmore, CA.)

The solution was collected by a gutter on the lower end which drained into the 3-inch PVC return pipe to the cistern (Fig. 6.105). The tables sloped 4 inches (10 cm) from the inlet to outlet ends. The table tops were covered with 6-mil thick black polyethylene to retain all moisture. The nutrient tank level was maintained by use of a float valve on the raw-water inlet (Fig. 6.104). Solution passed through a filter on the 1½-inch main inlet line when pumped to the beds. As it was found that excessive moisture buildup occurred in the beds once the crop became established, a time-clock controller was installed to operate the pumping cycles for 10 minutes each hour during the day and several times during the night.

Figure 6.104. Nutrient tank with circulation pump, filter on left and automatic-fill ball-valve in tank. (Courtesy of California Watercress, Inc., Fillmore, CA.)

Basil was seeded in 98-celled growing trays with perlite medium. Plants were thinned to two per cell after 4 weeks. They were transplanted to the growing beds when they were 5 weeks old. Spaced at 6-inch (15-cm) centers, 246 plants were transplanted per 12-ft (3.6-m) bed. Transplants grew vigorously into the capillary matting within 10 days (Fig. 6.103). The first harvest was done 3½ weeks after transplanting (2 months from seeding) (Fig. 6.106). Thereafter, the crop was harvested every 3 weeks.

Due to the continued presence of moisture in the capillary matting, algae grew on the matting, producing an ideal environment for fungus gnats. They had to be controlled by spraying Pyrenone and Safer's Soap (M-pede) weekly. A second problem arose due to the high moisture levels. Basil does not like excessive moisture at its crown, so as the plants

Figure 6.105. Collection gutters and return pipe to cistern from beds on one side. (Courtesy of California Watercress, Inc., Fillmore, CA.)

Figure 6.106. Basil (3 weeks after transplanting) in NFT capillary-mat system. (Courtesy of California Watercress, Inc., Fillmore, CA.)

matured and growth slowed, they began to die from puddling. After four harvests the plants were removed and replaced with mint which thrives in high moisture. The capillary mats were removed, washed, sterilized with a 10%-sodium-hypochlorite solution, rinsed and dried before replacing them in the beds.

Mint cuttings were taken from field-grown plants. They were washed to remove soil and weeds, then transported to the greenhouse for transplanting (Fig. 6.107). They were unifomly spread over the entire capillary matting surface of the beds. Overhead sprinklers moistened the plants for 20 minutes every hour during the day to prevent desiccation during their initial rooting for the first two weeks (Fig. 6.108). Nutrient solution was circulated after they rooted. The nutrient formulations used for most of the herbs were basically similar to that of the watercress.

Harvesting began within 5 weeks from transplanting and continued on a 5- to 7-week cycle depending upon the season (Fig. 6.109). The short days of winter months slowed growth despite the use of heating in the greenhouse. Similar to the basil, harvesting is done by hand. Depending upon market demand, 3 to 4 beds are harvested each day over a period of two weeks. Similar to watercress, mint and basil are bundled using rubber bands, washed and refrigerated in wax boxes containing ice-packing.

While fungus gnats were still a problem with the mint, their populations were not as high. There was less problem with algae once the mint was fully established, blocking out light from reaching the capillary matting. A large root-mass formed at the base of the capillary matting (Fig. 6.110). Puddling was not a problem, as mint prefers moisture at its crown, forming adventitious roots and additional plants. The mint crop was still healthy after two years.

6.17 Advantages of the NFT Systems

The advantages of the nutrient-film technique in glasshouse crop production are:

1. Low capital cost.
2. Elimination of soil sterilization and preparation.
3. Rapid turnaround between crops.
4. Precise control of nutrition.
5. Maintenance of optimal root temperatures by heating of the nutrient solution (77° F for tomatoes, 84° F for cucumbers).
6. Simplicity of installation and operation.
7. Reduction of transplanting shock by use of growing pots or cubes and preheating of nutrient solution to optimal root temperatures.
8. Easy adjustment of nutrient-solution formulation to control plant growth under changing light conditions.

Figure 6.107. Mint cuttings placed on capillary-matting NFT system. (Courtesy of California Watercress, Inc., Fillmore, CA.)

Figure 6.108. Mint cuttings 14 days after transplanting. Note the over-head sprinklers. (Courtesy of California Watercress, Inc., Fillmore, CA.)

Figure 6.109. Mint 5 weeks after transplanting, ready to harvest. (Courtesy of California Watercress, Inc., Fillmore, CA.)

Figure 6.110. Large root-mass of mint formed in capillary matting of NFT system. (Courtesy of California Watercress, Inc., Fillmore, CA.)

9. Use of systemic insecticides and fungicides in the nutrient solution to control insects and diseases of ornamental crops.
10. Possible energy savings by keeping the greenhouse air temperature at lower than normal levels due to maintenance of optimal root temperatures.
11. Elimination of plant water-stress between irrigation cycles by continuous watering.
12. Conservation of water by use of a cyclic system rather than an open system.

The future of successful crop production lies in a universal cropping system in which water and fertilizers can be used efficiently. This is particularly true in arid regions of the world, such as the Middle East, where land is nonarable and water is scarce. In such areas desalinated sea water can be used, but is very costly. Sand culture is used in many of these areas, but often it is of calcareous nature which causes rapid changes in *p*H and tying-up of essential elements such as iron and phosphorus. In these areas of high solar energy, efficient use of costly desalinated water is essential. The nutrient-film technique is the system which makes efficient use of water and fertilizers and at the same time does not rely on a suitable medium such as noncalcareous sand or gravel. I believe that the future of soilless culture in such harsh environments lies in the NFT systems.

References

Barrow, Joseph M. 1980. Hydroponic culture of grapes in the tropics. *Proc. 5th Int. Congress on Soilless Culture*, Wageningen, May 1980, pp. 443–451.

Cooper, A. J. 1973. Rapid crop turn-round is possible with experimental nutrient film technique. *The Grower*, May 5, 1973, pp. 1048–51.

———. 1974. Soil? Who needs it? Part I. *Am. Veg. Grower*, August 1974, pp. 18–20.

———. 1974. Soil? Who needs it? Part II. *Am. Veg. Grower*, September 1974, pp.13, 64.

———. 1975. Rapid progress through 1974 with nutrient film trials. *The Grower*, Jan. 25, 1975.

———. 1976. *Nutrient film technique of growing crops*. London: Grower Books.

———. 1985. New ABC's of NFT. *Hydroponics Worldwide: State of the Art in Soilless Crop Production*. Ed., Adam J. Savage. pp. 180–185, Int. Center for Special Studies, Honolulu, Hawaii.

———. 1987. NFT developments and hydroponic update. *Proc. 8th Ann. Conf. of the Hydroponic Society of America*, San Francisco, Calif., April 4, 1987, pp. 1–20.

———. 1987. Hydroponics in infertile areas: Problems and techniques. *Proc. 8th Ann. Conf. of the Hydroponic Society of America*, San Francisco, Calif., April 4, 1987, pp. 114–121.

Douglas, J. Sholto. 1976. Hydroponic layflats. *World Crops*, March/April 1976, pp. 82–87.

Edwards, Kenneth. 1985. New NFT breakthroughs and future directions. *Hydroponics Worldwide: State of the Art in Soilless Crop Production.* Ed., Adam J. Savage. pp. 42–50. Int. Center for Special Studies, Honolulu, Hawaii.

Goldman, Ron. 1993. Setting up a NFT vegetable production greenhouse. *Proc. 14th Ann. Conf. on Hydroponics.* Hydroponic Society of America, Portland, Oregon, April 8–11, 1993, pp. 21–23.

New twist for hydroponics. *Am. Veg. Grower*, November 1976, pp. 21–23.

NFT culture—incalculable potential. *The Grower*, Feb. 14, 1976.

Nutrient film technique: cropping with the hydrocanal commercial system. 1976. *World Crops* 28:212–18.

Resh, H.M. 1976. A comparison of tomato yields, using several hydroponic methods. *Proc. 4th Int. Congress on Soilless Culture*, Las Palmas, Oct. 25–Nov. 1, 1976.

———. 1990. A world of soilless culture. *Proc. 11th Ann. Conf. on Hydroponics.* Hydroponic Society of America, Vancouver, B.C., March 30–April 1, 1990, pp. 33–45.

———. 1993. Outdoor hydroponic watercress production. *Proc. 14th Ann. Conf. on Hydroponics.* Hydroponic Society of America, Portland, Oregon, April 8–11, 1993, pp. 25–32.

Robinson, Jean. 1976. Soil—a thing of the past. *Gardener's Chronicle*, July 23, 1976, pp. 22–25.

Schippers, P.A. 1977. Soilless culture update: Nutrient flow technique. *Am. Veg. Grower*, May 1977, pp. 19,20,66.

———. 1980. Hydroponic lettuce: the latest. *Am. Veg. Grower* 28(6):22, 23,50.

Tube Culture—a challenging idea. *Am. Veg. Grower*, November, 1974, pp. 46,47.

Chapter 7

Gravel Culture

7.1 Introduction

Gravel culture was the most widely used hydroponic technique from the 1940s through the 1960s. During the late 1960s and early 1970s one commercial operation, Hydroculture, near Phoenix, Arizona had almost 20 acres of greenhouses in gravel culture. Many smaller commercial greenhouses throughout the United States used gravel culture as it was well documented and several companies such as Hydroculture were selling "mom & pop" packages to people interested in setting up their own hydroponic greenhouse.

W.F. Gericke introduced hydroponics commercially using gravel culture. It was used in most outdoor operations established during World War II on nonarable islands, as mentioned in Chapter 1. Gravel culture is still useful in areas having an abundance of volcanic rock, such as the Canary Islands and Hawaii. Today, NFT, rockwool and perlite cultures are more widely accepted as they have more consistent properties, are easier to sterilize between crops and are less laborious to handle, maintain and manage.

7.2 Media Characteristics

Some of the general characteristics media should possess have been discussed in Chapter 4. The best choice of gravel for a subirrigation system is crushed granite of irregular shape, free of fine particles less than $\frac{1}{16}$ inch (1.6 mm) in diameter and coarse particles more than $\frac{3}{4}$ inch (1.9 cm) in diameter. Over half of the total volume of particles should be about $\frac{1}{2}$ inch (1.3 cm) in diameter. The particles must be hard enough so they do not break down, able to retain moisture in their void spaces, and drain well to allow root aeration.

The particles should not be of calcareous material—in order to avoid *p*H shifts. If only calcareous material is available, the amount of calcium and magnesium in the nutrient solution will have to be adjusted according to the levels of these elements released by the aggregate into the nutrient solution. The calcium carbonate in calcareous aggregates such as limestone and coral gravels reacts with the soluble phosphates of the nutrient solutions to produce the insoluble di- and tri-calcium phosphates. This process continues until the surfaces of the calcareous aggregate particles are coated with insoluble phosphates. After they are thoroughly coated, the reaction slows down to a point at which the rate of decrease of phosphates in the nutrient solutions is slow enough to maintain phosphate levels.

A new unused aggregate containing more than 10 percent of acid-soluble materials calculated as calcium carbonate should be pretreated with soluble phosphates to coat the particles with insoluble phosphates (Withrow and Withrow, 1948). The aggregate is treated with a solution containing from 500 to 5,000 grams of treble superphosphate per 1,000 liters or 5 to 50 pounds per 1,000 gallons. The new gravel should be soaked for several hours. The *p*H of the solution will rise as its phosphate content decreases. If the phosphate concentration drops below 300 ppm (100 ppm of P) after one to two hours of soaking, the solution should be drained to the reservoir and a second phosphate addition made to the solution, which is repumped into the beds. This should be repeated until the phosphate level stays above 100 ppm (30 ppm of P) after several hours of exposure to the gravel. When this occurs, it indicates that all of the carbonate particles have been coated with phosphates. The *p*H of the solution will then remain about 6.8 or less. The phosphate solution then is drained from the reservoir and the reservoir filled with fresh water. The beds are flushed with this fresh water several times and once again the reservoir drained and refilled with fresh water. The beds then may be planted.

Over time the *p*H will begin to rise as the phosphates become depleted and free calcium carbonate is exposed on the surface of the aggregate. The process of phosphate treatment will then have to be repeated.

Schwarz and Vaadia (1969) have demonstrated that pretreated calcareous gravel or washed calcareous gravel could not prevent lime-induced chlorosis. The high *p*H of calcareous gravel also makes iron unavailable to plants. Victor (1973) found that daily addition of phosphorus and iron at the rate of 50 milliliters of phosphoric acid and 12 grams of chelated iron (FeEDTA) per 1,000 gallons (3,785 liters) of nutrient solution prevented the occurrence of lime-induced chlorosis with tomato plants.

"Haydite" or "Herculite" fired shale, available in a variety of particle sizes, is often used by smaller backyard hydroponic units. It is porous and has given good results in many cases. However, after continued use, the adsorption of fertilizer salts on the surfaces of the particles may cause difficulty. The absorbed salts are not easily removed by washing. Plant roots become lodged in the small pores of the rock surface, making sterilization of the medium between crops difficult. Also, the material fractures and breaks up into small pieces which eventually form a fine sand and silt which plugs feeder lines and pipes.

If a spaghetti feeding system is used rather than a subirrigation system, a smaller medium must be used. For such a system, "pea" gravel between ⅛ inch and ⅜ inch (3 mm–10 mm) in diameter should be used. Over half of the total volume of gravel should have a particle size of 3⁄16–¼ inch (5 mm–6 mm) in diameter. No silt or particles larger than ⅜ inch (10 mm) in diameter should be present. The Haydite gravel is particularly suitable to a spaghetti feeding system because its capillary action moves the nutrient solution laterally around the plant root system. However, as plant roots grow laterally they also will intercept the water and cause it to flow laterally.

7.3 Subirrigation Gravel Culture

Almost all gravel culture uses a subirrigation system. That is, water is pumped into the beds and floods them within several inches of the surface, then drains back to the nutrient reservoir. Such a system is termed *closed* or *recycled* since the same nutrient solution is used each pump cycle over a period of from 2 to 6 weeks. Then the solution is disposed of and a new solution made up.

The frequency and length of time of irrigation cycles is important to the success of the system. Each irrigation cycle must provide adequate water, nutrients and aeration to the plant roots.

7.3.1 Frequency of Irrigation

The minimum frequency of irrigation depends upon:

1. The size of the aggregate particles.
2. The surfaces of the aggregate particles.
3. The nature of the crop.
4. The size of the crop.
5. Climatic factors.
6. Time of day.

Smooth, regularly shaped, coarse aggregates must be irrigated more frequently than porous, irregularly shaped, fine aggregates having large surface areas. Tall crops bearing fruit require more frequent irrigation than short-growing leafy crops such as lettuce, due to their greater surface area and therefore higher evapotranspirational losses. Hot, dry weather promotes rapid evaporation and makes more frequent irrigation necessary. During mid-day when light intensities and temperatures are highest, the period between irrigation cycles must be reduced.

For most crops the aggregate must be irrigated at least 3 to 4 times per day during dull winter months, and in summer it is often necessary to irrigate at least every hour during the day. Pumping at night is not necessary. In temperate zones during the summer, irrigation should be from 6:00 a.m. to 7:00 p.m., while during the winter an 8:00 a.m. to 4:00 p.m. period may be set up.

Water is absorbed by the plant from the nutrient solution much more rapidly than the inorganic elements. As a result, the films of nutrient solution on the aggregate become concentrated with inorganic salts as absorption takes place. The concentration of the nutrient solution in the aggregate becomes greater as the rate of transpiration of the plants increases, and the rate of water absorption thus increases. By increasing the number of irrigation cycles, the high water demand of the plant is met and the water content within the void spaces of the gravel is maintained at a more optimal level. In this way the nutrient solution does not become so concentrated between irrigation cycles. It is also necessary to irrigate frequently so that the films of nutrient solution over the aggregate particles in contact with the root tips are not depleted of any nutrient between irrigations.

Immediately following irrigation, the nutrient solution in the aggregate has nearly the same composition as the reservoir solution. As absorption proceeds, the composition of the nutrient solution in the bed continually changes, including the proportion of the various ions, the concentration of the solution and the *p*H. If the frequency of irrigation is not sufficient, nutrient deficiencies may develop, even though adequate quantities of the nutrient elements may be present in the reservoir solution. If the aggregate does not contain a large proportion of fine particles, it is unlikely that frequent irrigation will cause aeration problems so long as the beds are completely drained between irrigations and the irrigation period is not too long. The more often the beds are irrigated, the more nearly the composition of the aggregate solution approaches that of the reservoir solution.

7.3.2 Speed of Pumping and Drainage

The speed of pumping and draining of the nutrient solution from the gravel determines the aeration of the plant root system. Roots require oxygen to carry on respiration, which in turn provides energy needed in the uptake of water and nutrient ions. Insufficient oxygen around the plant roots retards their growth or may cause their death, which results in plant injury, reduced yields and eventually plant death.

In a subirrigation system, as the nutrient solution fills the gravel voids from below, it pushes out air which has a relatively low content of oxygen and a high content of carbon dioxide. Then, as the nutrient solution flows from the gravel, it sucks air into the medium. This new supply of air has a relatively high content of oxygen and a low content of carbon dioxide. The greater the speed of solution movement in the gravel medium, the greater the speed of air displacement. Also, since the solubility of oxygen in water is low, the period of low oxygen supply while the free water is in the gravel is shortened if the speed of filling and draining is rapid.

This factor of aeration is also related to the frequency of pumping. When free water is present too often in the gravel, the voids (air spaces between the particles) are filled with water rather than with moist air. Thus the oxygen concentration around the roots is lowered.

A 10- to 15-minute period for filling and draining respectively, or a total time of 20 to 30 minutes, is generally acceptable. In the removal of the free solution from the bed, nothing less than complete drainage is recommended. Only a film of moisture on the gravel particles is desired. If puddles of solution remain in the bottom on the plant bed, poor plant growth results. This rapid filling and draining of the growing beds can be achieved by use of large pipes in the bottom of the beds. In summary, proper irrigation cycles should: (a) fill the bed rapidly; (b) drain the bed rapidly and (c) get all the solution out.

7.3.3 Effect of the Irrigation Cycle on Plant Growth

By reducing the number of nutrient solution pumpings, the moisture content in the gravel medium is lowered. This has a concentrating effect on the nutrient ions contained in the water film on the gravel particles. Whenever the osmotic concentration of the nutrient solution is increased, the water-absorbing power of the plant is reduced, also reducing the nutrient ion uptake. Consequently, the rate of plant growth is retarded and a harder and firmer type of growth develops.

In the greenhouse during the dark, cloudy, short days of winter, reduction of the number of irrigation cycles per day will help keep the plants reasonably hard and sturdy.

7.3.4 Height of Irrigation

The nutrient solution level should rise to within about one inch of the surface of the aggregate. This practice keeps the aggregate surface dry, preventing the growth of algae, and reduces the loss of water and high humidity buildup at the base of the plants. It also prevents the growth of roots into the surface inch of the aggregate, which, under conditions of high light intensities, may become too high in temperature for satisfactory root growth. The nutrient level in the beds can be regulated by installation of overflow pipes in the plenum. These construction details are discussed in Section 7.3.6.

7.3.5 Nutrient Solution Temperature

In greenhouse culture, the temperature of the nutrient solution should not fall below the night air temperature of the house. The temperature of the solution in the reservoir sump may be raised by use of an immersion heater or electric heating cable. Be careful not to use any heating elements having lead or zinc sheathing as these can be toxic to the plants. Stainless steel or plastic coated cables are best. Electric heating lamps have also been used successfully. In many locations the temperature of outside water used to fill the sump may be as low as 45° to 50° F (7° to 10° C). If a large reservoir needs to be filled, it should be done later in the day after the last irrigation cycle has been completed. Then there will be sufficient time before the next irrigation cycle the following morning for the heating unit to bring the solution temperature to a more optimal level.

In no case should heating cables be placed in the growing beds themselves. This causes localized high temperatures around the heating elements, which injure the roots.

7.3.6 Greenhouse Subirrigation System

Construction Materials. Since fertilizer salts used in making up the nutrient solution are corrosive, any metal parts, such as pumps, pipes or valves, exposed to the nutrients will wear out in a very short time. Galvanized materials may release sufficient zinc to cause toxicity symptoms in the plants. Copper materials offer the same problem. Plastic pipes and fittings, pumps with plastic impellers, and plastic tanks are noncorrosive and should be utilized. Plant beds can be built of wood and lined with at least 6-mil plastic film, but preferably 20-mil vinyl.

During World War II, concrete was used in numerous commercial operations. It has the advantages of permanence and corrosion resistance, but the cost is the highest of any material. Cedar or redwood is the most often used construction material, and direct lining of compacted normal ground substrate is cheapest for bed construction.

Beds. The beds must be designed to provide rapid filling and draining—and complete drainage. The use of a 3-inch (7.6 cm) diameter PVC pipe and a V-shaped bed configuration will fulfill these watering requirements (Fig. 7.1). The beds should have a minimum width of 24 inches (61 cm), a depth of 12 inches to 14 inches (30.5 cm to 35.5 cm) and a maximum length of 120 to 130 feet (36.5 to 40 meters). The beds should slope from 1 to 2 inches (2.5 to 5 cm) per 100 feet (30.5 m). Water enters and drains from the beds through small ¼- to ½-inch (0.64- to 1.27-cm) diameter holes or ⅛-inch-thick (0.32-cm) sawcuts on the bottom one-third of the PVC pipe. These holes or sawcuts are made every 1 to 2 feet (30.5 to 61 cm) along the entire length of the pipe.

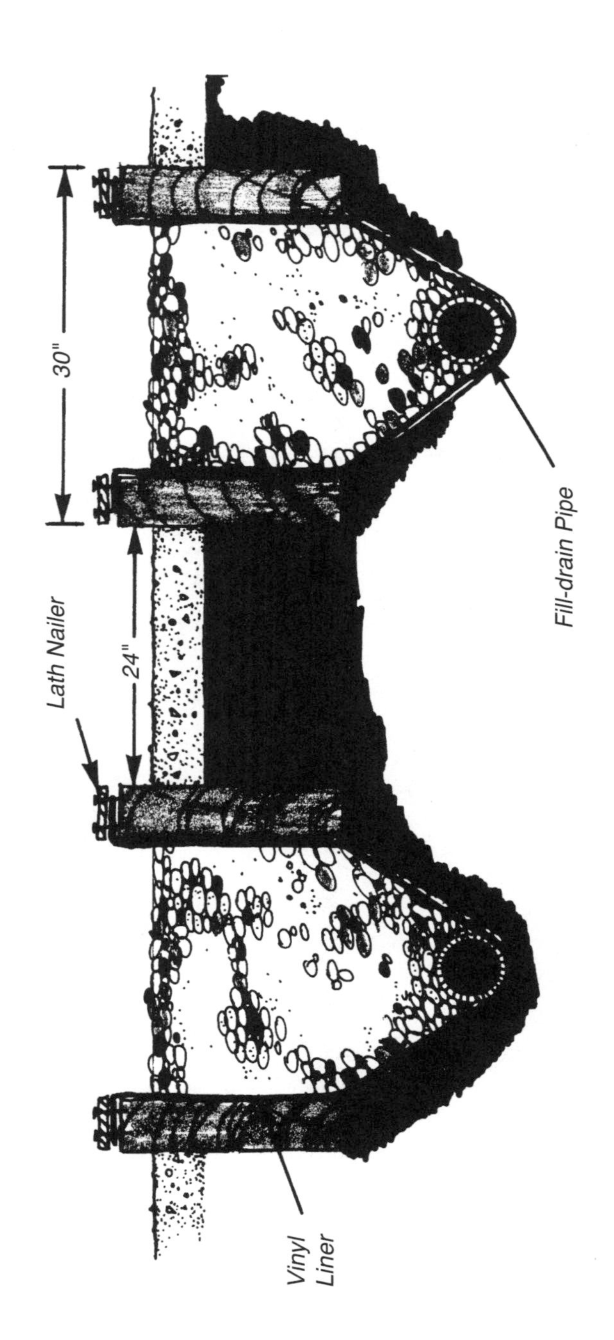

Figure 7.1. Cross section of a subirrigation gravel bed.

The slope can be achieved by sloping the side boards and staking them. In some cases growers may pour concrete walkways between these beds. The beds should be constructed in compacted river sand. A jig containing the desired configuration can be used to dig out the bed (Fig. 7.2). Once the proper configuration and slope has been compacted, the beds are lined with a vinyl liner of 20-mil thickness commonly used for swimming pools (Fig. 7.3). The 3-inch diameter PVC fill-drain pipe is then set in place with the holes or cuts facing downward to prevent roots readily growing into the pipe. The vinyl is held on the sides of the beds by folding it over the rough 2-inch-thick cedar side planks and nailing a wooden strip on top of it the entire length.

Figure 7.2. Digging bed configuration in compacted river-sand fill.

The PVC pipe allows water to flow rapidly along the bottom of the beds, and fills and drains the beds equally and vertically along the entire length. With this rapid filling and draining, old air is pushed out of the aggregate and new air is sucked in as discussed earlier.

The beds should be filled with gravel to within 1 inch of the top at the

Figure 7.3. Vinyl liner placed into beds and PVC drain pipes located in bottom groove.

end near the nutrient tank and within 2 inches at the far end. To prevent uneven near-surface moistening during the irrigation cycles, level the top surface of the gravel. Remember that the beds are not level, but water remains level; so, with the top surface of the gravel leveled, the water level in the flooded beds will be parallel to the gravel surface. (If the gravel were placed to within 1 inch of the top of the bed along the entire length, it really would have a 1-inch slope and therefore if during the irrigation cycle the beds were filled to within 1 inch of the gravel surface at the reservoir end of the beds, it would be 2 inches from the top at the far end. This would cause uneven watering of the plant roots from one end to the other end of the beds. Recently transplanted seedlings might suffer a water stress at the far end due to insufficient height of water in the bed. If the water level were within 1 inch at the distant end, it would come to the surface at the reservoir end, creating algae problems there.)

The 3-inch PVC pipe should have a 45-degree elbow, and project above the gravel surface with a cap at the far end from the tank to allow for cleaning of the pipe. Usually the pipes are cleaned of roots every year or so with a Roto-Rooter machine. The end of the pipe at the reservoir drains into a plenum.

Plenum. The filling and draining times can be greatly decreased by the use of a plenum rather than solid plumbing from the pump(s). The plenum is simply a trough into which the water is pumped from the tank. The

3-inch PVC pipes running along the bottom of the beds open into this plenum (Figs. 7.4, 7.5 and 7.6). The bed pipes must be sealed into the plenum or water will leak behind the tank, resulting in loss of nutrient solution and buildup of water under the greenhouse and tank.

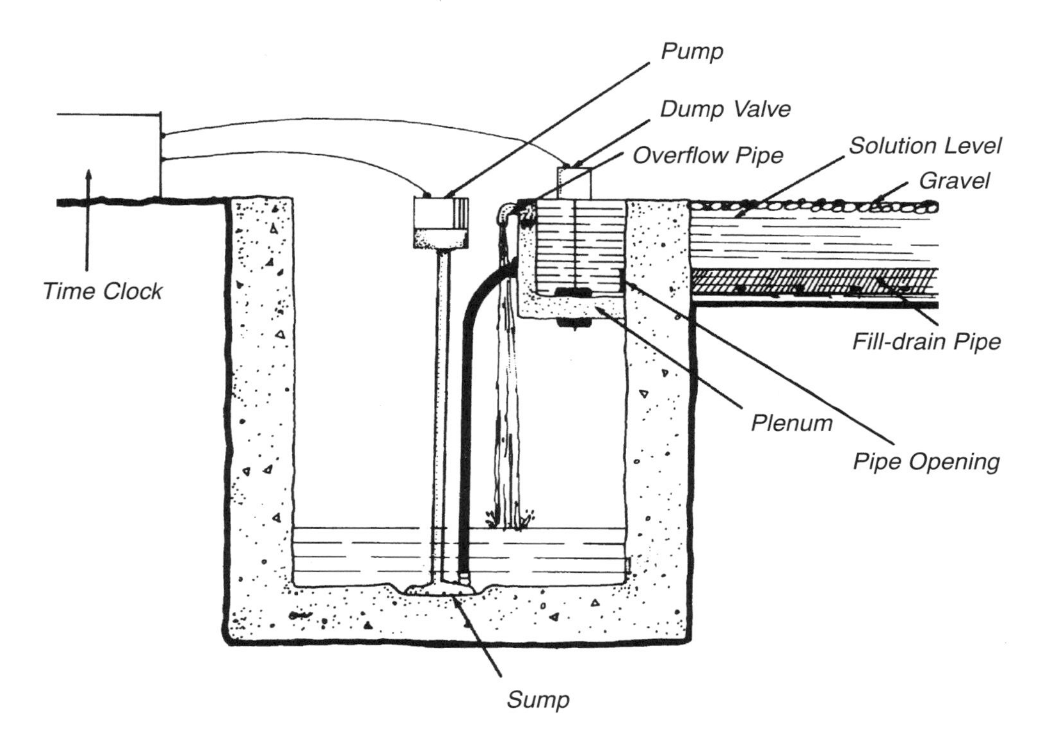

Figure 7.4. Cross section of plenum and nutrient tank.

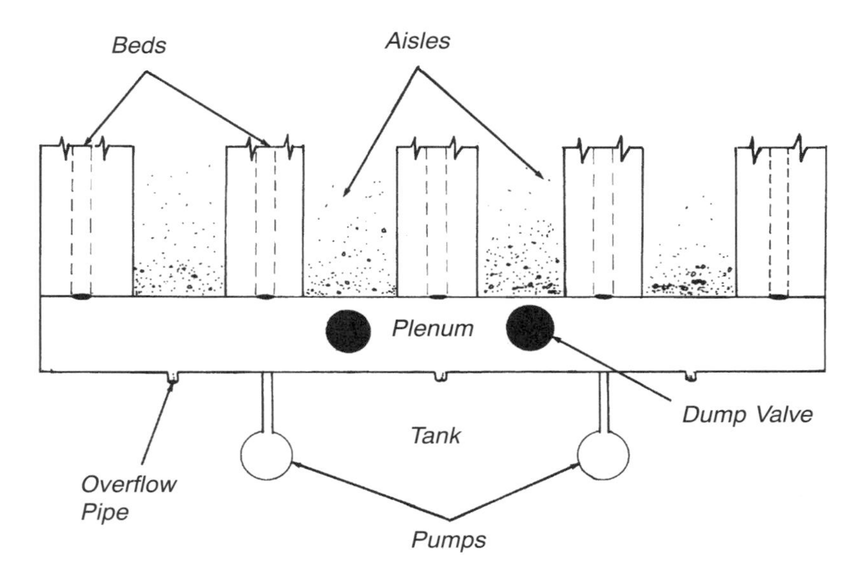

Figure 7.5. Plan view of plenum and nutrient tank.

Figure 7.6. Bed fill-drain pipe entering plenum.

Either sump or submersible pumps may be used to pump the nutrient solution from the reservoir into the plenum. The pump is activated by either a time clock or a feedback mechanism attached to a moisture sensor in the beds. The water level in the beds is regulated by an overflow pipe in the plenum. The plenum and beds will fill to that level which corresponds to 1 inch below the gravel surface in the beds. A dump valve, activated by the same time clock or feedback mechanism as the pumps, closes the drain holes of the plenum while the pumps are operating (Fig. 7.4). The operation is as follows:

1. The moisture sensor in the bed signals the feedback mechanism to turn on the pumps (or the preset time clock strikes an irrigation cycle) and activates the plenum valves.
2. The pumps are activated; the dump valves of the plenum close.
3. The plenum and beds begin to fill.
4. They continue filling until excess solution spills back into the tank via the overflow pipes.
5. The preset irrigation period terminates; the beds are full of solution.
6. The pumps stop, the dump valves fall open, nutrient solution drains from the beds back to the plenum and finally falls into the tank, being well aerated.
7. The whole cycle of pumping and draining should be completed within 20 minutes.

Nutrient Tank. The nutrient tank must be constructed of a watertight material. Steel-reinforced concrete 4 inches (10 cm) thick, coated with a bituminous paint to make the concrete watertight, is the most long-lasting material. The plenum should be a part of the tank so that leakage cannot occur between them. The volume of the tank must be sufficient to completely fill the gravel beds, based on the amount of void space in the gravel. This can be determined by taking a sample of the rock (about a cubic foot) and filling it with water, then measuring the volume of water required to fill the void spaces. Then extrapolate to calculate the total void space in all the beds. The tank should hold a volume 30 to 40 percent greater than the total volume required to fill the beds. For example, a tank of about 2,000 Imp. gallons (9,092 liters) would be required to supply 5 beds 2 feet (61 cm) wide by 12 inches (30.5 cm) deep by 120 feet (36.5 m) long (Fig. 7.7).

An automatic float valve could be attached to a water refill line in order to maintain the water level in the tank. In this way, water lost during each irrigation cycle through evapotranspiration by the plants would immediately be replaced.

The nutrient tank should have a small sump in which the pumps sit in order to facilitate complete drainage and cleaning of the tank during nutrient changes (Fig. 7.4).

Figure 7.7. Nutrient tank construction.

The tank-plenum could be changed somewhat in structure to reduce its size. This can be done (as shown in Fig. 7.8) by dividing the plenum into two parts, each servicing 3 beds. A three-way automatic valve would discharge the nutrient solution alternately into one plenum upon activation of the pumps and dump valve, filling 3 beds during each cycle (Fig. 7.9). In this case a time clock, rather than a moisture-sensor feedback system, must be used, with shorter intervals between irrigation cycles than that required for one larger plenum as described earlier.

The pump could also be piped into a discharge line which could be connected to the outside drainage system. Gate valves would be used to control the direction of flow of the nutrient. When the nutrient solution is changed, the gate valve in the waste line is opened and the solution is pumped out to waste. Normally during regular feeding cycles, this waste valve is closed and the valve between the pump and the three-way valve is open. The three-way valve is available commercially.

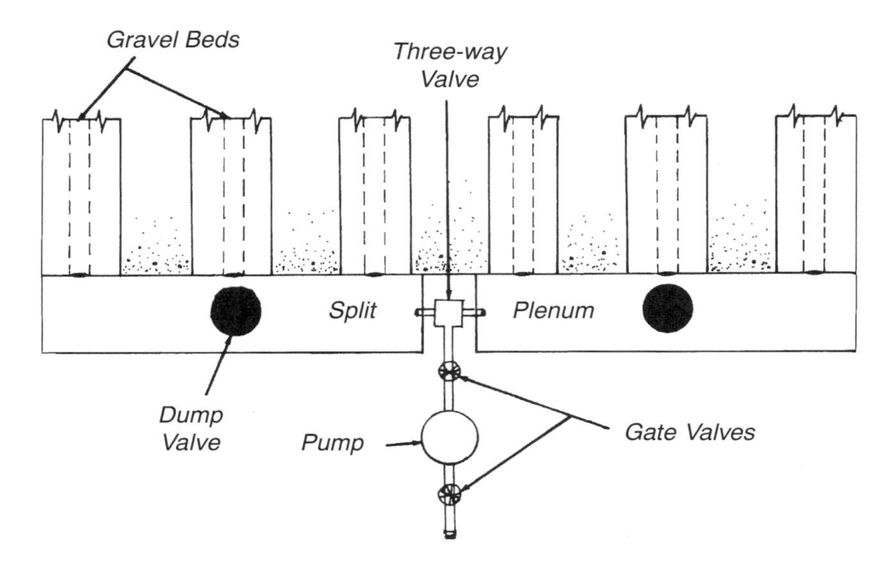

Figure 7.8. Plan view of nutrient tank with a split plenum.

Figure 7.9. Three-way automatic valve used with a split-plenum design.

By using such a split plenum, the tank capacity can be reduced to almost half the volume of one large enough to fill all 6 beds at once. Thus, a tank of 1,200 Imp. gallons (5,455 liters) should easily supply 3 gravel beds 24 inches (61 cm) wide by 12 inches (30.5 cm) deep by 120 feet (36.5 m) long.

An overall plan of a gravel culture system having 6 beds and the attached plenum and nutrient tank is illustrated in Figure 7.10. A profile sketch of an individual Quonset-style greenhouse which would contain 6 gravel beds is given in Figure 7.11. Successful crops of high productivity can be grown with a subirrigation gravel system, as illustrated in Figures 7.12 to 7.14.

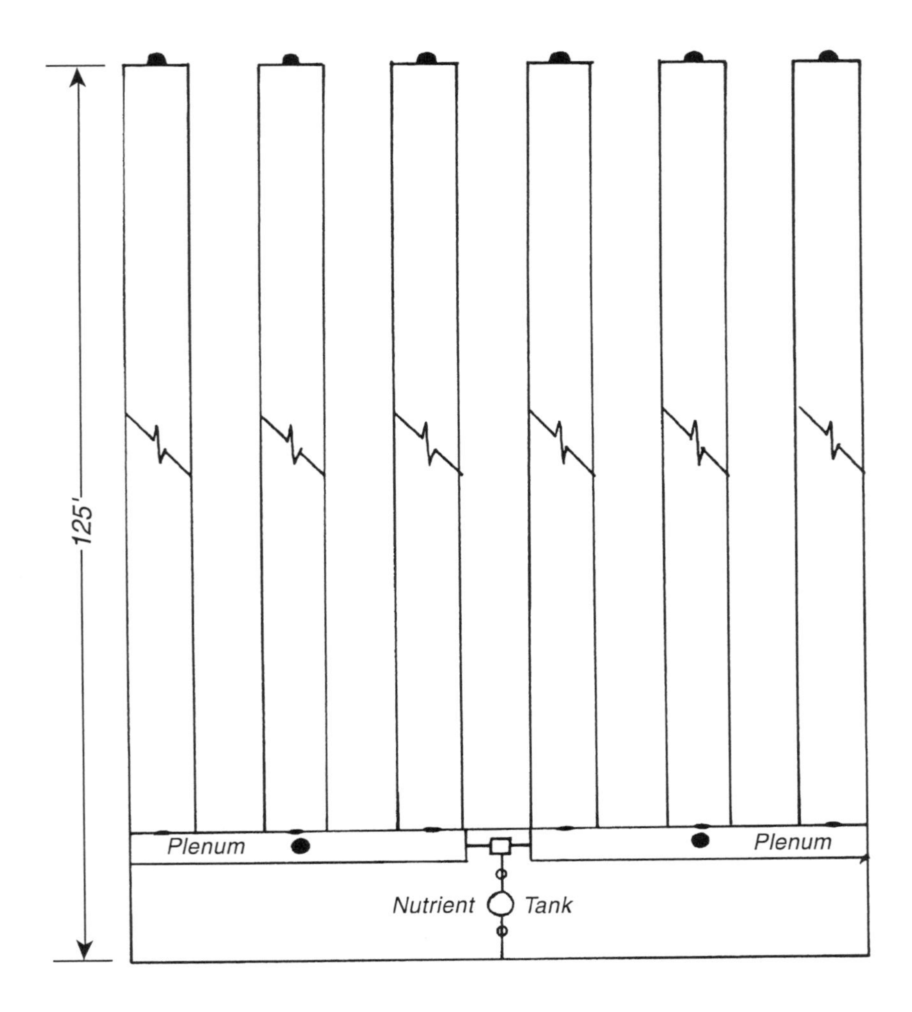

Figure 7.10. Plan of a greenhouse with six gravel beds and split-plenum nutrient tank.

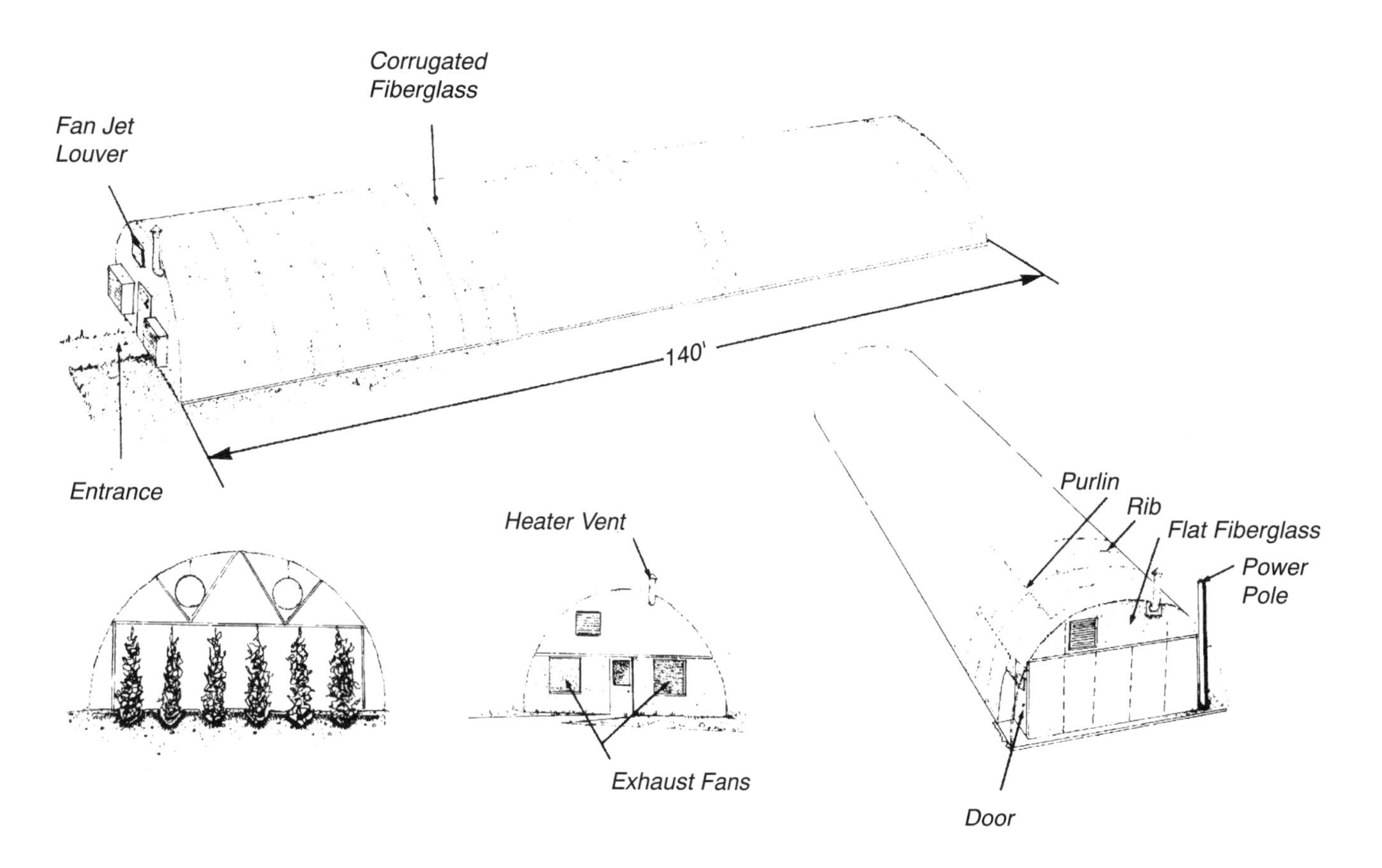

Figure 7.11. Schematic of an individual Quonset-style greenhouse using a subirrigation gravel-culture system.

Figure 7.12. Crop of tomatoes (about 6 weeks old) growing in a subirrigation gravel-culture system.

Figure 7.13. Crop of cucumbers (about 5 weeks old) growing in a subirrigation gravel-culture system.

Figure 7.14. Crop of mature tomatoes ready for harvesting.

7.3.7 Outdoor Subirrigation Gravel-Cultural Systems

Flume System. The flume system, commonly used outdoors, uses a series of beds built on exactly the same level, parallel with one another. A flume or channel (equivalent to a plenum) at least 10 to 12 inches (25 to 30 cm) deep and 18 to 20 inches (46 to 51 cm) wide runs perpendicularly to the beds. The length of the beds depends upon the size of the reservoir. Usually beds no longer than 100 to 120 feet (30.5 to 36.5 m) are built on both sides of the flume. They must be sloped several inches (5 cm) toward the flume to allow gravitational drainage (Fig. 7.15). A reservoir can be constructed above or below ground level into which the flume is connected. If the reservoir is located above ground, a sump must be located at the end of the flume into which nutrient solution will flow from the beds. A pump is then needed to lift the solution from the sump to the reservoir during the draining cycle. The nutrient solution enters the flume via a valve from the above-ground reservoir and flows along the flume into each bed. A valve or sluice-gate must be used to close the flume off from the sump while the beds are filling. When draining the beds this sluice-gate or valve must be opened.

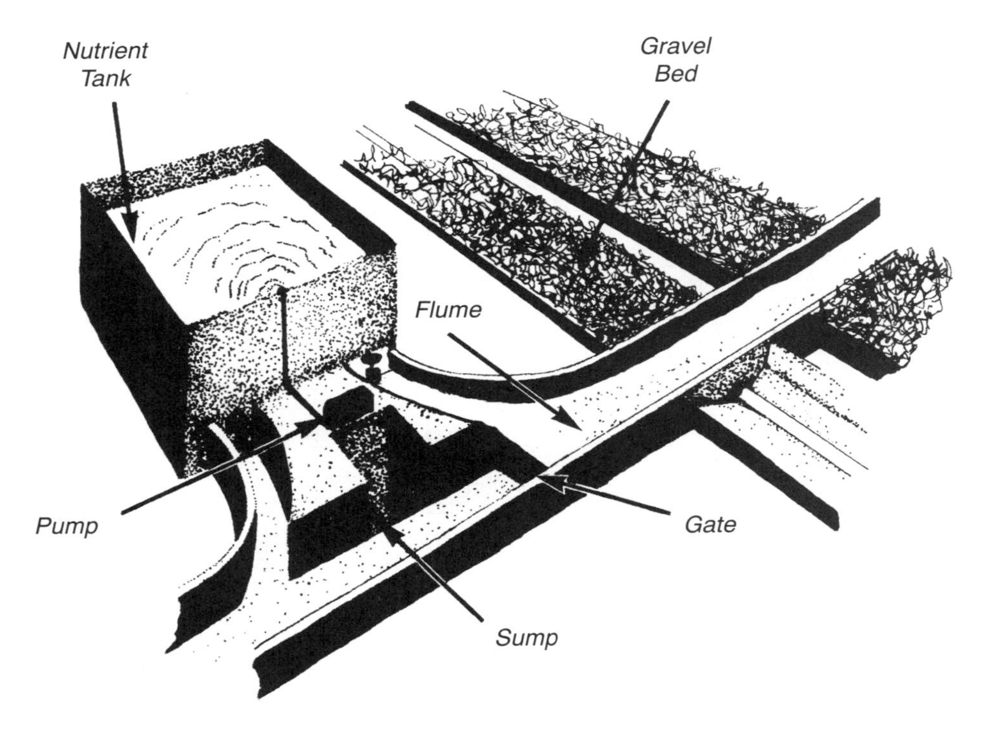

Figure 7.15. A flume system of subirrigation gravel culture using a sump and above-ground reservoir.

If a cistern reservoir is built to take the place of the sump, no other reservoir is required. Then a sump or submersible pump is used to pump the water into the flume. This method is very similar in construction to the one described earlier for greenhouses. A three-way valve can be used to alternately direct the flow of water into one of two flumes, and therefore twice as many beds can be serviced by the same volume of nutrient tank (Fig. 7.16). The cistern could be located underneath and at the ends of the flumes. A dump valve located at the end of each flume over the cistern would open during the drainage cycle.

Terrace System. In the terrace system, a series of beds are made over sloping or terraced ground, each set at a lower level than the previous one, as shown in Fig. 7.17. The bottom of a higher bed should be level with the top of the lower bed. Nutrient solution is stored in an above-ground header tank raised several feet above the level of the first bed. An automatic solenoid or manual gate valve is used to control the flow of solution. The beds are constructed on the same design as previously described for a subirrigation system, but a valve is located at each end of the beds to regulate the flow of solution from one bed to the next. The valves

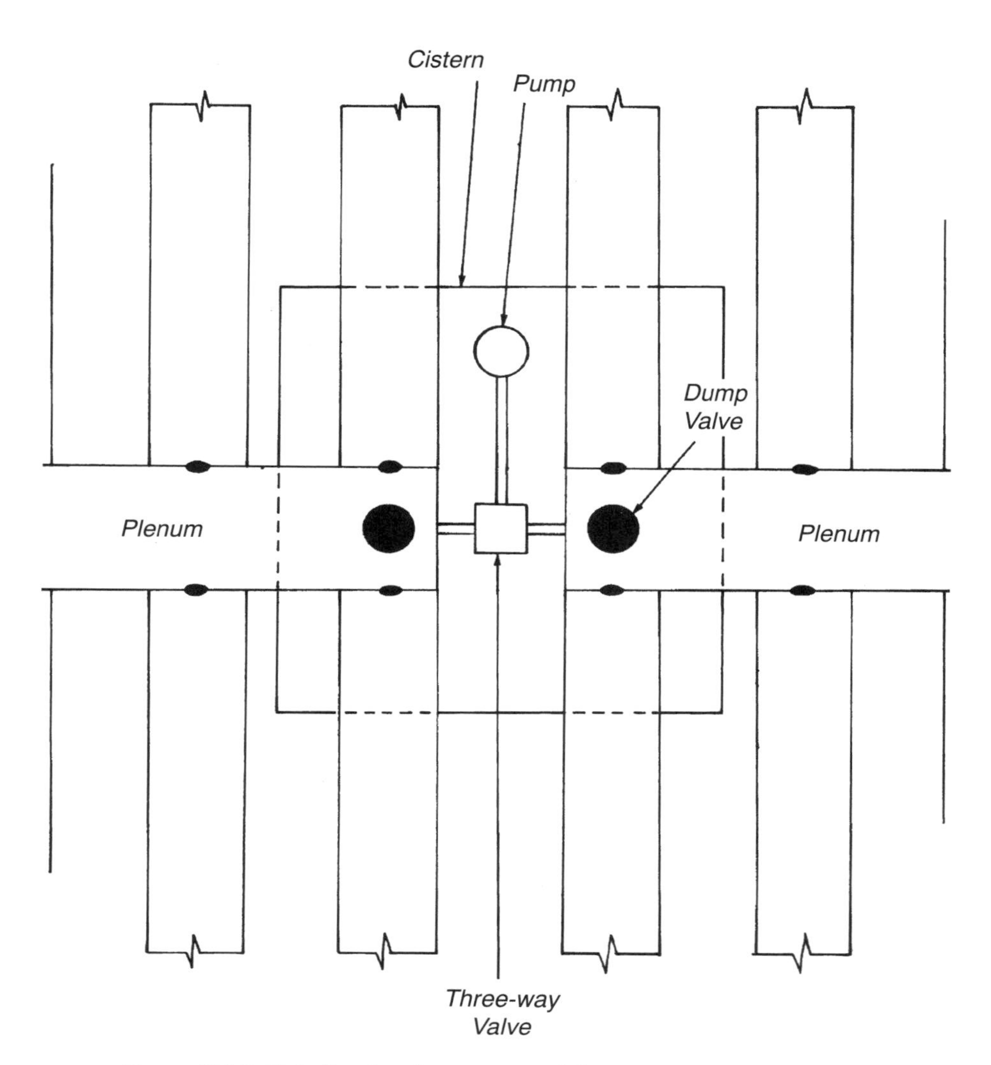

Figure 7.16. Details of a cistern type of flume system with a three-way valve to control water flow alternately to the flumes.

may be either manual or automatic. Sometimes an automatic siphon device replaces the valve, but such a system drains the beds very slowly. Finally, the nutrient solution flows into a sump from where it is pumped back into the raised header tank.

Each bed must be filled completely before the solution is allowed to drain into the following bed. Since there is some water loss due to evapotranspiration from one bed to the next, each subsequent bed should be 20 percent shorter than the preceding one. The first bed (nearest to the header tank) should not exceed 100 to 120 feet (30.5 to 36.5 m) in length; therefore, subsequent beds in a four-bed system would be 80 to 96 feet (24 to

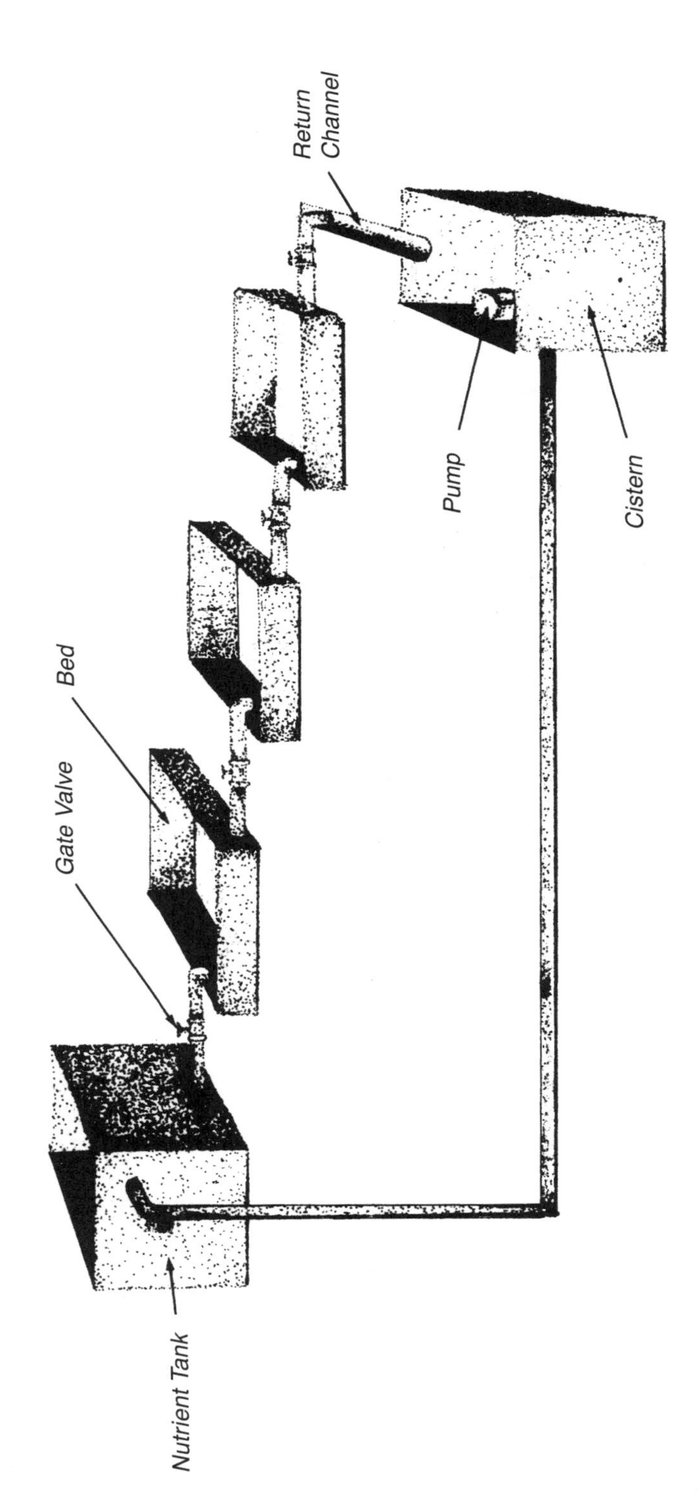

Figure 7.17. A terrace system of subirrigation gravel culture having a single series of bed levels.

29 m), 64 to 77 feet (19.5 to 23.5 m), and 51 to 62 feet (15.5 to 19 m) respectively. The number of beds running parallel to each other would be a function of the header tank and sump capacity. An automatic float valve could be installed in the sump to maintain the solution level. Automatic solenoid valves would be controlled by one or more time clocks. The valves for each series of beds at the same level should be synchronized to operate simultaneously.

If the sump and header tank were constructed so they did not extend the entire length of the series of beds, return channels (or flumes) could be used to conduct water from the header tank to the beds and from the beds to the sump (Fig. 7.18).

The sump is the main storage reservoir for the system. Therefore, the total capacity must be sufficient to fill all the first-level beds with an additional safety factor volume of about 20 percent. Such a gravity system saves reservoir and pump capacity, since only the requirements of the first-level beds need to be satisfied. By means of time clocks, solenoid valves and float valves, the whole system can be automated.

By a further modification, splitting the top flume into two parts and using a three-way valve connected to the header tank and flumes, alternate irrigation of these beds each time could be operated in a similar fashion as that described earlier. This would allow the reduction of tank and sump capacities with resultant savings in water, nutrients and construction costs.

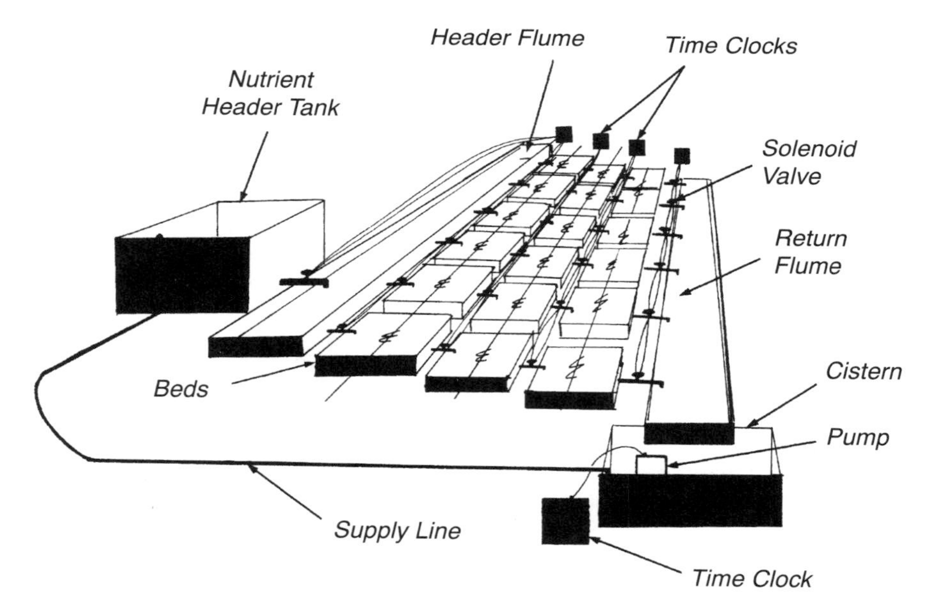

Figure 7.18. A terraced gravel-culture system with six series of bed levels fully automated.

7.3.8 Small-Scale Subirrigation Systems

One of the earliest systems of gravel culture used on a small scale was the bucket system (Fig.7.19). In this system, a plant was grown in a bucket of gravel set on top of a bench. A second bucket, containing the nutrient solution, was connected by a hose to the bucket containing the plant. Normally, the second bucket remained on the floor or on a bench below the bottom of the "growing" bucket. Irrigation was achieved by raising the bucket of nutrient solution above the growing one and attaching it to a hook. After all the nutrient solution drained into the growing bucket, the solution bucket was lowered to the floor and drainage occurred.

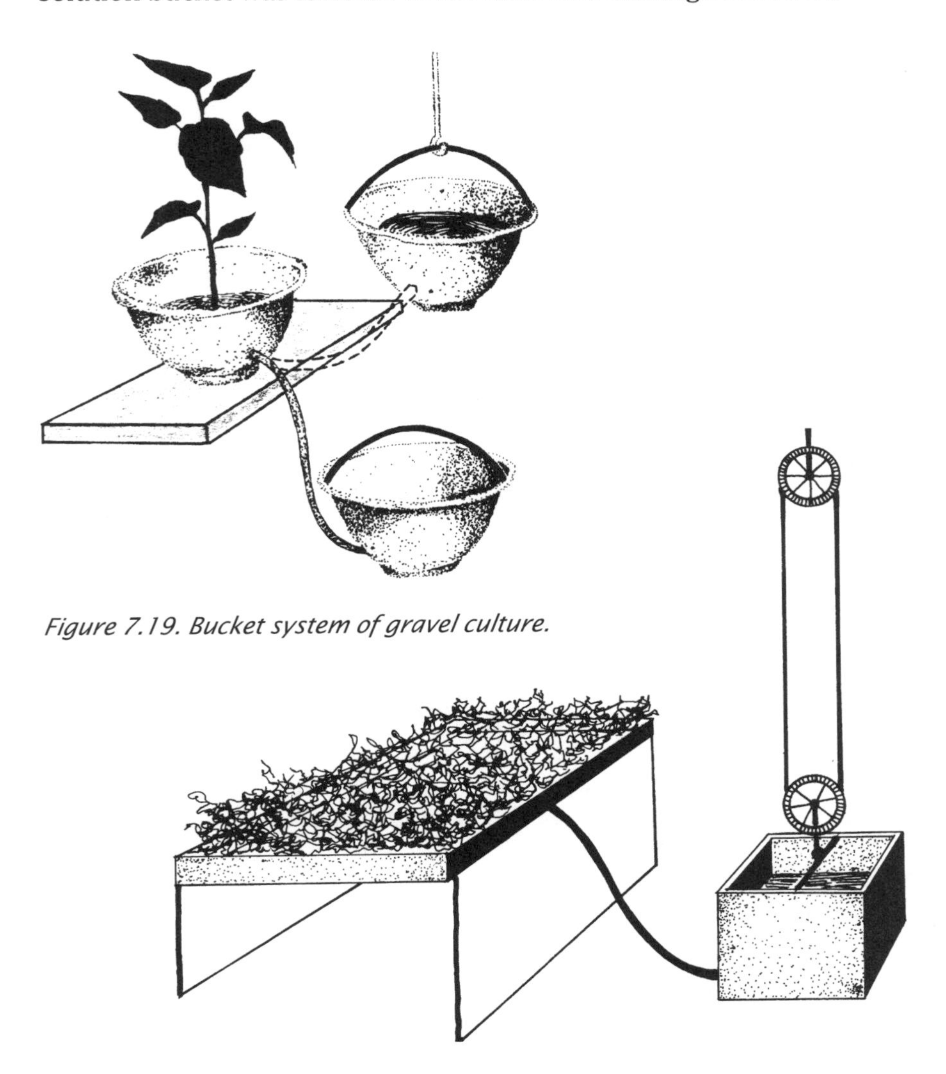

Figure 7.19. Bucket system of gravel culture.

Figure 7.20. Bucket and pulley gravel-culture system.

This system can be expanded to a series of buckets, each attached to a pulley which can be used to raise and lower the buckets during the feeding-irrigation period. On a larger scale, a bed can be constructed at bench level and a larger reservoir or barrel capable of filling the entire bed can be attached to a pulley system which facilitates the lifting of the reservoir during the irrigation cycle (Fig. 7.20). This system could be automated by connecting an electric motor and winch to the pulley ropes. The motor could be activated by a time clock having preset irrigation periods for the whole day.

This system could be simplified by use of a pump in the nutrient reservoir (Fig. 7.21). The system would require a tray to hold plants, a tank for the nutrient solution, a pump and control system, and a suitable connecting pipe. The solution is pumped into the plant tray and allowed to drain back to the nutrient reservoir through the pump. A time clock is used to control the pump and thus preset irrigation cycles are programmed for daily watering and feeding. All storage tanks and delivery lines should be opaque to reduce algal growth.

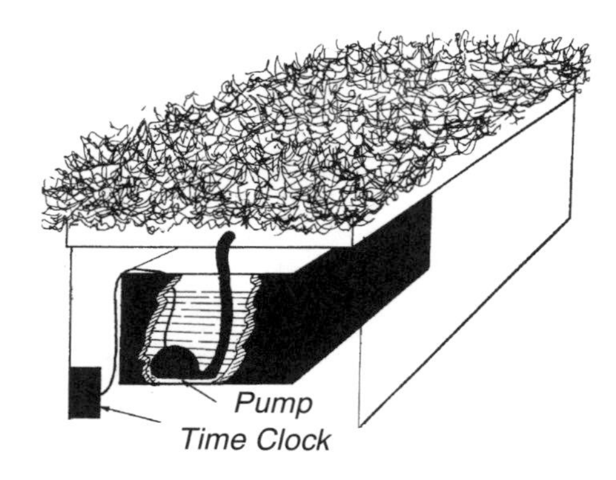

Figure 7.21. A simple small-scale subirrigation gravel bed with automatic pumping.

Another simple home unit is one using a trickle feeding system in which the nutrient solution is stored in a reservoir under the tray and is pumped along the gravel surface. Excess water percolates through the perforated bottom of the tray back to the reservoir (Fig. 7.22). The irrigation cycles can be operated by a time-clock-activated pump.

Several types of small units are available commercially. One is a tray with a water storage tank underneath which uses a common fish-aquarium pump to distribute the nutrient solution (Fig. 7.23). Self-watering and feeding planters for tropical foliage houseplants are available (Figs. 7.24 and 7.25). These work on capillary action of expanded shale for feeding.

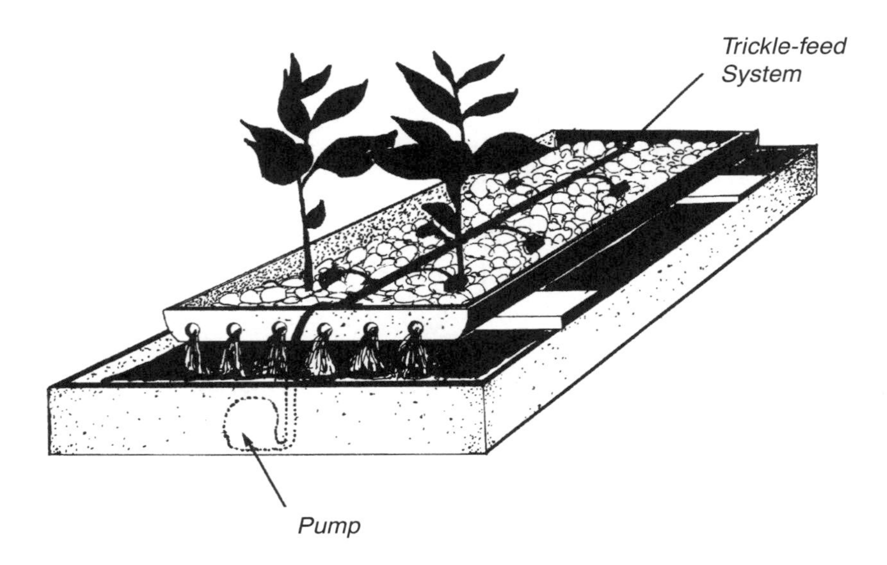

Figure 7.22. A simple small-scale trickle gravel bed with automatic pumping.

Figure 7.23. "City Green" home hydroponic units.

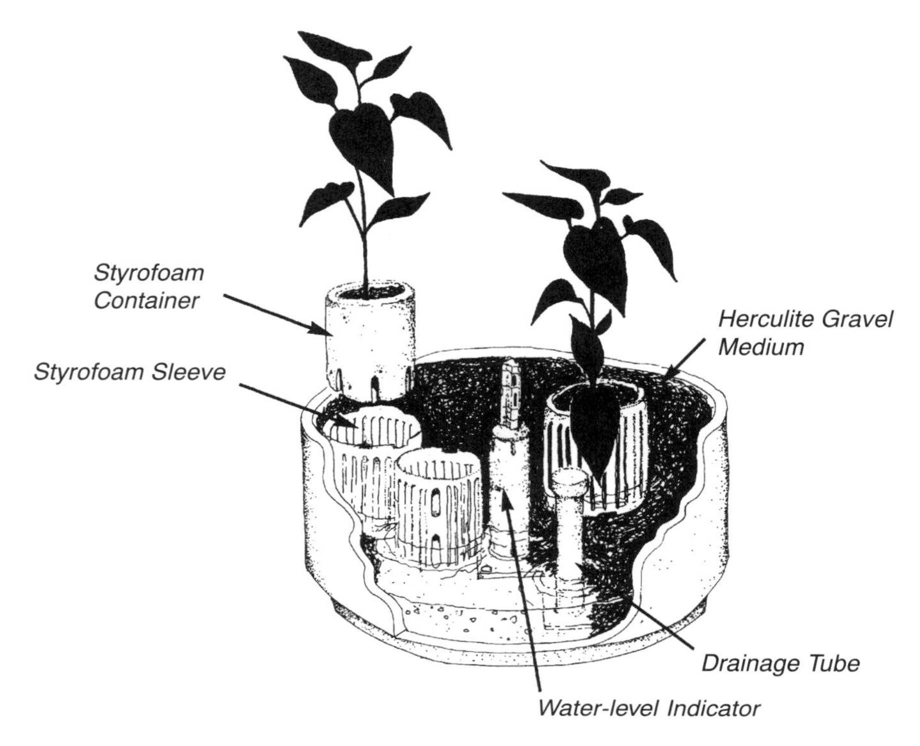

Figure 7.24. "Hydroculture Luwasa" hydroponic planters—cross-sectional sketch.

Figure 7.25. "Hydroculture Luwasa" planters in an office growing indoor tropical foliage plants. (Courtesy of Hydroculture Luwasa, Downsview, Ontario, Canada.)

7.4 Trickle-Irrigation Design

The bed design and construction of trickle-irrigation systems is similar to that of subirrigation systems, but can be somewhat simplified, as shown in Figure 7.26. The bed may be either round or V-shaped; both configurations will give proper drainage. In this system, the nutrient solution is applied to the base of every plant by either a spaghetti- or ooze-hose. The solution runs down through the plant roots. The use of smaller gravel (1/8 to 1/4 inch in diameter) with the trickle system is essential to facilitate lateral movement of the nutrient solution through the medium. Lateral distribution of the solution also occurs along the lateral roots of the plants.

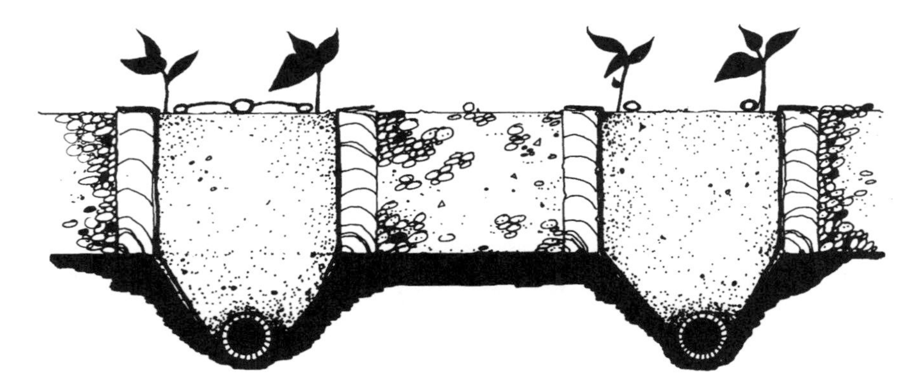

Figure 7.26. Cross section of a trickle-irrigation gravel-culture bed.

The spaghetti system is constructed of ½-inch (1.3-cm) outside diameter (O.D.) thin-wall black polyethylene tubing. Spaghetti tubes of 0.045-inch or 0.060-inch diameter are inserted into the ½-inch lateral feed lines by brass inserts (Fig. 7.27). These inserts are easily pushed in with a pointed insert tool similar to an awl. The inserts automatically seal themselves in the ½-inch pipe (Fig. 7.28). A 1-inch (2.5-cm) piece of the ½-inch O.D. pipe is placed on the end of each feeder line to disperse the jet of solution coming out of the line. At the same time, this end piece prevents clogging of the feeder line by salt buildup through evaporation between irrigation cycles. Lead end pieces are available commercially. They also serve to keep the spaghetti line in place, but are more costly than the homemade ones.

The spaghetti lines must be long enough to reach the base of the plant they are to feed. Excessively long lines greater than 2 feet (61 cm) should be avoided, as the internal friction against the flow of solution will lessen the outflow.

The alternative to a spaghetti system is the ooze-hose system. This thin-wall ½-inch O.D. tubing is available from several sources. Two

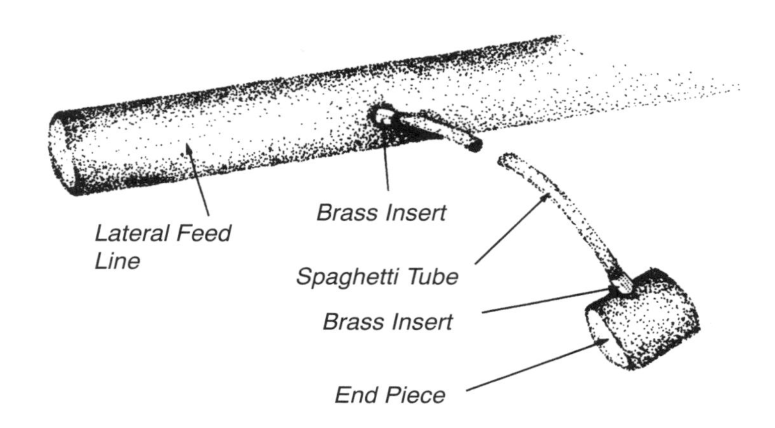

Figure 7.27. Spaghetti (trickle) feed system.

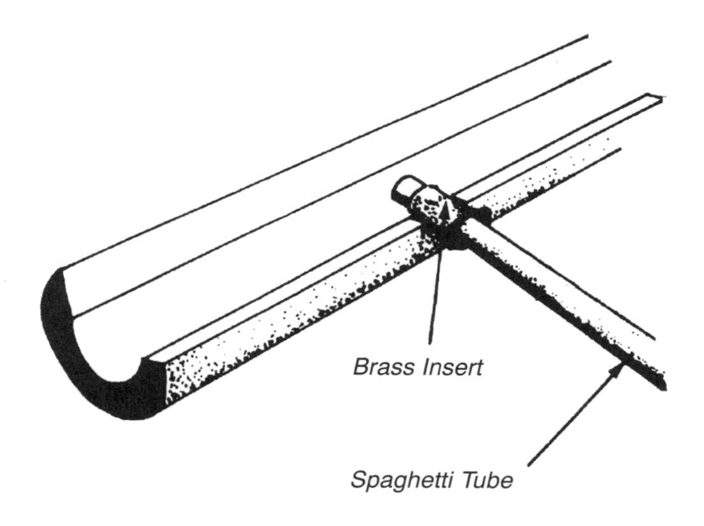

Figure 7.28. Brass inserts "seal" into the ½-inch lateral feed line.

commercial types are Chapin Watermatics' "Twin-Wall" and Dupont's "Via-Flow." Both "ooze" or "sweat" water slowly so that it drips from the hose along its entire length at intervals of 6 to 8 inches (15 to 20 cm) or whatever spacing you request from the manufacturer. Some complaints against these hoses are that they clog and that algae grows in them. Also, they must be replaced with each crop; however, they are relatively inexpensive (several cents a linear foot).

The disadvantages of the spaghetti system are that the tubes clog, roots grow into them, and they are in the way during the changing of crops. Also, they are often accidentally pulled out by workers passing down the rows. As a result, some plants may suffer from lack of water. Proper filters of 100 to 200 mesh placed in the feed lines after the fertilizer has been injected will overcome a lot of the plugging problems. Since the ooze hoses are thrown away between crops, there is no hindrance to changing of crops and lines are not in workers' way during normal cropping periods.

Another trickle-irrigation system common in commercial greenhouses involves the use of ½-inch polyethylene pipe for lateral lines and emitters inserted at each plant position along the entire lateral line. About 30 kinds of drip-irrigation water emitters, perforated hoses and porous pipes are available for use in drip-irrigation systems. A typical system will be discussed in detail in Chapter 8.

The nutrient solution is normally supplied to the plants via the trickle-irrigation system from either a storage tank or injectors. This will also be described in detail in Chapter 8. However, if a pea-gravel culture system is used, a storage tank will have to be used to collect the solution draining through the 3-inch PVC drain pipes in the bottom of the beds. These drain pipes must connect into a main return line which conducts all the solution back to the nutrient tank. The nutrient tank should therefore be placed fairly close to the growing area, preferably at a lower end of the greenhouse so that all drain pipes can slope to that point.

The use of trickle irrigation with a pea-gravel system is not commonly practiced on a commercial scale. Trickle irrigation is usually used with sawdust or sand culture. On a small backyard scale, I have found it to work well with pea gravel (Fig. 7.29). In fact, the drainage pipe is not even necessary if the beds are not much longer than 20 feet (6 meters). A lot of root buildup on the bottom of the beds will occur during cropping, and this should be cleaned out at least once a year by turning over the gravel with your hands or a shovel. In a backyard operation this labor is not excessive; however, in a commercial operation it would be prohibitive. Because of this root buildup and resistance of both roots and gravel to the lateral flow of water, in beds greater than 25 feet (7.6 m) a drainpipe is absolutely essential.

Figure 7.29. A backyard greenhouse with a trickle-feeding system.

7.5 Advantages and Disadvantages of Trickle Irrigation

The advantages of the trickle system over the subirrigation system are:

1. Fewer problems of roots plugging the drain pipes.
2. Better aeration to the roots since at no time are they completely submerged in water. Also, water trickles down past the roots, carrying fresh air with it.
3. Lower construction costs since smaller nutrient tanks are needed and no valves or plenums are required.
4. A much simpler system with fewer chances of failure. Coordination of valves and pumps, etc., is not involved. It is simple to install, repair and operate.
5. The nutrient solution is fed directly to each plant.

Some disadvantages are also present:

1. Sometimes "coning" of water movement occurs due to the relatively coarse particles of gravel. That is, water does not move laterally in the root zone but flows straight down. This results in water shortage to the plants and roots growing along the bottom

of the beds where most water is present, eventually plugging drainage pipes. A subirrigation system uniformly moistens all plant roots and the medium.

2. The trickle lines sometimes get clogged or pulled out by workers. The use of filters in the main header lines will reduce clogging. The use of sweat hoses or emitters can reduce problems of workers accidentally pulling out lines.

7.6 Sterilization of Gravel Between Crops

The sterilization of gravel between crops can easily be done with household bleach (calcium or sodium hypochlorite) or hydrochloric (muriatic) acid used for swimming pools. A 10,000 ppm of available chlorine solution is made up in the nutrient tank and the beds are flooded several times for 20 minutes each time. The chlorine solution is then pumped to waste, and the beds rinsed several times with clean water until all bleach residues are eliminated. The greenhouse then should be allowed to air out for 1 to 2 days before planting the next crop.

In a trickle system, the drain-pipe exit must be plugged to allow the beds to fill up and the sterilizing solution can be pumped through the trickle lines. This will take time, so it is helpful to use an auxiliary pump and hose to flush the beds from above until they fill up. The same procedure is followed in rinsing the beds with clean water.

With each crop, some roots will be left in the medium. Over the years chlorine sterilization will become less effective unless the roots are removed. Removal would be very costly, therefore eventually a more powerful sterilant will have to be used. Steam sterilization or such chemicals as Vapam, chloropicrin or methyl bromide should be used, observing cautions of the manufacturers as mentioned earlier.

7.7 Advantages and Disadvantages of Gravel Culture

Gravel culture initially is blessed with many advantages, but over time some of these advantages are lost.

Here are the advantages:

1. Uniform watering and feeding of plants.
2. Can be fully automated.
3. Gives good plant root aeration.
4. Adaptable to many types of crops.
5. Has proven to be successful on many commercial crops grown both outdoors and in greenhouses.
6. Can be used in nonarable areas where only gravel is available.
7. Efficient use of water and nutrients by use of a recycle system.

Here are the disadvantages:

1. Costly to construct, maintain and repair.
2. With automatic valves, etc., failures occur often.
3. One of the biggest problems is the root buildup in the gravel, which plugs drainage pipes. Each crop leaves some roots behind and the moisture-holding capacity of the medium increases. Consequently, watering frequency may be reduced each year. Watering and aeration stresses occur. Over the years this root buildup results in a graveled soil and the advantages over a soil system will be lost. Eventually the gravel will have to be cleaned of roots, if not completely changed. Sterilization between crops by use of chlorine alone becomes ineffective.
4. Some diseases such as *Fusarium* and *Verticillium* wilts can spread through a cyclic system very rapidly.

References

Schwarz, M., and V. Vaadia. 1969. Limestone gravel as growth medium in hydroponics. *Plant and Soil* 31:122–28.

Victor, Roger S. 1973. Growing tomatoes using calcareous gravel and neutral gravel with high saline water in the Bahamas. *Proc. of Int. Working Group in Soilless Culture Congress,* Las Palmas, 1973.

Withrow, R. B., and A. P. Withrow. 1948. *Nutriculture.* Lafayette, IN: Purdue Univ. Agr. Expt. Stn. Publ. S.C. 328.

Chapter 8

Sand Culture

8.1 Introduction

Sand culture was the most commonly used hydroponic method in areas of the world having an abundance of sand. It has been particularly well suited to desert regions of the Middle East and North Africa. Now, however, NFT and rockwool systems are replacing sand culture due to their ability to recirculate the nutrient solution and to automatically control nutrition through the use of computerization.

Since high quality water is rare in most of these desert locations, some form of purification through distillation or reverse osmosis is imperative for success. Recirculating hydroponic systems that efficiently utilize the costly purified water are essential from an economic standpoint.

Some of the larger sand-culture operations have been:

Superior Farming Company, Tucson, Arizona (11 acres) (4.4 hectares) (Fig. 8.1)
Quechan Environmental Farms, Fort Yuma Indian Reservation, California (5 acres) (2 hectares)
Kharg Environmental Farms, Kharg Island, Iran (2 acres) (0.8 hectares) (Fig. 8.2)
Arid Lands Research Institute, Sadiyat, Abu Dhabi, United Arab Emirates (5 acres) (2 hectares) (Fig. 8.3)
Sun Valley Hydroponics, Fabens, Texas (10 acres) (4 hectares).

The Environmental Research Laboratory, a branch of the University of Arizona, worked closely with many of these projects. They were instrumental in proving sand culture in these areas. In 1966 the laboratory began a pilot project in Puerto Peñasco, Mexico, to test the feasibility of sand culture and later based the establishment of commercial operations in the U.S. Southwest and in the Middle East on the findings from that project.

Figure 8.1. Aerial view of 11-acre sand-culture greenhouse complex of Superior Farming Company, Tucson, Arizona. At the time of the photograph, the air-inflated greenhouses in the foreground were being covered with polyethylene. (Courtesy of Superior Farming Company, The Environmental Research Laboratory and Manley, Inc., Tucson, Arizona.)

Figure 8.2. Aerial view of 2-acre sand-culture greenhouse complex on Kharg Island, Iran. (Courtesy of The Environmental Research Laboratory and Manley Inc., Tucson, Arizona.)

Figure 8.3. Aerial view of 5-acre greenhouse complex of Abu Dhabi. Note the barren dunes and proximity to the sea from which water is desalted for use in the growing crops. At the time of the photograph, a fire had damaged the packing facilities to the rear of the greenhouses. (Courtesy of the government of Abu Dhabi, The Environmental Research Laboratory, Tucson, Arizona, and Gulf Aviation, Bahrain.)

8.2 Medium Characteristics

In Mexico and the Middle Eastern countries where greenhouse projects were established on the seacoast, the normal beach sand was used as a medium. Once the growing medium (sand) was leached free of excess salts, vegetables were either seeded directly in the sand or planted as transplants. In the U.S. Southwest, a concrete river-wash sand is used, not a mortar sand, as it is too fine and will puddle. Puddling is indicated by water coming to the surface upon vibration of the sand. It is a result of a high percentage of silt and fine sand. The aggregate must be washed free of fine silt and clay. It also should be relatively free of particles over 1⁄16 inch (2 mm) in diameter and under 1⁄40 inch (0.6 mm). A properly screened sand culture aggregate will drain freely and not puddle after an application of a large amount of water.

Aggregates which are soft and tend to disintegrate should be avoided. However, it may not be possible to avoid using soft particles in areas where only limestone sand is available. In such cases nutrients should be added and the *p*H adjusted daily, as discussed earlier in Chapter 7.

8.3 Structural Details

Two methods of utilizing sand as a growing medium have proven satisfactory. One is the use of plastic-lined beds; the other involves spreading sand over the entire greenhouse floor.

8.3.1 Beds with Plastic Liner

Growing beds may be built as above-ground troughs with wooden sides (Fig. 8.4) similar to those described for gravel culture in Chapter 7. Six-mil black polyethylene can be used for the liner, but 20-mil vinyl is more

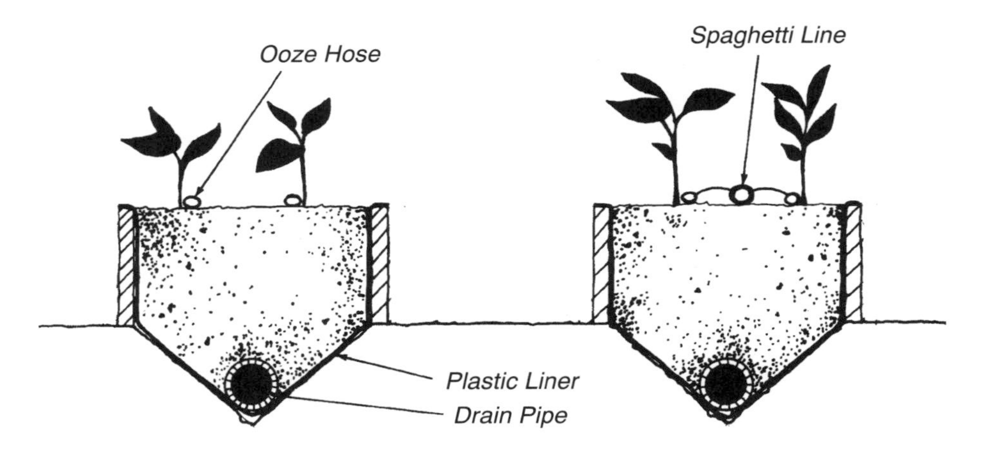

Figure 8.4. Cross section of sand-culture beds.

durable. The bottom of the trough should have a slight slope of 6 inches (15 cm) per 200 feet (61 meters), so that it can be drained or leached when necessary. The drain pipe should be placed in the entire length of the bed. A 2-inch (5 cm) diameter pipe is large enough since in sand culture only excess solution (about 10 percent of that added) is drained. Connect the drain pipes from all beds to a main at one end which collects the waste water and conducts it away from the greenhouses.

Similarly to gravel culture, drainage holes are cross-cut with a saw, one-third the distance through the pipe every 18 inches (46 cm). The cuts must be against the bottom of the bed so that plant roots are discouraged from entering the pipes. As in gravel culture, one end of each pipe should be left above the ground so that a Roto-Rooter can be used to clean them. An alternative to making these drain pipes, which is very laborious, is to purchase the black coiled plastic drainage pipe from an irrigation supplier. This drain pipe is prepunched with holes.

The width of the beds may be 24 to 30 inches (61 to 76 cm) and the depth 12 to 16 inches (30.5 to 40.6 cm). The bottom of the bed may be level, round or V-shaped with the drainage pipe in the middle.

An alternative to above-ground beds would be a wire strung 2 inches (5 cm) off the ground on either side of the bed. A 6-mil polyethylene tube is strung over the two wires to form a double-layer barrier between the sand medium and soil (Fig. 8.5). Elevating the edges of the bed above ground level prevents soil from being kicked into the sand beds. In combination,

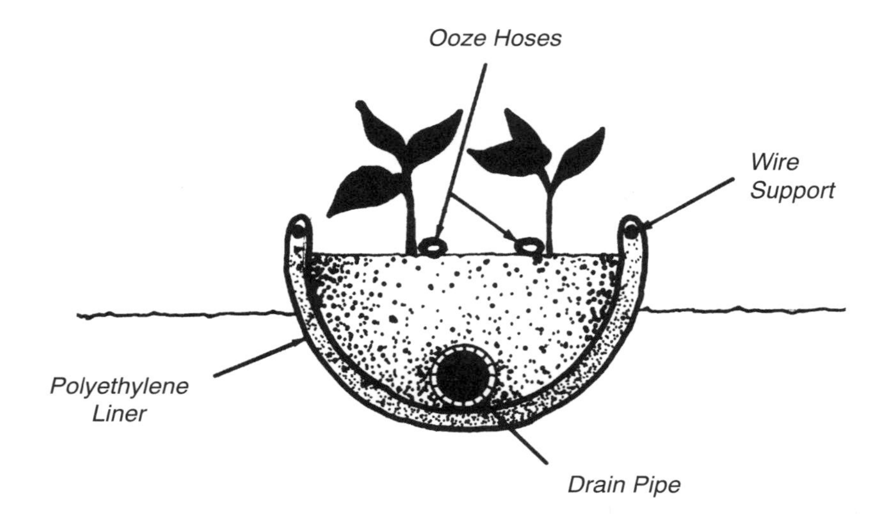

Figure 8.5. Cross section of bed using wire side supports.

the sides of the trench and the wire support the sides of the polyethylene bed. This replaces wood, which is expensive in most desert regions.

8.3.2 Greenhouse Floors Lined with Polyethylene

Construction costs can be reduced in areas where lumber is expensive or difficult to obtain by lining the greenhouse floor with black 6-mil polyethylene and filling it with 12 to 16 inches (30.5 to 40.6 cm) of sand. The floor should have a slight grade of 6 inches (15 cm) per 100 feet (30.5 m), so that the area may be drained or leached when necessary. Generally, two layers of 6-mil plastic are used to cover the entire floor.

Prior to installing the polyethylene, the floor should be graded and packed. Polyethylene sheets should be overlapped several feet (0.6 m) when more than one sheet is required in wide houses. The drain pipe, 1¼ to 2 inches (3.1 to 5 cm) in diameter, or black drainage pipe as mentioned above, is then placed on top of the polyethylene at a uniform spacing of 4 to 6 feet (1.2 to 1.8 m) between pipes, depending upon the nature of the sand. The finer the particles, the closer the pipes will have to be spaced. These drain lines must run parallel with the slope down into a main drain running across the low end of the greenhouse grade. Once the drain pipes are in place, sand is spread over the entire area to a depth of 12 inches (30.5 cm) (Figs. 8.6–8.8).

If the medium is spread shallower than 12 inches (30.5 cm), there will be a problem in obtaining uniform moisture conditions and a greater chance of roots growing into the drain pipes. The surface of the bed should be graded to the same slope as the floor.

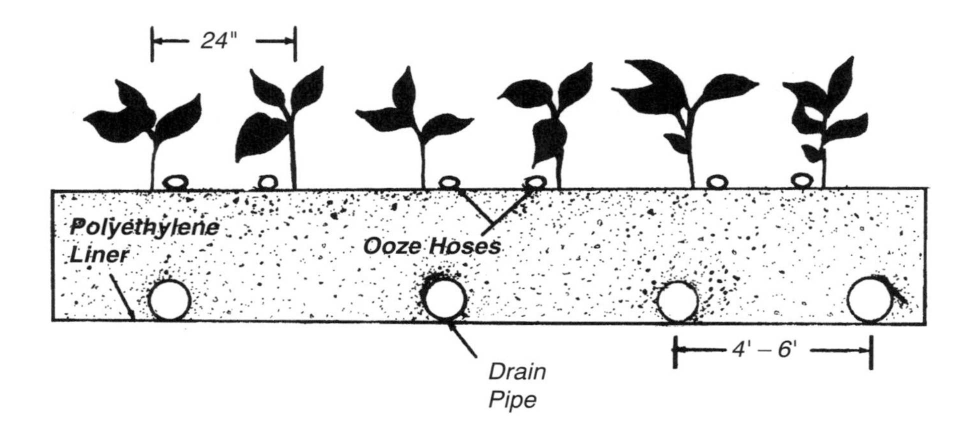

Figure 8.6. Cross section of greenhouse floor design of sand culture.

Figure 8.7. The laying of a polyethylene liner and drain pipes. (Courtesy of Superior Farming Company and The Environmental Research Laboratory, Tucson, Arizona.)

Figure 8.8. Back filling with 12 inches of sand. (Courtesy of Superior Farming Company and The Environmental Research Laboratory, Tucson, Arizona.)

8.4 Drip (Trickle) Irrigation System

A drip-irrigation system must be used with sand culture. Waste nutrients (approximately 10 percent of that applied) are not recycled. Such a system is termed an open system, as opposed to the recycled or closed system of gravel culture. The drip-irrigation system feeds each plant individually by use of spaghetti feed lines, sweat (ooze) hoses or emitters (Fig. 8.9). If the Chapin Twin-Wall ooze hose is used, a 4-inch (10-cm) spacing between outlets is recommended. If the surface of the bed is level, the tube should not be over 50 feet (15 m) long. If the surface is sloped 6 inches (15 cm), the tube can be 100 feet (30.5 m) long, with the supply manifold at the high end.

Figure 8.9. The installation of ooze hoses for an automated drip-irrigation-feeding system. (Courtesy of Superior Farming Company and The Environmental Research Laboratory, Tucson, Arizona.)

On a level bed, the supply manifold may run down the center of a 100-foot (30.5-meter) bed, with 50-foot (15-meter) lines off either side. The objective of a greenhouse drip-irrigation system is to apply uniform water at optimum levels to all plants.

8.4.1 Planning a Drip-Irrigation System

Divide the total greenhouse area into equal or similar crop sections or into individual houses. Plan irrigation systems so that each house or section can be irrigated independently (Fig. 8.10). Piping to each section should be capable of distributing 1.6 to 2.4 gallons (6 to 9 liters) per minute for each 1,000 square feet (93 square meters), or 8 to 12 gallons (30 to 45 liters) per minute for each 5,000 square feet (465 square meters) of growing area.

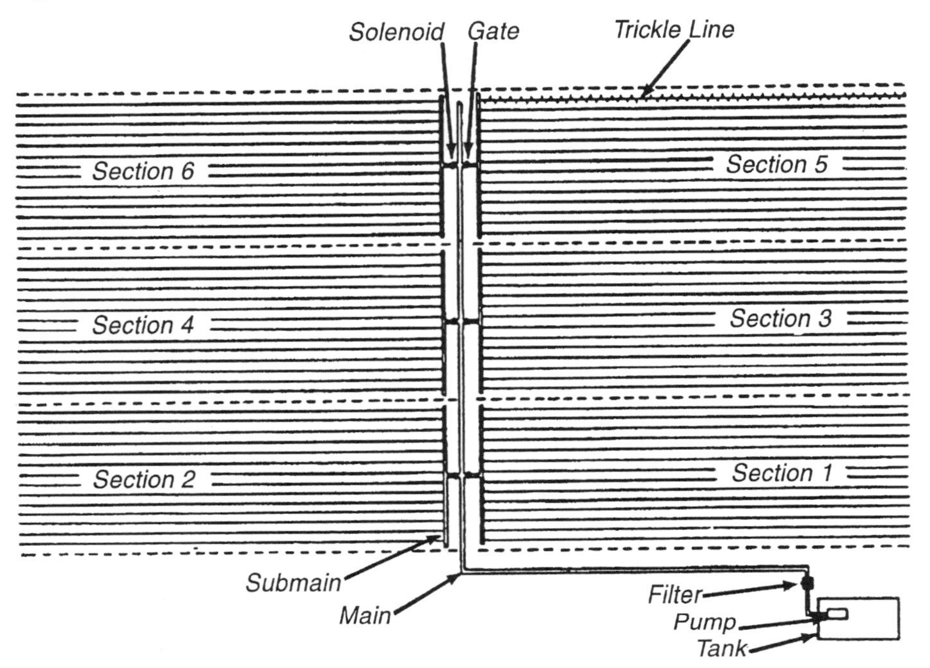

Figure 8.10. A typical drip (trickle) irrigation system for greenhouse sections which can be automatically irrigated independently.

The rate and length of time for each irrigation cycle will be a function of the type of plant, its maturity, weather conditions and time of day. In all cases a tensiometer system should be set so that no more than 8 to 10 percent of the nutrient solution applied at any irrigation cycle is wasted. This can be determined by measuring the amount of water passing through the main supply line and that flowing out of the main collector drain line.

The volume of water that can enter each greenhouse section should be regulated by a flow-control valve which is sized and selected according to plant water requirements of that greenhouse section. Flow valves are usually available in 1- and 2-gallons-per-minute (3.8- and 7.6-liters-per-minute) size increments and fit ¾- and 1-inch (1.9- and 2.5-centimeter) pipe connections. The flow valve should be located upstream from the solenoid valve where the irrigation cycle is automatically controlled. While a minimum water supply pressure of 15 pounds per square inch (103.5 kiloPascals or kPa) is required for proper operation of most flow control valves, for best performance the main water supply line should be maintained at a pressure between 20 and 40 pounds per square inch (138 and 276 kPa). The flow control valve assures a constant quantity of water and reduces water pressure in the irrigation system piping to 2 to 4 pounds per square inch (13.8 to 27.6 kPa) which is correct for low-pressure drip- irrigation emitters.

Main line piping should be 2- to 3-inch (5- to 7.5-cm) diameter PVC, depending upon the area of the largest greenhouse section or number of sections watered at any one time. Since all sections are not irrigated at the same time, the total volume capacity of the main line need satisfy only the maximum number of sections operating at a time. Header lines should be 1-inch (2.5-cm) diameter PVC pipe for a 5,000-square foot (465-square meter) section. Larger sections should use larger pipe. The header line is connected to the main line with a tee at its center to equally divide the water supply (Fig. 8.10). A gate valve with a solenoid valve downstream on this connection between the main and submain (header) operates the irrigation cycle. From this header line, ½-inch black polyethylene pipe is run along the inside of each plant row. The emitters are placed in these lateral lines at the base of each plant (Fig. 8.11). Flexible polyethylene pipe (usually 80 to 100 pounds per square inch) is normally used for these emitter laterals. Where manufactured emitters are used, ½-inch pipe will provide equal water distribution and uniform water application throughout 100- to 150-foot (30.5- to 46-meter) greenhouse irrigation runs.

Most drip emitters can deliver water at a rate of ½ to 3 gallons (2 to 11 liters) per hour, depending upon the water pressure in the lateral line. A greenhouse drip-irrigation system should be designed so that each emitter applies 1 to 1½ gallons (4 to 6 liters) per hour.

While spaghetti tubing is more economical than manufactured emitters, more labor is required for its installation and maintenance. Perforated (ooze) hose is installed more easily but is not as durable (it usually must be replaced between crops).

Emitters, pipes and fittings should be black to prevent algae growth inside the piping system.

Water must be filtered before flowing into a drip-irrigation system. Y-type, in-line strainers, containing at least 100 mesh screens and equipped with clean-out faucets, should be installed with the screen housing and flush valve down. The filter(s) should be installed downstream from the fertilizer injector in the main supply line. It is wise to also install a filter in the main supply line upstream from the fertilizer injector.

The fertilizer injector or proportioner automatically proportions the right amount of stock solution into the main supply line during every irrigation cycle. Positive-displacement pump injectors, venturi propor-

Figure 8.11. Location of ooze hose beside cucumbers. (Courtesy of Superior Farming Company, Tucson, Arizona.)

tioners and forced flow batch tanks are some of the types of fertilizer injectors available commercially (Fig. 8.12).

Alternatively, a nutrient solution can be pumped into the drip-irrigation system directly from a large storage tank. Many growers prefer such a method since they then know the exact formulation of the nutrient solution rather than depend on an automatic proportioner which may sometimes break down or make an error. However, manufacturers of fertilizer injectors claim they are very reliable.

Figure 8.12. Automatic-proportioner fertilizer-injector system used by Superior Farming Company, Tucson, Arizona.

All irrigation may be controlled by time clocks or by a tensiometer feedback system, as described earlier. These control the solenoid valves and the activation of the fertilizer injector or tank pump which allow only one section of the greenhouse to be irrigated at one time.

If calcareous sand is used, the amount of chelated iron going to the plants must be increased, as discussed earlier. If fertilizer proportioners are used, two stock solutions are prepared. One is a calcium nitrate and iron solution and the other contains magnesium sulfate, monopotassium phosphate, potassium nitrate, potassium sulfate, and the micronutrients. The proportioner must be a twin-head type. For example, if each head injects one gallon (3.785 l) of stock solution into each 200 gallons (757 l) of water passing through the water line, then the stock solution will be 200 times the final concentration to reach the plants. In this case, the proportioners would have a ratio of 1:200. Two manufacturers of proportioners in the United States are Anderson and Smith.

8.5 Watering

If a time clock is used, the crop should be watered two to five times per day, depending on the age of the plants, weather and time of the year. As mentioned earlier, enough water is added during each cycle to allow 8–10 percent of the water applied to drain off. Twice a week a sample of this drainage should be tested for total dissolved salts. If the total dissolved salts reaches 2,000 ppm, the entire bed should be leached free of salts by using pure water. However, if no extraneous salts such as sodium are found, pure water may be irrigated onto the crop until the plants themselves, over a few days, lower the salt level down to a point at which nutrients can again be added to the irrigation water.

When using a proportioner, the total dissolved salts of the nutrient solution going onto the plants should be checked twice each week to be sure that the injector is functioning properly. Also check whether each injector pump is injecting the proper amount of stock solution to the irrigation water.

If a nutrient tank is used to store the nutrient solution, it must be large enough to supply water to all the plants in the greenhouse for at least one week. Its size therefore will depend upon the total greenhouse area. If several crops having very different nutrient requirements are grown, two storage tanks should be used, each having its specific formulation suitable to the crop it is feeding. The entire irrigation system connected to one tank must be independent of the other.

Since a sand-culture system is an open system in which excess nutrient solution goes to waste, there will be little change in the formulation of the nutrient solution in the storage tank. The *p*H, however, should be checked daily, especially in areas having highly alkaline waters.

The nutrient solution in the storage tank does not have to be changed regularly as was necessary in gravel culture. It need only be cleaned out periodically of any sludge and sediment due to inert carriers in the fertilizer salts. When the volume of solution in the tank is almost depleted, a new batch is mixed up.

Many growers prefer such a storage-tank system over the proportioner because they make up their own solution and know exactly what its formulation is. Nonetheless, fertilizer injectors have several advantages over a storage tank: (1) they require less space; (2) the initial capital outlay for an injector is less than for large storage tanks; (3) they are capable of making rapid changes in the nutrient solution formulation to compensate for changes in plant requirements under changing weather conditions. For instance, during a period of dull weather, the rate of nitrogen can be easily reduced, whereas in a storage-tank system the entire volume of nutrient (usually at least one week's supply) would have to be altered.

8.6 Sterilization of Sand Beds Between Crops

While fumigation can rid the sand of any soil-borne disease, plus any nematodes which may have been introduced, it will not rid the sand of tobacco mosaic virus (TMV) or cucumber mosaic virus II (CMV II). One of two fumigants may be used—Vapam, which is added through the irrigation system, or methyl bromide, which is put through the drain system under pressure. In either case, the entire floor must be covered with polyethylene before the fumigant is applied. Methyl bromide is also available in canisters which come with a special dispenser hose to inject the fumes into the polyethylene cover. In this case the polyethylene should be inflated with air and sealed with a layer of sand around the edges to hold the

Figure 8.13. Injecting methyl bromide under an inflated polyethylene cover.

air inside before injecting the gas (Fig. 8.13). After 48 hours, the cover is removed and the area leached with water. This removes any remaining fumigant and salts. If Vapam is put through the watering system, the system must be flushed with water after application. The beds may be replanted four to five days after fumigation.

To rid sand of TMV or CMV II, steam sterilization must be used. If the greenhouse is heated with a hot water boiler, the boiler should be assembled with a steam converter which can produce sufficient steam for sterilization of the growing beds. The steam may be put through the drain system provided the drain lines will not be damaged by high temperature. Alternatively, pipe can be placed within the top few inches of the sand and the beds covered with heavy canvas or polyethylene before releasing the steam. Pipes are moved along the beds as sterilization is completed for each section of bed.

8.7 Operation and Productivity of Sand-Culture Greenhouses in Arid Lands

An example of how sand culture can be utilized to produce food on a coastal desert is the commercial-scale power-water-food facility of Abu Dhabi (Fig. 8.3). The University of Arizona Environmental Research Laboratory (ERL), with a grant from the ruler of Abu Dhabi in 1972, installed a power-water-food complex which utilizes desalted seawater for the irrigation of vegetable crops in large, controlled-environment greenhouses. With rainfall averaging less than 5 centimeters per year and strong winds occurring frequently, outdoor cultivation is severely limited. Air-inflated polyethylene and structured greenhouses with a desalination plant could meet the needs for producing crops. Half of the 5-acre complex was constructed of air-inflated polyethylene greenhouses connected to two central structured corridors, as shown in Figures 8.3 and 8.14. Low-growing crops that need no overhead support such as turnips (Fig. 8.15), lettuce, peppers, eggplant, radishes, etc. are grown in the air-inflated greenhouses. Crops such as tomatoes, cucumbers and string beans, which are trained vertically on a cordon system, are grown in the structured houses.

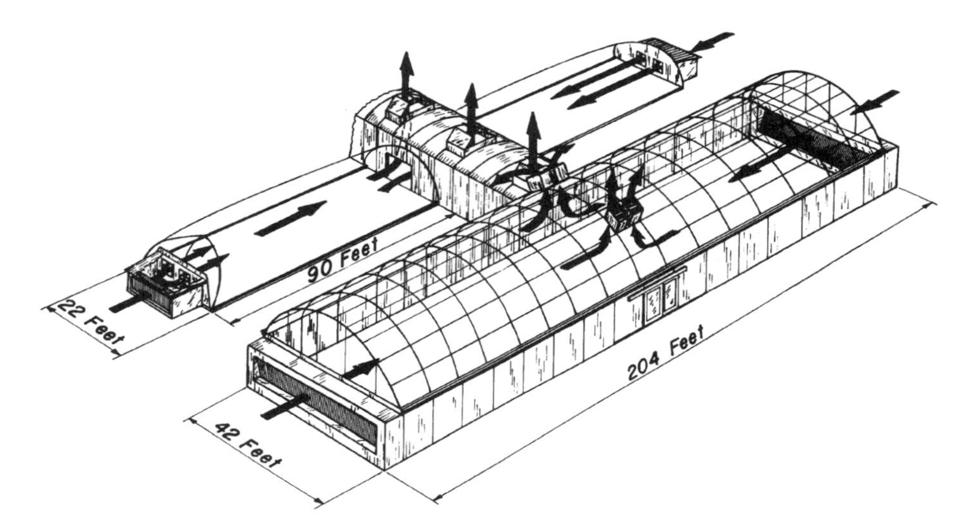

Figure 8.14. Schematic drawing of air-inflated and structured greenhouses shows air flow pattern. (Courtesy of The Environmental Research Laboratory and Manley, Inc., Tucson, Arizona.)

The greenhouses are cooled by use of evaporative cooling pads at the ends of each unit. Seawater, rather than fresh water, is distributed over the top of the pad and runs through it as centrally located exhaust fans pull air through the cooling pads and across the houses, exhausting the air in the middle of the structured houses, as shown in Figure 8.14. In the inflated greenhouses, air is sucked through the cooling pads by fans mounted in the end wall. The air is forced through the house and exhausted through the roof of the central corridor.

The greenhouses protect the plants from blowing sands, drying winds and high and low temperatures, and the high humidity created by seawater-cooling systems reduces water consumption.

Crops are grown in the existing sands which is composed of essentially fine calcium carbonate. The *p*H of the medium consequently is 8.3. Water and fertilizers are applied at regular intervals by dissolving commercial-

Figure 8.15. Air-inflated greenhouses are especially suited to low-growing crops such as turnips. (Courtesy of The Environmental Research Laboratory and Manley, Inc., Tucson, Arizona.)

grade fertilizers in the distilled seawater and distributing them individually to each plant by use of a drip-irrigation system and fertilizer injector similar to those shown in Figure 8.12.

Plastic-lined beds and a drip-irrigation system similar to that described earlier in this chapter are used to grow the crops.

Fontes (1973) points out that production projections for the 2 hectares of greenhouses in Abu Dhabi were an average of 1 ton per day of vegetables, which was accomplished by the middle of 1972. Yields in tons/ha/day for various vegetable crops in 1972 are given in Table 8.1.

TABLE 8.1 Comparison of Yields of Various Vegetable Crops Grown in Abu Dhabi Greenhouses

Type of Vegetable	*Tons/ha/day*
Cabbage	1.4
Cucumber	2.8
Eggplant	1.3
Lettuce	1.5
Okra	0.4
Tomato	1.2
Turnip	2.4

For the first six months of 1973, 230 tons of vegetables were shipped to market. Table 8.2 shows the yields of various crops grown in the Abu Dhabi greenhouses.

TABLE 8.2 Yield of Crops Grown in Abu Dhabi Greenhouses

Type of Vegetable	*Yield/ Crop (in tons)*	*Crops/ Year*	*Total Yield per Acre per Year (in tons)*
Broccoli	13.0	3	39.0
Bush Beans	4.6	4	18.4
Cabbage	23.0	3	69.0
Chinese Cabbage	20.0	4	80.0
Cucumber	70.0	3	210.0
Radish	9.0	8	72.0
Tomato	45.0	2	90.0

The 1972 production of tomatoes alone exceeded 150,000 kilograms, enough to supply almost 29,000 persons at U.S. levels of consumption.

8.8 Sand Culture of Herbs

Chives, basil, sage and mint have been grown successfully in sand-culture beds at California Watercress, Inc. in Fillmore, California. In one greenhouse of dimensions 30 ft × 156 ft (9 m × 47.5 m), seventeen beds 8 ft × 12 ft (2.4 m × 3.65 m) on one side and seventeen beds 8 ft × 14 ft (2.4 m × 4.27 m) on the other side were constructed using 1" × 6" treated lumber. The 6-inch to 8-inch (15–20-cm) deep beds were lined with 6-mil black polyethylene. The bottom of the beds were sloped 2 inches (5 cm) from one side to the other to assist in drainage.

At a second location, a greenhouse of one-half acre (0.2 hectare) was erected to grow herbs. In that greenhouse, one range 30 ft by 165 ft (9 m by 50 m) was used to grow sage and mint in sand. Prior to building the beds the irrigation system was installed. A 2-inch PVC main with a ¾-inch riser to each bed was set in place. A solenoid valve on the main line operated by an irrigation controller activated the feeding cycles from an injector system. The entire floor of the greenhouse was covered with a nursery weed-mat to prevent weed growth (Fig. 8.16). The underlying soil was very sandy with many rocks which provided excellent drainage. The beds, of the same dimensions as in the other greenhouse, were placed on top of the weed matting and staked in position. However, unlike the other beds,

Figure 8.16. Greenhouse floor covered with a weed mat. (Courtesy of California Watercress, Inc., Fillmore, CA.)

these were not lined with polyethylene. The treated boards making up the beds were simply set on top of the weed-mat covering as the weed mat allows water to move through it freely while roots cannot easily penetrate.

A coarse river wash sand of granitic origin was used as the medium. In the larger greenhouse, the beds of sand were covered with 1 inch (2.5 cm) of a peat-perlite mix to improve lateral movement of the solution. It was later found that the sand had too much silt in it as crusting occurred with the hard water of the area. To improve it, approximately one-third was removed and replaced with the peat-perlite mix which was incorporated with the sand. The other beds in the smaller greenhouse had a better quality of sand from a different source and grew chives and basil well.

Drip-irrigation lines were placed along the length of the beds from the ¾-inch header at 12-inch (30.5-cm) centers (Fig. 8.17). "T-tape" having small holes every 12 inches (30.5 cm) provided 38 gallons per hour (gph) per 100 feet at 10 psi pressure, which is equivalent to 144 liters per hour per 30.5 meters at 69 kPa (kilopascals) pressure. The T-tape was secured with an adapter attached to a ¼-inch black poly drip emitter line which in turn was inserted into the ¾-inch header via a hole and sealed with silicone rubber (Fig. 8.18). The ends of the T-tape were sealed by folding them several times and taping them with pipe tape.

Figure 8.17. Drip-irrigation system of sand-culture beds growing herbs.(Courtesy of California Watercress, Inc., Fillmore, CA.)

Figure 8.18. Attaching T-tape drip line to poly-tube adapter which in turn is attached to a ¾-inch header. (Courtesy of California Watercress, Inc., Fillmore, CA.)

Sage was seeded in a peat-lite medium in 98-celled trays and later transplanted to the beds (Fig. 8.19). Mint cuttings were rooted in 2¼-inch × 3-inch deep (6 cm × 7.5 cm) plastic pots with a peat-lite medium before transplanting. Chives were transplanted from the field after washing the soil from their roots. The basil was grown from seed. The mint was ready for first harvest after 6 weeks (Fig. 8.20), with continued harvests every 4 to 5 weeks depending upon the season. The basil was harvested every three weeks (Fig. 8.21). Chives were harvested on a 30- to 35-day cycle depending upon daylength and light conditions (Figs. 8.22–8.24). Two to four beds were cut each day over a two-week harvesting period. Similar to watercress, these herbs are bundled with elastic bands or twist ties and sold as dozens of bunches.

Figure 8.19. Transplanting herbs into sand-culture beds. (Courtesy of California Watercress, Inc., Fillmore, CA.)

Figure 8.20. Mint ready to harvest 6 weeks after transplanting seedlings. (Courtesy of California Watercress, Inc., Fillmore, CA.)

Figure 8.21. Basil in sand culture, harvesting every 3 weeks. (Courtesy of California Watercress, Inc., Fillmore, CA.)

Figure 8.22. Chives 7 days after harvesting.
(Courtesy of California Watercress, Inc., Fillmore, CA.)

Figure 8.23. Chives 33 days after cutting, ready for another harvest. Note the space heater in the foreground. (Courtesy of California Watercress, Inc., Fillmore, CA.)

Figure 8.24. Chives just harvested from four beds on the right. The rest of the beds on the right are ready to harvest at 23 to 25 days after a previous harvest. Beds on the left from front to back are 2 to 4 days after harvesting. (Courtesy of California Watercress, Inc., Fillmore, CA.)

8.9 Sand Culture in the Tropics

Growing temperate crops in tropical countries presents some unique problems. Certain tropical countries, such as Venezuela, which have a resource-based industrial economy are developing an expanding middle class of people whose demands are similar to those in other industrialized countries throughout the world. They demand similar diets which include fresh fruits, vegetables and similar esthetically pleasing surroundings with the use of flowers.

Such crops, which are not native to the tropics, require specific growing conditions and culture. Strawberries, sweet peppers, eggplants, tomatoes, cucumbers, lettuce, cabbage, cauliflower, carnations, chrysanthemums and foliage plants are in demand. In many regions of the country the climate is not suitable for growing such crops due to excessive temperatures and high humidity. However, in the mountainous areas with elevations above 1,500 meters (4,900 feet), the temperatures drop sufficiently to allow the growing of cool-season temperate crops such as lettuce, strawberries, carnations and chrysanthemums. The temperatures in these areas range from 22 to 28 degrees Celsius (72–82° F) during the day, and from 16 to 20 degrees Celsius (61–68° F) during the night. The daytime temperatures are at the upper limit of the tolerable range for the growing of cool-season crops such as lettuce, cabbage, cauliflower, carnations and chrysanthemums. Any prolonged variations in daytime temperatures can cause crop failures.

The higher in elevation crops can be located, the less chance of experiencing excessive daytime temperatures which may cause bolting of lettuce or breaking of flower blossoms. By seeking higher elevations, generally steeper terrain is encountered (Fig. 8.25). As a result, conventional farming is very difficult. Small plots of land are cultivated by hand repeatedly year after year. Production declines with fertility and soil erosion increases.

Figure 8.25. Typical steep terrain of mountainous regions of the tropics with crops grown by hand cultivation in soil. (Courtesy of Hidroponias Venezolanas, S.A., Caracas, Venezuela.)

Where arable land is very limited and difficult to work, as in this case, hydroponic culture for intensive production is the answer. But it is often impossible to obtain local people with the necessary skills for this more technical growing of crops. Fertilizers are often unavailable for months; therefore, large inventories must be maintained. Equipment and materials such as pumps, plastic piping, pesticides and greenhouse supplies are difficult to obtain locally, so they must often be imported from the United States or Europe. All of these factors lead to frustrations and slow progress in a highly technical agriculture such as hydroponics. On the other hand, if technical agriculture is introduced successfully, with time and patience, a vast market awaits to be tapped because presently most fresh vegetables are imported from the United States.

The tropics have an ever present abundance of insects and soil-borne pests such as nematodes. Water is often high in dissolved salts, and soils are shallow and infertile. As a result, growing crops by conventional soil culture produces low yields, with major pest problems.

Hydroponics can overcome these problems of high-salt-content water, soil-borne pests and diseases, shallow infertile soils and limited arable land. Intensive culture under controlled conditions utilizing qualified technologists can open a new era of agricultural production in the tropics.

To demonstrate the problems, possible solutions and the potential of hydroponics in the tropics, an example of a sand-culture operation in Venezuela is presented. The company involved has been engaged in sand-culture production of lettuce for over 20 years. Most of this time has been devoted to the development of a suitable sand-culture system for the local conditions.

Figure 8.26. Hydroponic sand-culture beds of head lettuce on terraces in mountainous terrain of Venezuela. (Courtesy of Hidroponias Venezolanas, S.A., Caracas, Venezuela.)

Their farm is located on very steep terrain near Caracas. Terraces have been cut to maximize the available level ground (Fig. 8.26). Each terrace is about one-third of a hectare. As the soil is very rocky and the use of heavy equipment to level the sites is costly, it was decided to construct raised beds of metal frames. The cost of steel and the labor of welding it is relatively low.

The frames of the beds were welded on-site and set in concrete footings as shown in Figure 8.27. The bottoms were constructed of clay bricks and concrete, common building materials of the country (Fig. 8.28). The beds were sealed with a bituminous paint (Fig. 8.29). The system constructed was basically sand culture with some modifications. Since trickle-

Figure 8.27. Metal frames for raised sand-culture beds. (Courtesy of Hidroponias Venezolanas, S.A., Caracas, Venezuela.)

Figure 8.28. Bottoms of beds constructed of clay bricks. (Courtesy of Hidroponias Venezolanas, S.A., Caracas, Venezuela.)

Figure 8.29. Beds sealed with bituminous paint and drainage provided by clay tiles. (Courtesy of Hidroponias Venezolanas, S.A., Caracas, Venezuela.)

or drip-irrigation systems were not readily available in Venezuela at the time of construction, a subirrigation method was used. Large below-ground cisterns of concrete were built for storage of the nutrient solution (Fig. 8.30). Water is pumped from several wells to a storage tank above the hydroponic farm. The water from the wells is low in total dissolved salts, a rare and fortunate situation in the tropics. A complex system of piping and plumbing was installed to move the well water to the storage tank, and to distribute nutrient solution to each bed from the nutrient-solution reservoir (Fig. 8.31).

Figure 8.30. Large concrete cisterns for storage of the nutrient solution.

Figure 8.31. Complex pumping and piping system for distribution of nutrient solution to growing beds.

(Courtesy of Hidroponias Venezolanas, S.A., Caracas, Venezuela.)

The beds are flooded from one end by spraying the nutrient solution into a plenum (Fig. 8.32). Clay tiles cut in half placed along the bottom of the bed (Fig. 8.29) join the plenum to the opposite drain end. The drain is plugged manually to allow the water to rise in the bed to within an inch of the sand surface (Fig. 8.33). This generally takes about 15–20 minutes. The plug is then removed and the water flows out of the bed within 10 minutes, providing fairly good aeration.

Figure 8.32. Distribution of nutrient solution into plenum of each sand bed.

Figure 8.33. Water regulation and drainage of subirrigated sand-culture bed.

(Courtesy of Hidroponias Venezolanas, S.A., Caracas, Venezuela.)

Suitable sand is not available locally and therefore must be hauled by truck a distance of 500 miles. The sand is pure silica of a coarse texture normally used in the glass industry. Its cost is therefore very high, about $50 a cubic meter (slightly greater than a cubic yard). Due to the high cost of sand, the sand requirements were reduced by first putting a 3- to 4-inch (7.5–10-cm) layer of crushed clay bricks on the bottom of the beds over the drain tile lines (Fig. 8.29). Then about 4 to 5 inches (10–12.5 cm) of silica sand was placed on top of the crushed bricks (Fig. 8.29). The use of the clay tiles as a lower medium has proved unsatisfactory since over a period of six months to a year it breaks down into a fine powder causing excessive moisture in the roots of the lettuce plants. It contributes to a bacterial soft-rot problem from the high moisture content always present in the medium (Fig. 8.34). The solution would be to remove the lower clay chard layer and/or convert to a trickle-feed system. This would

Figure 8.34. Bacterial soft-rot of head lettuce. (Courtesy of Hidroponias Venezolanas, S.A., Caracas, Venezuela.)

conserve water, reduce puddling, improve root aeration, and minimize the spread of diseases and insects by lowering the relative humidity at the base of the lettuce. With continued use of beds containing clay chard, the clay breakdown caused excessive puddling, and it had to be replaced with a layer of coarse granitic rock followed by a layer of smaller pebbles before the top sand layer. Results using this sequence of media have been quite successful with good drainage and aeration (Fig. 8.35).

The availability and cost of construction materials, medium, and fertilizers in a specific location will determine the type of hydroponic system that should be used. In many tropical countries, lumber is scarce and very costly. Generally, steel products, concrete and clay bricks are readily available and inexpensive. Such is the case in Venezuela where hydroponic beds are constructed of steel frames, clay bricks and concrete as outlined above.

An alternative to this type of construction is to use an asbestos-concrete roofing channel ("Canal 90") fabricated by Industrias Eternit, S.A., Caracas, Venezuela. These channels are 1 meter (39 inches) wide by up to

Figure 8.35. Beds containing coarse gravel, smaller pebbles and coarse sand. (Courtesy of Hidroponias Venezolanas, S.A., Caracas, Venezuela.)

8 meters (26 feet) long. Depth of the channel is 24 centimeters (10 inches) which is sufficient for most crops. The "Canal 90" can be lined with heavy (12-mil) black polyethylene, or coated with bituminous paint to seal it from chemical reactions with the nutrient solution. It has a shallow "W" shape which is ideal for drainage. Drainage is provided by placing clay tiles end-to-end with several centimeters overlap in each depression of the channel (Figs. 8.36, 8.37). A screen is placed immediately over the tiles to prevent sand entering the drainage spaces. The drain tiles are covered with 7-8 centimeters (3 inches) of coarse rocks (3–4 cm diameter) followed by 10–12 centimeters (4–5 inches) of pea gravel (0.5 cm diameter) and the remainder is filled with coarse sand. All rocks and sand must be of granitic origin, preferably quartz. Subirrigation or trickle may be used. If a subirrigation system is used, a nutrient-solution tank must be located below level unless the beds are raised.

Each bed can grow two rows of tomatoes, cucumbers, eggplants or peppers. If lettuce is grown, two channels should be placed within 5 centimeters (2 in) of each other. Up to four rows of head lettuce can be planted in each bed. The distance between plants within rows should be a minimum of 25 centimeters (10 in) and spacing between rows should be 10 centimeters, 25 centimeters, 25 centimeters and 10 centimeters (4 inches,

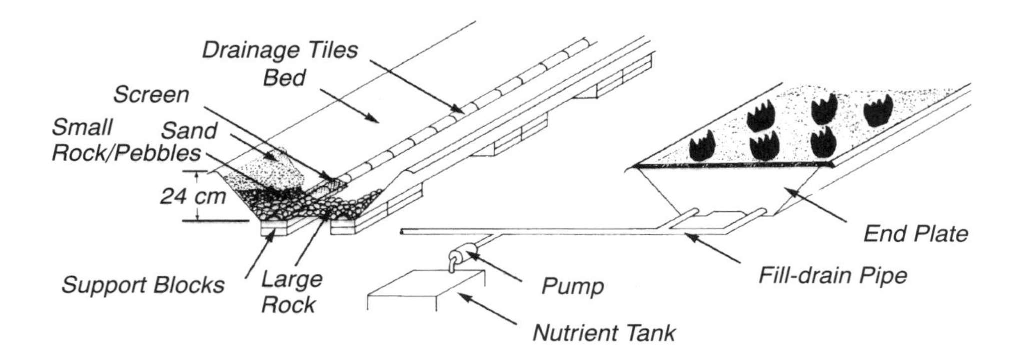

Figure 8.36. Sketch of "Canal 90" bed.

10 inches, 10 inches, and 4 inches). With two channels positioned 5 centimeters (2 inches) apart, the spacing between all rows of plants will be 25 centimeters (10 inches).

Tomatoes, eggplants, and peppers should be spaced 40 centimeters (16 inches) apart within the rows and 50 centimeters (20 inches) between rows in each bed. Beds should be individually situated with pathways of 65 centimeters (25 inches) in width to allow sufficient light penetration in the crop. Cucumbers should also be grown in individual beds placed 65 centimeters (25 in) apart. Spacing within rows should be at least 65 centimeters (25 inches), and the distance between rows a minimum of 50 centimeters (20 inches).

Watercress was grown successively in "Canal 90" beds of 6-meter (20-ft) length. (Fig. 8.37) The beds were filled with coarse gravel to just above the level of the drain tiles. A layer (2 cm) of coarse sand was placed on top of the coarse gravel. The watercress seedlings or cuttings were placed on top of the medium and the beds flooded with nutrient solution to a level of 10 centimeters (4 inches) above the medium. This level of solution was maintained constant. Additional solution was pumped up to recycle back to the nutrient tank every hour for 15 minutes during the daylight hours to increase oxygenation to the plants.

Due to increased cost of the "Canal 90," new beds have been constructed of steel supports with bricks, as described for the lettuce. No medium is used for the watercress. The solution level is maintained at 1–2 cm depth and is circulated every 15 minutes for 5 minutes. This is an ebb-and-flow system (Fig. 8.38).

Cuttings are taken from existing plants to transplant into new beds. Within 4 weeks they are ready for harvest. Plants are harvested every 25–28 days for bulk sales of 0.5 kg bundles and every 14–16 days for packaged pre-mixes.

Figure 8.37. Watercress growing in "Canal 90" bed. (Courtesy of Hidroponias Venezolanas, S.A., Caracas, Venezuela.)

Figure 8.38. Watercress growing in ebb-and-flow beds. (Courtesy of Hidroponias Venezolanas, S.A., Caracas, Venezuela.)

Every 6–8 months, as the root-mass becomes too thick to allow good oxygenation, the plants are changed.

Watercress sold as 0.5-kilogram (1.1-lb) bundles to restaurants in Caracas are placed in plastic tote bins having several centimeters of water on the bottom (Fig. 8.39). This maintains freshness and in fact, allows the watercress to take up water and increase its weight. The entire plastic bin is delivered to a restaurant, similar to a case. The client keeps the watercress fresh in the bin until it is used and then returns the bin to the grower at the next delivery.

Figure 8.39. Packing of watercress in plastic tote bins. (Courtesy of Hidroponias Venezolanas, S.A., Caracas, Venezuela.)

In the tropics where daylength and the hours of sunlight vary only slightly from one month to the next, an accurate correlation should exist among electrical conductivity (EC), total dissolved solutes (TDS), concentration of various nutrients and plant age as discussed earlier (Sec. 3.8.4). There are two seasons in the tropics, a wet season from June through December and a dry season from January through May. While there is more precipitation during the wet season, rainfall is very heavy over a short period of time each day. Rapid clearing of the sky follows with full sunlight. Therefore, the number of hours of sunlight for any month during the wet season does not differ greatly from that of the dry season. Consequently, the most significant effect upon nutrient solutions is that of plant growth stage, rather than the influence of sunlight. As a result, electrical conductivity can be used to accurately monitor the status of the nutrient solution.

In the tropics, year-round warm temperatures generate large insect populations. When crops are grown hydroponically outside without the protection of closed greenhouses, insects quickly invade the crops and rapidly spread with few or no natural predators. The use of pesticides is imperative. Without skilled people knowledgeable in the identification of insects and the use of pesticides, crops infested by one or several pests will soon be destroyed. Such was the case with the lettuce crop in early 1980 when a leaf-miner infestation spread throughout the crop causing heavy losses, with 70–80 percent of the crop non-marketable (Fig. 8.40). Some pesticides were used in an attempt to control the leaf miners but the chemicals were not effective. Often pesticides are not available and those which are may have become ineffective due to the buildup of insect resistance. Without the introduction of new pesticides to overcome such resistance, control diminishes until complete crop losses occur.

Figure 8.40. Leaf-miner damage to lettuce.

In the past, leaf miners were controlled by pesticides applied with a backpack sprayer (Fig. 8.41). Now by use of sticky traps, spot spraying of infested areas and removal of infected leaves, adequate control is achieved. Assisting this control is a natural predatory wasp.

Leaf miners infest the leaves of plants at an early stage. As larvae, they eat the leaf tissue tunnelling between the upper and lower leaf epidermal layers. This produces a characteristic reticulate pattern (Fig. 8.40). If the upper and lower leaf surfaces are pulled apart the larvae will be exposed.

The adult female, about 2 mm long, punctures the leaf surface with a tubelike appendage on the abdomen known as an ovipositor and inserts eggs through it. The injection of the eggs into the leaves causes small white blister-like spots on the leaf. The adult males and females feed on

Figure 8.41. Application of insecticide by use of "Solo" backpack sprayer. (Courtesy of Hidroponias Venezolanas, S.A., Caracas, Venezuela.)

the sap which oozes from the punctures. Each female lays about 100 eggs in its two- to three-week life span (see life-cycle in Section 12.14 *Diseases and Insects*). The eggs hatch in 5 to 6 days into soft white maggots (larvae) which reach about 2.5 millimeters in length when mature. The larvae tunnel for up to two weeks before pupating for two weeks to emerge as adult flies. The female flies then infest new leaves and the cycle is repeated. About five weeks is required for the completion of the life cycle from egg to adult. They infest many crops such as lettuce, tomatoes, carnations and chrysanthemums.

With such a short life cycle the insect can spread rapidly under favorable conditions of warm tropical climates. Populations increase to epidemic proportions in a short time if no effective control is practiced; crops can be completely destroyed.

While tropical climates have generally an abundance of sunshine favorable to the growing of crops, they have a wet and dry season. Extremes in rainfall are commonplace. For instance, in the plains of central Venezuela during the dry season, precipitation is so low it closely resembles desert conditions with extremely high temperatures over 30° C (86° F), while during the wet season, rainfall is so heavy the entire region becomes inundated with several feet of water. In such areas crops can only be grown during the dry season with use of irrigation.

Even in the mountainous regions the wet and dry seasons are distinct. During the wet season, temperatures are generally a few degrees warmer and rainfall occurs for several hours almost every day. As a result, crops that cannot tolerate continual moistening soon are damaged and infected with diseases. Such is the case with lettuce, especially head lettuce. Within several weeks of maturity, head lettuce is very sensitive to moisture. Any moisture that penetrates the head causes rotting (Fig. 8.34) which spreads rapidly with high temperatures. Losses of up to 40 to 60 percent are not uncommon.

Since many crops cannot be cultivated during the wet season, short supply raises prices dramatically. Head lettuce sells as high as $3.00–$4.00 each during this period of low production. Therefore, an excellent market can be established for hydroponically-grown lettuce and other produce if the market can be reliably supplied year-round.

Using sand culture in raised beds keeps the rainfall from upsetting the growing of crops through excessive root moisture, but moisture onto the plants still causes crop failures. To overcome this, a greenhouse type of structure was erected over the beds to prevent the rain from entering the crop as it matures. A closed greenhouse would not be suitable since daily rainfall occurs only for short periods and the sun shines for the majority of the daylight hours. Closed greenhouses would result in high temperature buildup due to the "greenhouse effect" of trapped heat radiation. Unless a cooling system were installed, temperatures would quickly exceed 40 degrees Celsius (104° F) within minutes of the appearance of the sun. While such cooling systems may be the answer to complete environmental control, they are very costly to import and difficult to maintain.

Since the only function of a greenhouse cover was protection from the rain, a better approach was to install a sloped fiberglass roof supported by a steel structure that could be made on-site by local workers. With no sides, free wind movement over the crop allowed natural ventilation for cooling.

The structure was designed in a saw-tooth fashion so that the wind could freely pass through the roof at each "tooth" (Fig. 8.42). Initially, the high points of the structures were oriented away from the prevailing wind, assuming that as the wind blew over the top a vacuum effect would suck air outside from underneath through the openings giving good air exchange. However, this was not the case, so the structures were re-oriented in the other direction with the high sides facing the prevailing wind.

At the low edges the roofs were less than one meter (39 inches) above the top of the growing beds and the distance between openings was less than one-half of a meter (19 inches). Air exchange was restricted and lack of sufficient ventilation caused temperatures to rise 6 degrees Celsius (11° F) above ambient. With ambient temperatures between 25 and 28 degrees Celsius (77–82° F), the temperatures within the "greenhouse" ranged between 30 and 34 degrees Celsius (86–93° F), exceeding the tolerable range for head lettuce. These high temperatures caused "bolting" of the lettuce (Fig. 8.43) so they were not marketable.

Several modifications were made to the structures to overcome excessive temperatures within the growing area. First, the structures were raised

Figure 8.42. Compare raised and reoriented structures (left) with original structures (right). (Courtesy of Hidroponias Venezolanas, S.A., Caracas, Ven.)

Figure 8.43. "Bolting" of head lettuce caused by excessive heat. (Courtesy of Hidroponias Venezolanas, S.A., Caracas, Venezuela.)

one meter (39 inches). Second, the openings were increased to one meter in height and oriented facing the prevailing winds (opposite to what they were) (Figs. 8.42, 8.44).

Ventilation was greatly improved. At the same time when little, if any, wind existed, the heated air would rise above the crop and escape through the wider openings creating natural ventilation. Cooler air would enter

Figure 8.44. Roofs were oriented toward the prevailing wind and raised one meter in height. Compare the raised roofs of foreground to the original lower ones in the background. (Courtesy of Hidroponias Venezolanas, S.A., Caracas, Venezuela).

from the sides setting up good air exchange. By orienting the openings toward the prevailing wind, air is caught and forced down onto the crop moving across the beds. With a prevailing wind, rapid air changes of at least one every 10 to 15 seconds kept the temperatures within one degree of ambient. Bolting of the lettuce was largely overcome.

The fiberglass coverings have been used successfully, but within several years the material discolors. High humidity and temperatures cause heavy growth of algae on the fiberglass panels. As a result, light transmissibility is greatly reduced. Washing the panels every 3 to 4 weeks to remove the algae became essential. However, such washing with a brush damages the panels by exposing the fibers. Consequently, the panels deteriorate rapidly through wear of the acrylic surface and light passage through the panels is reduced sufficiently to affect crop production.

An alternative to fiberglass is polyethylene. While dust and algae growth are also problems with polyethylene, washing is easier since no corrugations exist. Ultraviolet light from sunlight in the tropics is higher than in temperate regions and breakdown of the polyethylene is more rapid, necessitating changing the polyethylene every year. A more durable grade of polyethylene developed for greenhouses such as "Monsanto 602" has lasted up to two years in some tropical regions, but often if a plastics industry exists in a country, it is difficult or costly to import other polyethylene due to high tariffs protecting the local industry

With the use of polyethylene instead of fiberglass to cover the protective structures, some modifications in the basic sawtooth design must be made. Its use with roses has been successful in Venezuela (Fig. 8.45).

Figure 8.45. Polyethylene structures used for the growing of roses in the tropics.

Growing of head lettuce in areas of high temperatures, 25 to 28° C (77 to 82° F), is difficult due to "bolting" which occurs above these temperatures as discussed earlier. With a slight reduction in light from the use of a fiberglass or polyethylene cover, bolting occurs at lower temperatures (25° C to 27° C). For this reason, a temporary cover which could easily be removed would be beneficial to the growing of many cool-season crops. A polyethylene-covered greenhouse is now available having automated roll-up roofs. The motorized system rolls up the poly roof within one minute. This type of greenhouse combined with a fog-cooling system should greatly assist the growing of cool-season crops in the tropics. If the roll-up covers could be housed in an opaque cover or box while not in use, ultraviolet light could be excluded, extending the longevity of the polyethylene.

Head lettuce is very susceptible to bacterial soft rot during the last 15 to 20 days of maturity. It is during that stage that the polyethylene covers would have to be used to prevent rain from entering. Also, the covers would be used during the night and early morning when heavy dew forms.

Since ambient temperatures at the elevation of the hydroponics farm are very close to the upper tolerable limit for heading of lettuce, heat tolerant varieties must be used. Some 20 to 25 varieties were tested to find that "Great Lakes 659" was most resistant to bolting.

A final approach to solving the problem of bolting is to locate additional growing beds for lettuce production at higher elevation terraces, and use the existing terraces for growing carnations, chrysanthemums and warm-season crops such as tomatoes and peppers.

In many tropical countries, human sewage is used as fertilizer for field crops. This has resulted in endemic infestations of amoeba causing dysentery. Consequently, many people will not eat fresh salads for fear of becoming ill with dysentery. Hydroponically-grown leafy crops such as lettuce and many herbs such as watercress, which are very popular in salads, make amoeba-free crops possible. Such sterile products can be easily sold to exclusive restaurants and produce stores for a premium price.

Nematodes infest most soils in tropical countries. These are very small microscopic round worms, sometimes called eelworms, that infest the roots of plants. Depending upon the species involved, they may cause death of roots, injuries to roots which act as ports of entry for fungal diseases or swelling of roots so that they cannot function in normal water and mineral uptake (Fig. 8.46). This root damage often causes plants to wilt during the day when water uptake cannot meet evapotranspirational losses. If plants do survive the stress, they become stunted and non-marketable.

Soil temperature is critical in the development of nematodes. Females fail to reach maturity at temperatures above 33° C (92° F) or below 15° C

Figure 8.46. Swelling lettuce roots caused by nematode infection. (Courtesy of Hidroponias Venezolanas, S.A., Caracas, Venezuela).

(59° F). Soil temperatures in the tropics are near optimum levels for nematode development. It takes about 17 days at 29°C (85°F) for females to develop from infective larvae to egglaying adults. Spreading within the field occurs through movement of infested soil or plant debris by man, water, and wind.

Even with hydroponic culture, nematodes can easily be introduced into a crop by movement through water or wind. Preplanting treatments of steam or chemicals effectively eliminate nematodes from soil and other media such as sand, sawdust or gravel used in hydroponics. Nematodes are killed with steam sterilization by heating the medium under moist conditions for 30 minutes to 49° C (120° F). This and chemical methods of sterilization are described in Chapter 4. Basamid, a soil fumigant, may be used for the control of unencysted soil-borne nematodes, soil fungi, and germinating weed seeds. It is applied at a rate of 325 to 500 kilograms per hectare, or 3 to 5 kilograms per 100 square meters. It must be evenly distributed on the soil surface, then worked into the upper 15–23 centimeters (6–9 inches) of the medium. The medium is then covered with polyethylene sealing the edges with medium or weights (Fig. 8.47). The interval between treatment and planting depends on the temperature, the moisture and nature of the medium. For example, at warm temperatures of above 18° C (64° F) the medium may be opened up 5 to 7 days after the Basamid application. The medium should then be rototilled to release all

Figure 8.47. Basamid sterilization of sand in bed for elimination of nematodes. Polyethylene cover edges sealed with pipes and sand. (Courtesy of Hidroponias Venezolanas, S.A., Caracas, Venezuela).

residual Basamid. A waiting period of 7 to 10 days followed by germination tests must be carried out before planting. The period between treatment and planting depends upon the medium temperature at 10 cm (3.9 in.). If it is over 18° C (64° F), a period of 10–12 days is required. Caution must be exercised as Basamid is toxic to all growing plants.

Basamid can be removed from the sand beds within 24 hours by washing the beds with a spreader solution normally used in the application of pesticides. The overall sterilization cycle has been reduced to 3 days: One day of fumigation; one day for cleaning with the spreader and one day to wash with fresh water (at least 4–6 washings are required). No harmful effects of this short fumigation cycle have been found on lettuce production, and it has been very effective in the elimination of nematodes from the sand medium.

Steam sterilization is equally effective as chemicals and can be carried out in a shorter period of time, especially if the medium is not porous. Steam sterilization can be done with a small portable steam sterilizer (Fig. 8.48). Perforated pipes are placed several inches below the surface of the medium in the beds. Then the entire bed is covered with a heavy vinyl or canvas tarpaulin (Fig. 8.49).

Figure 8.48—Portable steam sterilizer. (Courtesy of Hidroponias Venezolanas, S.A., Caracas, Ven.)

Generally, it takes several hours to raise the temperature of the medium throughout the bed to 60° to 82° C (140°–180° F). It is best to moisten the medium prior to sterilizing as the moisture carries the heat uniformly throughout the medium. This process is really pasteurization, not sterilization, as it is best not to use as high a temperature as would be required for sterilization, since only the detrimental organisms must be killed. If

Figure 8.49. Steam sterilization of sand-culture beds. (Courtesy of Hidroponias Venezolanas, S.A., Caracas, Ven.)

the lower temperatures between 60° and 80° C (140°–180° F) are used, many beneficial organisms will remain alive. It may be a disadvantage to sterilize the medium and kill all organisms, as after sterilization any organisms can be easily introduced into the medium. Whereas, with pasteurization at slightly lower temperatures only the pest organisms will be killed. Beneficial organisms can then resist re-inoculation by detrimental organisms.

Hidroponias Venezolanas no longer uses Basamid or steam sterilization to control nematodes. Between lettuce crops they flood the beds with water for several hours to float the old roots to the surface, removing them by hand. After draining, the beds sit empty for several days. Sunlight raises the temperature of the sand sufficiently to kill any nematodes.

Once this multitude of problems can be resolved, excellent crops can be grown under tropical conditions using hydroponics (Figs. 8.50, 8.51).

Figure 8.50. Carnations grow well in sand culture in the tropics. (Courtesy of Hidroponias Venezolanas, S.A., Caracas, Venezuela.)

Figure 8.51. A healthy crop of head lettuce grown in sand culture in the tropics. (Courtesy of Hidroponias Venezolanas, S.A., Caracas, Venezuela.)

The disadvantages of unavailability and high costs of materials and equipment can be overcome by use of water cultural and NFT systems of hydroponics. While little commercial application of the nutrient film technique has been carried out in the tropics, it does offer a vast potential by reducing capital costs and dependency upon suitable media such as granitic sand or gravel which is often rare. Hydroponics offers the answer to combating the pests and diseases of the tropics, the lack of good quality water, the presence of steep terrain and relatively small areas for growing at altitudes which have suitable climatic conditions for growing temperate crops.

With high population centers creating strong demand for high quality products normally not obtained from conventional soil culture, superior quality, clean, hydroponically-grown crops find the marketplace willing to pay a premium price. This higher return for products grown hydroponically can justify the high initial capital costs of establishing a hydroponic farm in tropical countries. It will support imported technology required for successful operation of hydroponics and training of local workers.

The intense solar radiation in tropical countries produces high-yielding crops as long as they can be protected from torrential rainstorms. By choosing the correct altitudes, optimum temperatures can be encountered for growth of the crops. Since any structures need function only for protection of crops from rainstorms and not for heating or cooling, very simple inexpensive covers can be used from locally available materials as much as possible to reduce overall capital costs. With natural ventilation for cooling and no heating system, the use of energy from electricity or fossil fuels is minimal. Low-cost labor combined with low energy costs of operating a hydroponics farm in the tropics and high prices for produce all contribute to high profits. For this reason, while the initial struggle to overcome many problems and frustrations in establishing a hydroponic operation seem immense, eventual success will lead to handsome returns on the investment.

When all problems were resolved, each hydroponic bed 2½ meters by 20 meters by 25 centimeters deep (8.4 ft by 65.6 ft by 9.8 inches deep) was capable of producing 310 to 330 head of lettuce, each averaging 1 kilogram in weight (Fig. 8.52). In a continuous cropping system, new plants were sown every day so that harvesting could be done at least three times a week. In this way the market can be supplied year-round. Consistent supply to the marketplace is essential if long-term contracts are to be met. This reliable servicing of the market is the basis for establishing a strong market, with a constant premium price to be received for the hydroponic products.

Lettuce takes between 80 and 85 days from seeding to harvest. Therefore, yearly production of one hydroponic bed would be five crops of an

Figure 8.52. High quality head lettuce produced from sand culture in the tropics. (Courtesy of Hidroponias Venezolanas, S.A., Caracas.)

average 320 heads or 1,670 heads of lettuce annually. With an average wholesale price paid year-round for high quality head lettuce in Caracas about $2.00, the annual gross sales per bed would be $3,340. The return per square foot of growing area would be $6.89.

The return per square foot for lettuce in North American greenhouses is about $1.50 to $2.00. The gross sales revenue in Venezuela can be three to four times that obtained in North American greenhouses. In addition, since no heating is required and labor is much cheaper in Venezuela, net profits would be six to eight times that of a similar crop grown under greenhouses in North America.

This profit differential between North American and South American growing of vegetable and flower crops hydroponically is the impetus needed to cope with the economic, political and agricultural problems that must be contended with in South America. Once established, hydroponic growing of certain crops offers an attractive return on investment in many tropical countries where arable land is scarce and strong markets exist.

8.10 Small-Scale Sand-Culture System

A simple home unit can be designed similarly to a commercial unit but on a much smaller scale. Basically, it should consist of a bed or growing tray, a nutrient reservoir, and a trickle-feeding system operated by a pump energized by a time clock (Fig. 8.53).

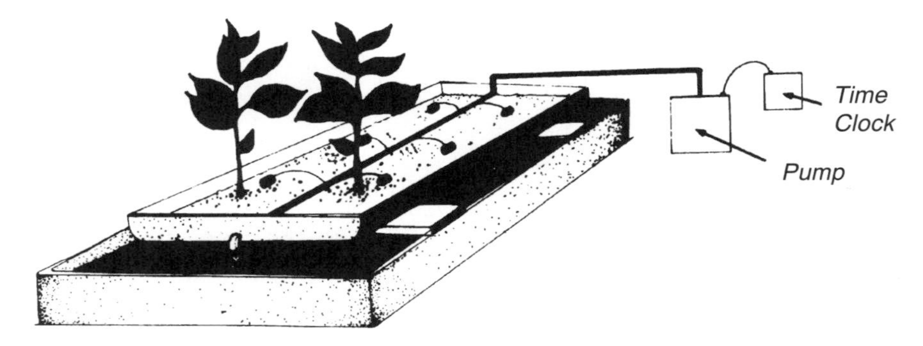

Figure 8.53. A simple small-scale trickle sand-culture system.

The growing tray could have small holes in the bottom of the plastic liner, or use a perforated plastic pipe for drainage. One of the oldest and still popular hydroponic methods for individual potted plants is the wick system. Such a system consists of a double pot, one containing the medium and plant and the other the nutrient solution. A fibrous wick is set into the growing pot about one-third of the way with the other end suspended in the nutrient solution below (Fig. 8.54). As water evaporates from the plant and moves from the medium to the plant, capillary action moves the solution from the nutrient reservoir through the wick to the plant root zone. The ends of the wick would be teased out so that a tuft of untwined fiber is present at both ends. Be sure that at no time is the solution of the reservoir in contact with the bottom of the growing pot or excess water will be drawn into the root zone, causing puddling. Several pots with wicks may be set up over a window-box reservoir. The reservoir can be covered to prevent evaporation losses, but holes are needed at the location of each pot to allow the wick to pass through (Fig. 8.55). Though not essential, it is useful to have a floating marker to indicate the level of liquid in the reservoir.

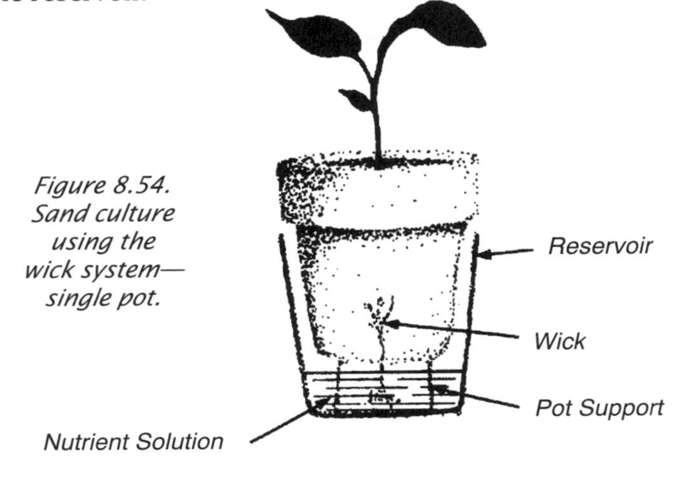

Figure 8.54. Sand culture using the wick system—single pot.

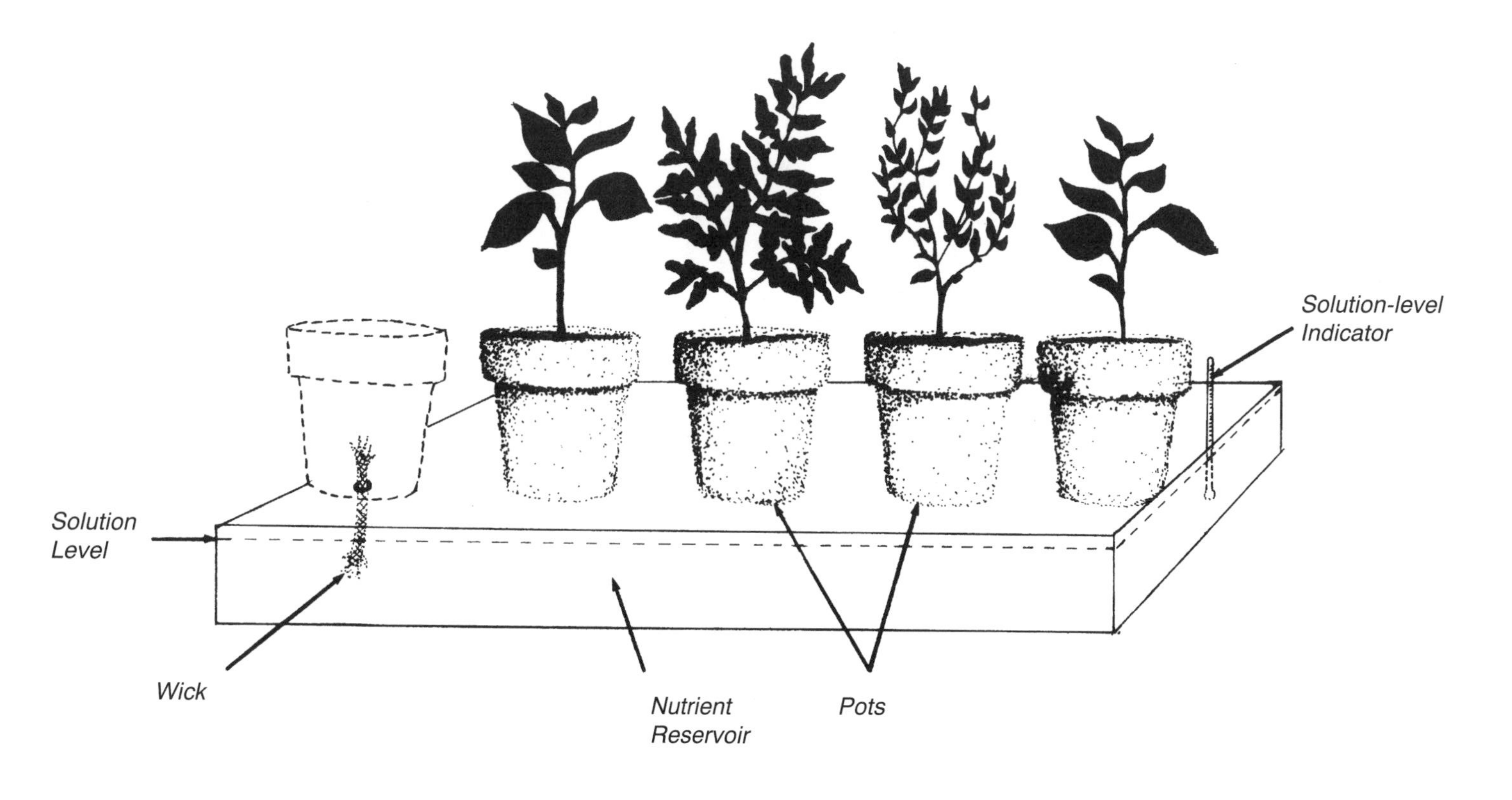

Figure 8.55. Sand culture using the wick system—several pots on a window-box reservoir.

8.11 Advantages and Disadvantages of Sand Culture

Advantages over gravel culture:

1. It is an open system, that is, the nutrient solution is not recycled, so the chances of diseases such as *Fusarium* or *Verticillium* spreading in the medium are greatly reduced.
2. There is less problem with drain pipes getting plugged with roots since the more dense medium of sand favors lateral root growth.
3. The finer sand particles allow lateral movement of water through capillary action so that solution applied at each plant becomes evenly distributed throughout the root zone.
4. With the right choice of sand combined with a drip-irrigation system, adequate root aeration is achieved.
5. Each plant is fed individually with a new complete nutrient solution during each irrigation cycle—no nutrient imbalance occurs.
6. Construction costs are lower than for a subirrigation gravel system.
7. The system is simpler, easier to maintain and service, and more foolproof than a subirrigation gravel-culture system.
8. Due to the smaller particle size of sand, water retention is high and only several irrigations are required each day. If a failure occurs, there is more time available to repair the system before the plants will use up the existing water in the medium and begin to experience water stress.
9. Smaller, centrally located nutrient reservoirs or injectors can be constructed away from the actual growing area of the greenhouse.
10. Sand is readily available in most locations. When using calcareous sand, the formulation can be adjusted to compensate for daily *p*H changes and shortages of iron and/or other elements

.

Disadvantages of sand culture compared with gravel culture:

1. One of the major disadvantages is that either chemical or steam sterilization methods must be used to fumigate between crops. However, such methods are thorough, even if they are a little more time-consuming than the use of bleach with gravel culture.
2. Drip-irrigation lines can become plugged with sediment. This, however, can be overcome by the use of in-line 100- to 200-mesh filters which can be easily cleaned daily.
3. Some claims are made that sand culture uses more fertilizer and water than a cyclic gravel-culture system. Again, this can be overcome by good management. The waste should be monitored and feeding adjusted so that no more than 7 to 8 percent of the solution

added is actually drained out. Even in gravel culture the waste can be equally as great, if not greater, due to the need to change the nutrient solution periodically.

4. Salt buildup may occur in the sand during the growing period. This can be corrected, however, by flushing the medium periodically with pure water. Again, proper management with monitoring of salt accumulation from the drainage water is important to prevent excess salt problems.

References

Fontes, M.R. 1973. Controlled-environment horticulture in the Arabian Desert at Abu Dhabi. *HortScience* 8:13–16.

Hodges, C. N. and C. 0. Hodge. 1971. An integrated system for providing power, water and food for desert coasts. *HortScience* 6:30–33.

Jensen, M. H. 1971. The use of polyethylene barriers between soil and growing medium in greenhouse vegetable production. *Proc. 10th National Agr. Plastics Conf.*, Chicago, IL., Nov. 2–4, 1971. Ed. J. W. Courter, pp. 144–50.

Jensen, M. H. and N. G. Hicks. 1973. Exciting future for sand culture. *Am. Veg. Grower,* November 1973, pp. 33, 34, 72, 74.

Jensen, M. H., H. M. Eisa and M. Fontes. 1973. The pride of Abu Dhabi. *Am. Veg. Grower*, November 1973, pp. 35, 68, 70.

Jensen, M. H. and Marco Antanio Teran R. 1971. Use of controlled environment for vegetable production in desert regions of the world. *HortScience* 6:33–36.

Massey, P. H., Jr., and Yasin Kamal. 1974. Kuwait's greenhouse oasis. *Am. Veg. Grower*, June 1974, pp. 28, 30,

New, L. and R. E. Roberts. 1973. *Automatic drip irrigation for greenhouse tomato production.* Texas A & M Univ. Ext. bulletin MP-1082.

Chapter 9

Sawdust Culture

9.1 Introduction

Sawdust culture is especially popular in areas having a large forest industry such as the West Coast of Canada and the Pacific Northwest of the United States. In British Columbia, Canada, the Canada Department of Agriculture Research Station at Saanichton carried out extensive research for a number of years to develop a sawdust-culture system for greenhouse crops (Maas and Adamson 1971). The need for a soilless-culture system became evident with increased soil-borne nematode infestations and diseases coupled with poor soil structure which made the profits from greenhouse crops very marginal. Today, in British Columbia, over 90 percent of all greenhouses use some form of soilless culture for vegetable and flower production. Vegetable growers usually use sawdust culture, while flower producers use a peat-sand-pumice mixture.

9.2 The Growing Medium

Sawdust was adopted as a growing medium in the coastal region of British Columbia because of its low cost, light weight and availability. A moderately fine sawdust or one with a good proportion of planer shavings is preferred, because moisture spreads better laterally through these than in coarse sawdust.

Sawdust from Douglas fir (*Pseudotsuga menziesii* [Mirb.] Franco) and western hemlock (*Tsuga heterophylla* [Raf.] Sarg.) were found to give best results (Maas and Adamson 1971). Western red cedar (*Thuja plicata* D.) is toxic and should never be used.

While other media such as sphagnum peat, ground fir bark, and mixtures of sawdust with sand and/or peat were tested successfully, they are more expensive and therefore might be used if sawdust is unavailable.

One precaution that should be taken with sawdust is that of determining its sodium chloride content. Logs are floated in barges on the ocean and often remain in the water for many months before going to the sawmill. They will absorb sea water over this time and thus acquire salt (sodium chloride) levels toxic to plants. Therefore as soon as the sawdust is received, samples should be taken and the sodium chloride content tested. If any significant amount of sodium chloride is found (greater than 10 ppm), the sawdust should be thoroughly leached with pure fresh water once it is placed in the beds, but before planting. This leaching process may take up to a week in order to reduce the sodium chloride to an acceptable level.

9.3 Bed System

The growing beds are usually constructed of rough cedar and lined with 6-mil black polyethylene or 20-mil vinyl similar to those designs discussed under sand culture (Fig. 9.1). Rough cedar boards 1 by 8 inches (2.5 by 20 cm) can be used for the sides. Either a V-bottom or round-bottom bed configuration is suitable (Fig. 9.1). The depth of the beds should be 10 to 12 inches (25 to 30.5 cm). A 2-inch (5-cm) diameter drain pipe is

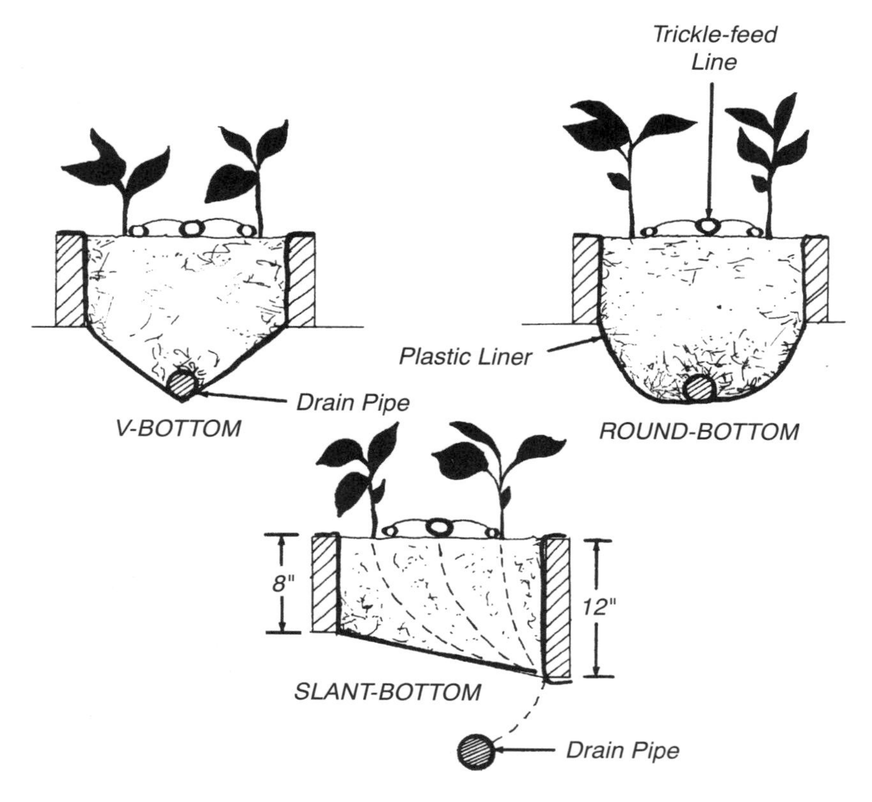

Figure 9.1. Cross sections of sawdust-culture beds.

placed on the bottom of the bed. Beds are usually 24 inches (61 cm) wide, however, 20-inch (51-cm) wide beds (inside dimension) are adequate with 32-inch (81-cm) pathways between the beds (Fig. 9.2). Studies by Maas and Adamson (1971) have shown that even somewhat narrower and shallower beds, with a volume of ⅓ cubic foot (0.009 cubic meter) of medium per plant, are satisfactory. If narrower beds are used, the pathways should be widened to provide the same total amount of greenhouse space for each plant, as light requirements for the plants are the same regardless of the ability of the volume of medium required to provide adequate nutrition.

Figure 9.2. Tomato crop in sawdust beds. Note the location of heating pipes and trickle-irrigation feed lines.

An alternative to these standard bed designs is the use of sloped-bottom beds (Fig. 9.1). In this case the beds are constructed of rough cedar 1 × 8 inches (2.5 × 20 cm) on one side and 1 × 12 inches (2.5 × 30.5 cm) on the other side. The 1-inch × 12-inch lumber is covered with the liner on the inside face and around the top and bottom edges. The 1-inch × 8-inch lumber is covered with the liner on one face and extends on a slope toward the wide side. But a small gap (¼ inch) (0.6 cm) between the end of the poly liner and the wide side allows drainage of excess solution to a drainage pipe located several feet (61 cm) below the beds (Fig. 9.1). The drain pipes are installed when the site is leveled prior to the construction of the greenhouse. In this case, standard perforated plastic drain pipe is used. It is placed in a ditch dug by an automatic ditch-digging machine (Ditch Witch). When a fine-texture (clay loam) soil is present, a filtering material (blinding) placed above and around the pipe will prevent clogging of the holes in the pipe (Fig. 9.3). The best material is a composition of varied particle sizes: coarse sand, to fine sand, pea gravel. The material should be placed to a depth of approximately 6 to 8 inches (15 to 20 cm) above and around the tile.

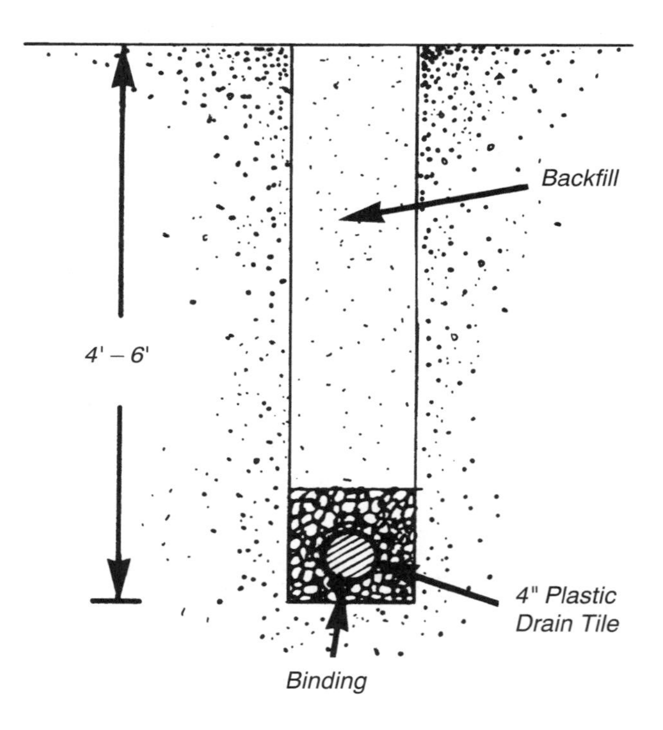

Figure 9.3. Cross section of a drainage ditch and 4-inch perforated drain pipe.

9.4 Bag System

An alternative to beds is the use of polyethylene bags filled with sawdust (Fig. 9.4). Regular "kitchen" garbage bags of 1.25 mil thickness, measuring 20 by 22 inches (51 by 56 cm) can be used. Drainage holes must be punched in the bottom to allow good drainage. The bags are often set on top of a poly sheet to prevent any roots growing out of the drainage holes from contacting the underlying soil. Depending upon the bag size, up to three plants may be grown in each bag. The three-plant bags are placed in a single row but the plants are positioned in the bags and trained vertically (e.g., tomatoes) so as to make two rows.

In the growing of European cucumbers, bags are set in a single row, then every other bag is planted with one plant (Fig. 9.5). These are trained on a sloping trellis. Once this first crop nears completion, new transplants are placed in the alternate empty bags and the process of using alternate bags is repeated after replacing or sterilizing the medium of those plants just removed. In this way up to six short cucumber crops can be grown annually.

A modification of the bag system involves the use of five-gallon (20-liter) plastic nursery pots in place of the polyethylene garbage bags. The

Figure 9.4. Bag system of sawdust culture growing tomatoes.

Figure 9.5. European cucumbers growing in a bag system of sawdust culture.

use of pots eliminates the cost of replacing the bags annually. Sawdust, sand, pea gravel or a mixture of peat-sand-sawdust may be used as medium in pots. A trickle-feeding system must be used with pot culture. If sawdust is used, it is important to spread about ½ inch (1 cm) of sand over the surface to achieve better lateral movement of nutrient solution as it is applied to the surface by the trickle-feed system. The use of a shallow layer of sand on top of sawdust applies to bag culture also, to prevent coning of the nutrient solution through the plant roots.

Pots have several advantages over bags. Plastic patio pots, commonly used in the nursery container industry, can be re-used for three to four years. If sawdust is used, it can be emptied from the pots into a pile and easily sterilized by use of steam or chemicals between crops. Even if the medium is discarded annually, pots lend themselves to efficient handling when filling them with new medium. They can be filled by use of mechanized potting machines such as those used in the nursery industry. With such equipment, six people can fill and plant over 5,000 pots per day. Filling plastic bags is much slower and more difficult. Plastic bags, unlike pots, cannot be easily transported to the greenhouse facilities. Plastic pots of 5-gallon size, having volume of growing medium similar to plastic bags, will produce equally healthy plants (Figs. 9.6, 9.7).

Since plastic pots have drainage holes, it is important to closely monitor the amount of nutrient solution applied to be certain that no more than 10 percent waste occurs. Pots, like bags, should be placed on plastic sheets so that roots will not grow into the underlying medium. If pea gravel is used as the medium, the plastic underlay sheets could be arranged to conduct excess nutrient solution back to the nutrient reservoir. Shallow troughs may be built to house the pots. This arrangement is particularly useful for small systems as used in backyard greenhouses (Fig. 9.7). The plastic underlay sheets should be black in color in order to prevent algae growth.

More recently, sawdust bags are made from 6-mil white polyethylene "layflat" plastic heat-sealed on one end, then filled with the medium before heat-sealing the other end. Filling is done with a hopper and chute that fits into the top of the bag. After filling, the bag is heat-sealed with a heat-sealing machine. The process can be fully mechanized to reduce labor. Finished bags measure approximately 8–10 inches (20–25 cm) wide by 2 feet (60 cm) long by 3–4 inches (8–10 cm) thick (Fig. 9.8). Bags are manufactured by a local company at a cost of about CAN$0.13 (US$0.10). Holes are punched in the bags by the manufacturer for drainage. These bags are particularly suitable for the growing of tomatoes. Each bag contains two tomato plants.

Seedlings are started in rockwool cubes, which are transplanted into rockwool blocks, which are then placed into the sawdust bags, similar to

Figure 9.6. Five-gallon plastic patio pots filled with pea gravel. Trickle-feed system.

Figure 9.7. Plastic patio pots in vinyl-lined bed for lettuce and tomatoes in a backyard greenhouse.

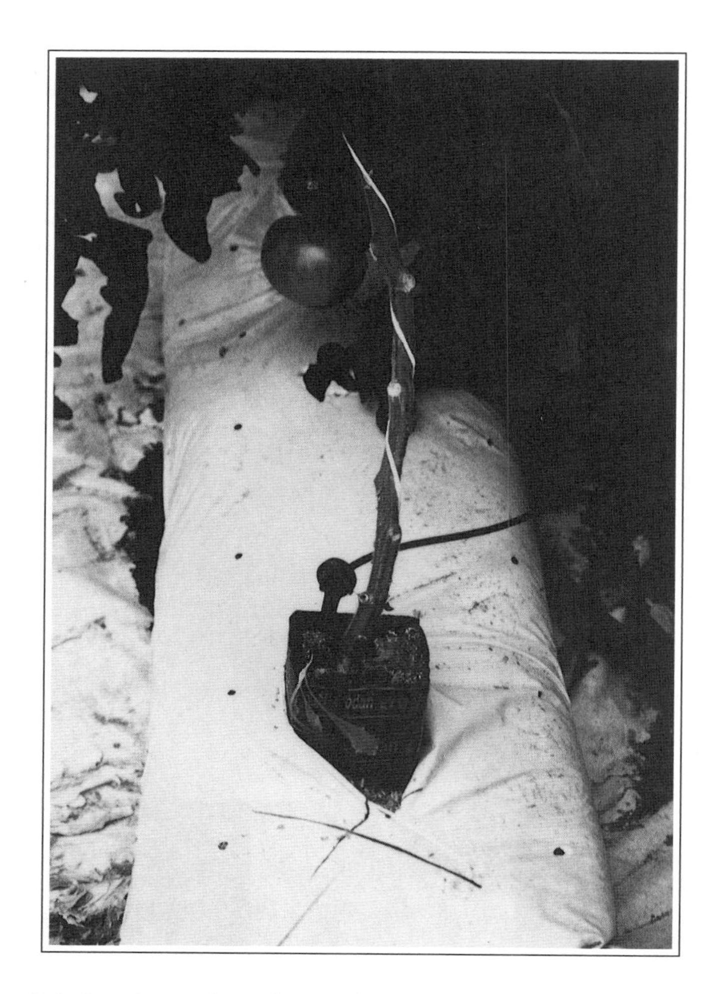

Figure 9.8. Sawdust-culture bag with rockwool block in which tomato plant seedling was grown. Note the drip emitter on the rockwool block at the base of the plant. (Courtesy of Gipaanda Greenhouses Ltd., Surrey, B.C., Canada.)

rockwool culture. For an early crop (crops seeded by mid-December), the growing blocks are placed on top of the sawdust bags, but not allowed to root out until one flower truss sets fruit. Then holes are cut below the blocks in the plastic bag and the blocks set into the holes.

In British Columbia, where light is limited during the winter months, when tomato seedlings are started it is advisable to use high-intensity discharge (HID) sodium-vapor 400-watt lights to give 5,500 lux (510 foot-candles) intensity at plant surface with a 20-hour photo-period. Seedlings should be grown under this propagation section of the greenhouse until they are ready for transplanting into the bags. Seedlings are generally sown in mid-December and transplanted in mid-January.

The greenhouse floor is lined with white polyethylene to prevent contact of the roots growing from the bags with the underlying soil. The white polyethylene also serves to reflect light (much needed during the winter months) and seal out any insect and disease organisms present in the underlying soil. Similar to rockwool culture, the floor is sloped to create a swale between each set of two rows of bags for drainage of excess solution out of the greenhouse. Hot water heating pipes conducting heat from a central boiler are laid in the aisles between the double rows of bags (Fig. 9.9). The heating pipes also serve as a track for a mobile harvesting plant maintenance cart to run on (Fig. 9.10).

Figure 9.9. Sawdust culture with hot-water heating pipes in aisle next to rows of tomato plants. Note the white poly ground cover on the floor.

Figure 9.10. Mobile harvesting working cart runs on heating pipes. (Courtesy of Houweling Nurseries Ltd., Delta, B.C., Canada).

Each plant is fed individually using a spaghetti trickle-feed line placed on the rockwool block in the sawdust bag. The irrigation line runs the length of the rows between each set of two rows of bags (Fig. 9.11). The nutrient solution is pumped from a central injector system using stock solution tanks. Carbon dioxide enrichment is also distributed to the plants by a small polyethylene conduction tube running the length of the beds between each set of two rows of plants (Fig. 9.11). The carbon dioxide is generated as a by-product of combustion of natural gas from the central boilers and is piped to each greenhouse (Fig. 9.12).

Figure 9.11. Sawdust culture with trickle-irrigation lines to each plant. Carbon dioxide is released through a small poly-ethylene tube between the rows.

Figure 9.12. Carbon dioxide recovery unit attached to central boiler. (Courtesy of Gipaanda Greenhouses Ltd., Surrey, B.C., Canada.)

Tomatoes are harvested into 25 pound (11 kg) plastic tote bins (Fig. 9.13) using the mobile carts running on the heating pipes. Pollination, suckering, tying string supports, etc., are all made easier by use of a mechanically raised platform running on the heating pipes. Harvested tomatoes in tote bins are moved using a pallet jack (Fig. 9.14). Peppers have also been grown successfully using sawdust bags.

In British Columbia, the most common tomato varieties used are Dombito and Trust. With the sawdust-culture system, generally one crop a year of tomatoes is grown. Seeding is in mid-December with production beginning by late March and continuing until November.

Plant density of about 8,000 plants per acre (20,000 plants/hectare) produces an average of 110–130 tons/acre (255–300 tonnes/hectare). This plant density is 5.45 square feet of greenhouse area per plant, producing 5 to 6 pounds/square foot (25–30 kg/sq m). The best growers produce up to 160 tons/acre (370 tonnes/hectare) or 37 kilograms/square meter. This is equivalent to 42 pounds (19 kg) per plant per year. Tomatoes and cucumbers are marketed through a greenhouse vegetable cooperative in British Columbia. The cooperative sells the boxes to the growers, grades, packs, and markets the produce (Fig. 9.15).

Figure 9.13. Tomatoes are harvested into plastic tote bins.

Figure 9.14. A pallet jack is used to transport palletized tote bins to the packing area of the greenhouses.

Figure 9.15. Greenhouse tomatoes packaged by the B.C. Greenhouse Vegetable Growers Cooperative.

9.5 Nutrient Solution Distribution System

In both the bed and bag systems of sawdust culture, a drip-irrigation system is used to supply the water and nutrient requirements of the plants. As pointed out in Chapter 8, adequately sized main headers and valves are needed to balance the flow to the row headers. Row headers are usually made of ¾-inch plastic hose, which can handle 200 (0.045 inch inside diameter) feeding tubes. Row headers of ½-inch and 1-inch sizes can supply 100 and 300 feeding tubes, respectively.

In the bed system, spaghetti feeders, ooze hoses or emitters may be used, but in the bag system spaghetti feeders or emitters are best adapted, due to the distances between bags.

The plants are supplied with nutrient solution directly from a dilute-solution storage tank or through a fertilizer proportioner from containers of concentrated stock solutions, as described for sand culture (see Section 8.4). The dilute-solution system needs a storage tank, a pump, and a distribution system.

Determine the capacity of the tank from the number of plants to be fed at one time. The tank should be able to supply at least 1 quart (1 liter) of solution per feeding for each plant for 1 week's total feeding requirements. The total volume needed depends upon the number of irrigation cycles required, which in turn is a function of weather conditions, plant maturity and the nature of the plant. Some growers install more than one tank so that they can prepare the nutrient solution at least a day in advance of the estimated time of depletion of the other tank. In this way, the recently made nutrient solution can be heated for at least 12 hours by an immersion heater to bring its temperature to an optimum level of 65° to 70° F (18° to 21° C) before it is applied to the plants.

Wooden tanks may be constructed of ¾-inch plywood and lined with vinyl or 6-mil polyethylene. Vinyl is a better liner than polyethylene since it does not puncture easily and can be easily repaired with a common swimming pool repair kit should a hole develop.

Fuel-oil or gas storage tanks may also be used, but with caution because of the possibility of an explosion if any gases remain in used tanks. These tanks must have a manhole opening cut and holes drilled and tapped for pipe connections. Several coats of epoxy resin or asphaltum should be painted on the inside and one coat on the outside of steel tanks to prevent corrosion and/or rust by the nutrient solution. Also, tanks may be made of concrete and lined with asphalt paint. Large fiberglass or plastic tanks, with no corrosion problem, are available commercially. It is a good idea to install an agitator inside the tank to facilitate the mixing of the fertilizers and their dissolution in water. The tank should be supplied with a 50-mesh screen filter at the inlet end to the feeding system. Such tanks should be located in headerhouses near boilers and water supply.

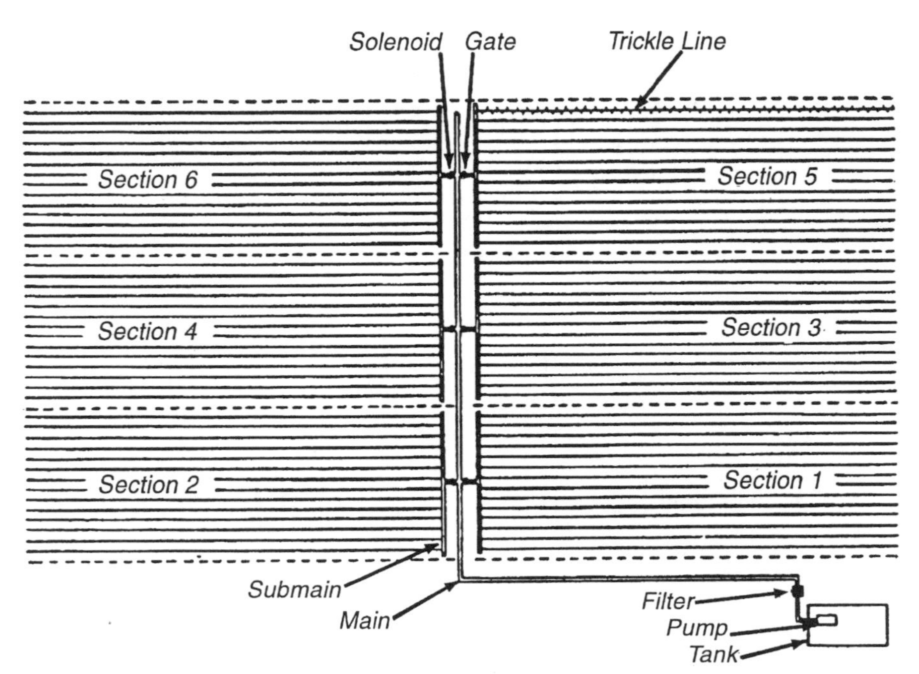

Figure 9.16. A typical drip (trickle) irrigation system for greenhouse sections which can be automatically irrigated independently.

The distribution system consists of a main supply line, submain headers, laterals, leader tubes (spaghetti lines) or emitters, fittings, and controls (Fig. 9.16). Gate valves and solenoid valves should be installed in the submain lines so that you have the option of controlling the feed supply to each greenhouse section. In this way, you can vary the flow to each section if there are differences in the volume requirements of the plants.

The submain header is connected to the main header and located underground close to the main header. It is best to position the main supply line so that the walkway is unobstructed (Fig. 9.17).

Select the main supply pipe size carefully, taking into consideration future expansion. For efficient operation of medium-volume low-pressure pumps, a total friction loss of 25 psi (172.5 kPa) should not be exceeded. For example, the friction loss in 100 feet (31 meters) of 2-inch (5-centimeter) pipe with a flow rate of 60 Imp. gallons (273 liters)/minute is 3.8 psi (26.2 kPa), whereas in the same length of 1½-inch (3.8-centimeter) pipe at the same flow rate, it is 13 psi (89.7 kPa), which is impractical, since in 200 feet (61 meters) the total friction loss of 25 psi would be exceeded. A nomograph such as the one in Figure 9.18, available from most plastic pipe suppliers, can be used to determine pipe size. By calculating the total friction loss in the supply lines and the submain headers, including fittings, and adding the pressure required at the row laterals (usually 5 to 10

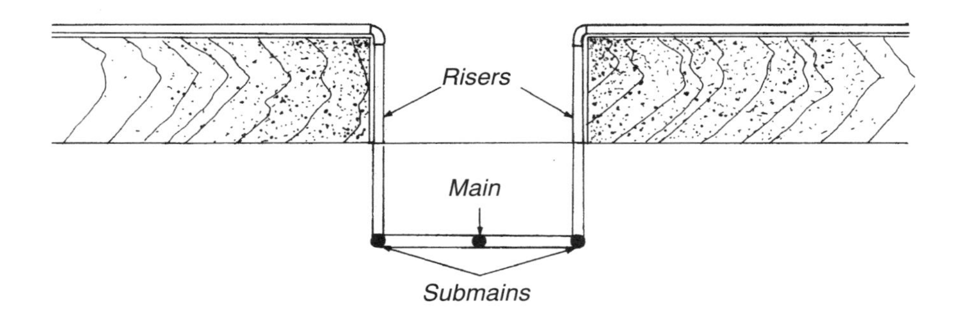

Figure 9.17. Main header with submain headers placed underground, with risers coming up from submains to laterals along beds.

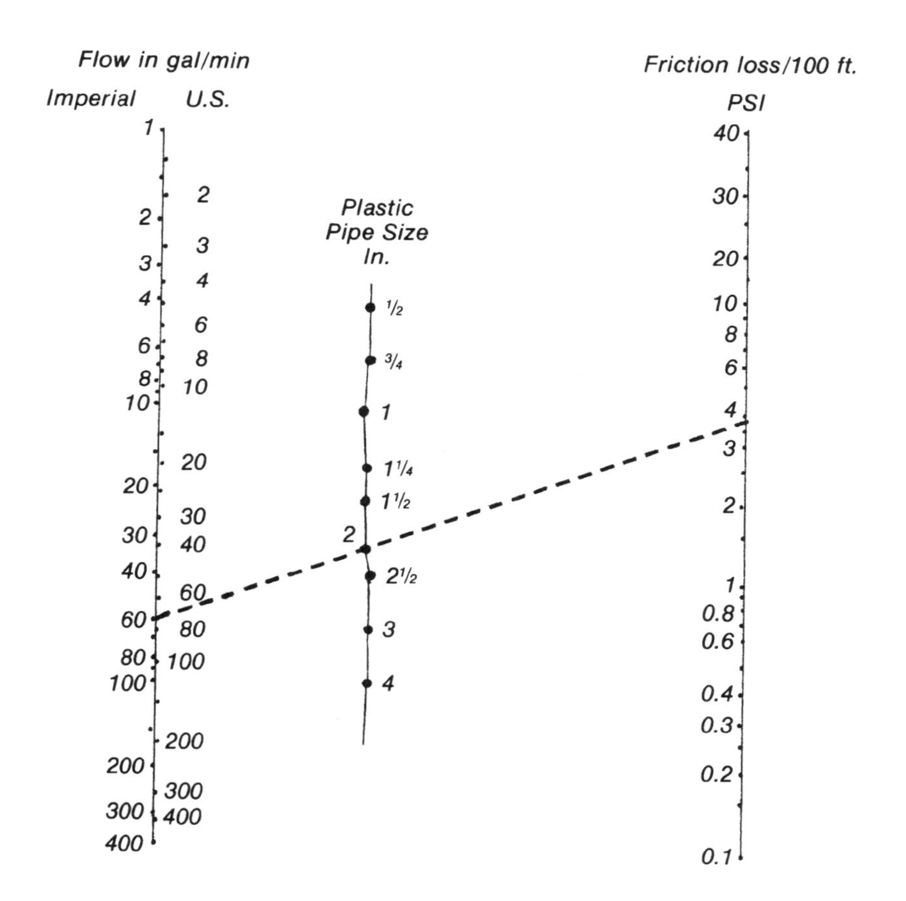

Figure 9.18. Nomograph for determining pipe size. For a "flow" of 60 Imperial gallons/minute, and a total friction loss of 3.8 psi/100 feet, a suitable pipe diameter is 2 inches. (Adapted from Mason and Adamson, Trickle Watering and Liquid Feeding System.*)*

psi) and using the total flow requirements in gallons per minute, the suitable pipe diameter may be determined from the nomograph.

High spots in the lateral lines should be avoided because trapped air will reduce the flow rate in the system. Install threaded plugs at the ends of the submains for easy flushing. Lateral lines of ½-inch (1.3-centimeter) black polyethylene pipe can supply 150 tubes. A ¾-inch (1.9-centimeter) pipe is the most suitable size for up to 300 leader tubes. This pipe delivers 20 fluid ounces (0.59 liters) per leader tube at a pressure of 2 psi in 10 minutes or at 1 psi in 13 minutes. Often higher pressures of 10 to 15 psi are used. To correctly measure the output at any given emitter, simply irrigate for a specific time, for example 10 minutes, while collecting the solution in a flask. Then measure the volume of solution collected in the flask using a graduated cylinder or other measuring container. To provide for flushing out the laterals, install a figure-8 fitting at the end of each lateral.

Spaghetti tubes of 0.045-inch (1.1 millimeter) inside diameter are usually cut in lengths of 12 to 15 inches (31 to 38 centimeters). Sloping cuts should be made on the ends to prevent sealing off the tubes if they are inserted too far into the row header. For uniform distribution, cut them all the same length. Install brass inserts in the row lateral opposite each plant as shown in Figures 7.27 and 7.28. An important advantage of this system is the almost instant flow of nutrient solution as soon as the pump is started, since the headers remain full of solution.

Many growers prefer to use emitters of ½- to 1-gallon (2- to 4-liter) per hour volume instead of spaghetti lines alone. The emitters produce a uniform volume of solution and they are also available as "pressure compensating" should there be differences in the level of the greenhouse floor. Pressure-compensating emitters are more expensive than the regular ones, but, their output is constant regardless of elevation differences.

At the end of the season, spot check the terminal end of the spaghetti tubes. If there has been a buildup of salts or root growth into them, cut off enough of the tip to clear the inside of the tube (usually ¼ inch). Thoroughly flush the whole system with clean water. If salt deposits are difficult to remove, flush the system with an acid solution to dissolve the deposits.

9.6 Feeding Methods

There are two methods of feeding the plants: One in which a complete nutrient solution is supplied each day, and one in which most of the nutrients are mixed with the growing medium before planting.

In the complete nutrient solution method all nutrients are supplied either directly from a storage tank or from concentrated stock solutions passed through an injector. The Saanichton Research Station has prepared formulae to make 600 Imp. gallons (2,728 liters) of diluted nutrient solution (Table 9.1). Their formulae are especially suited to Canadian West Coast and U.S. Pacific Northwest conditions.

In the fertilizer premix method the phosphorus, calcium, magnesium, and minor elements are mixed with the medium before each crop is planted. Only nitrogen and potassium are applied throughout the season in a nutrient solution (Table 9.2). They state that this method avoids the hazard of precipitation of incompatible fertilizers and reduces the labor required in preparing the nutrient solution. For the spring crop (tomatoes) they recommend incorporating 2.4 ounces (68 grams) of 19 percent superphosphate (0-19-0), 4 ounces (113 grams) of dolomite limestone and 1 fluid ounce (28.4 ml) of minor element solution into each cubic foot (0.0283 cubic meter) of sawdust.

TABLE 9.1 Fertilizers Used in Preparing 600 Imp. Gallons (2,728 l) of Complete Nutrient Solution at Three Nitrogen Levels, with Phosphorus at 37 ppm of P (84 ppm P_2O_5), and Potassium at 208 ppm of K (252 ppm of K_2O) for Tomatoes

	Nitrogen Levels		
	126 ppm (up to first truss set)	168 ppm (first to third truss set)	210 ppm (after third truss set)
A			
Potassium chloride (0-0-60)	40 oz (1.13 kg)	23 oz (0.65 kg)	nil
Potassium nitrate (13-0-44)	nil	23 0z (0.65 kg)	55 oz (1.56 kg)
Magnesium sulfate (Epsom salts)	48 oz (1.36 kg)	48 oz (1.36 kg)	48 oz (1.36 kg)
B			
Diammonium phosphate (21-53-0)	15 oz (0.425 kg)	15 oz (0.425 kg)	15 oz (0.425 kg)
C			
Calcium nitrate (15.5-0-0)	58 oz (1.64 kg)	64 oz (1.81 kg)	64 oz (1.81 kg)
Minor element solution *	20 fl oz (568 g)	20 fl oz (568 g)	20 fl oz (568 g)

* To prepare the minor element solution, dissolve 12 ounces (340 g) of dry minor element mix in 1 Imperial gallon (4.54 l) of boiling water and store in a dark bottle. Twelve ounces of minor element mix contains 57 g boric acid, 72 g manganese sulfate, 9 g zinc sulfate, 3 g copper sulfate, 1 g molybdic acid, and 200 g ferric citrate.

Source: Maas, E. F. and R. M. Adamson. 1971. *Soilless culture of commercial greenhouse tomatoes.* Can. Dept. Agric. Publ. 1460.

TABLE 9.2 Fertilizers Used in Preparing 600 Imp. Gal. (2,728 l) of Nitrogen-Potassium Solution at Three Nitrogen Levels and Potassium at 208 ppm of K (252 ppm of K_2O) for Tomatoes

	Nitrogen Levels		
	126 ppm (up to first truss set)	168 ppm (first to third truss set)	210 ppm (after third truss set)
Potassium nitrate (13-0-44)	55 oz (1.56 kg)	55 oz (1.56 kg)	55 oz (1.56 kg)
Ammonium nitrate (34-0-0)	14 oz (397 g)	26 oz (737 g)	38 oz (1,077 g)

Source: Maas, E. F. and R. M. Adamson. 1971. *Soilless culture of commercial greenhouse tomatoes.* Can. Dept. Agric. Publ. 1460.

To obtain a more uniform mixture of the minor elements, dilute 1 fluid ounce (28.4 ml) of minor element solution with 3 fluid ounces (85.2 ml) of water and blend with 1 pint (568 ml) of dry sawdust before incorporating into the medium. The directions for preparing the minor element solution are given in the footnote to Table 9.1. If the sawdust is to be reused for a fall crop of tomatoes, apply the premix fertilizers and minor elements at one-half the spring rate. The use of a complete nutrient solution with each irrigation provides more uniform feeding than the incorporation of dry premixes, which decline in nutrients through the cropping period.

9.7 Watering and Salt Buildup

When conductivity tests show salt levels above 4 millimhos/cm, plant growth may be suppressed. Salt buildup can be prevented by applying the nutrient solution at a volume 15 percent greater than the requirements of the plants, and allowing the excess solution to drain freely from the bottom of the bed. If salt levels are found to be too high, dilute the nutrient solution by one-quarter and increase the volume by one-third or apply pure water for several days until the conductivity level is reduced sufficiently. The foregoing information has been adapted from several publications by Maas and Adamson (1971) and Mason and Adamson (1973). Further details should be obtained from these publications, which are listed in the bibliography.

9.8 Small-Scale Sawdust-Culture Systems

A single home unit can be constructed similar to sand cultural home units. It should consist of a growing tray, a nutrient reservoir, and trickle-feeding system operated by a pump on a time clock (Fig. 9.19).

A small-scale operation can be set up by the use of a trickle-feeding system and the bag system of sawdust culture (Fig. 9.20). This can be

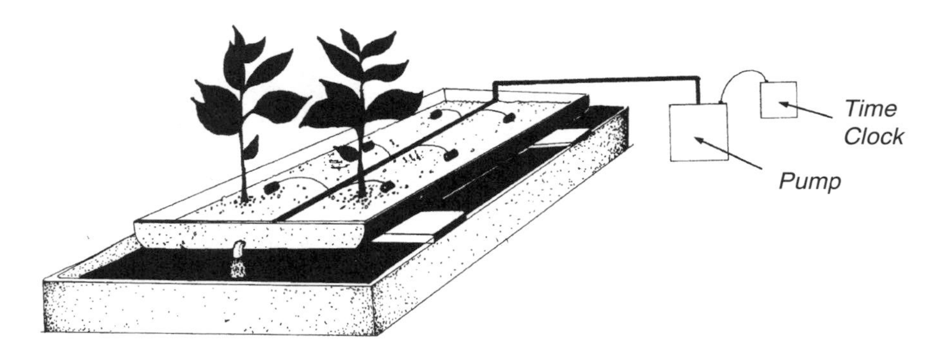

Figure 9.19. A simple small-scale trickle sawdust-culture system.

made any size by simply increasing the number of growing bags. It could even be used outdoors for growing vegetables during the summer months in the northerly latitudes. If facilities are not available for sterilizing the sawdust between crops, simply put it in the compost pile and the next season use new sawdust.

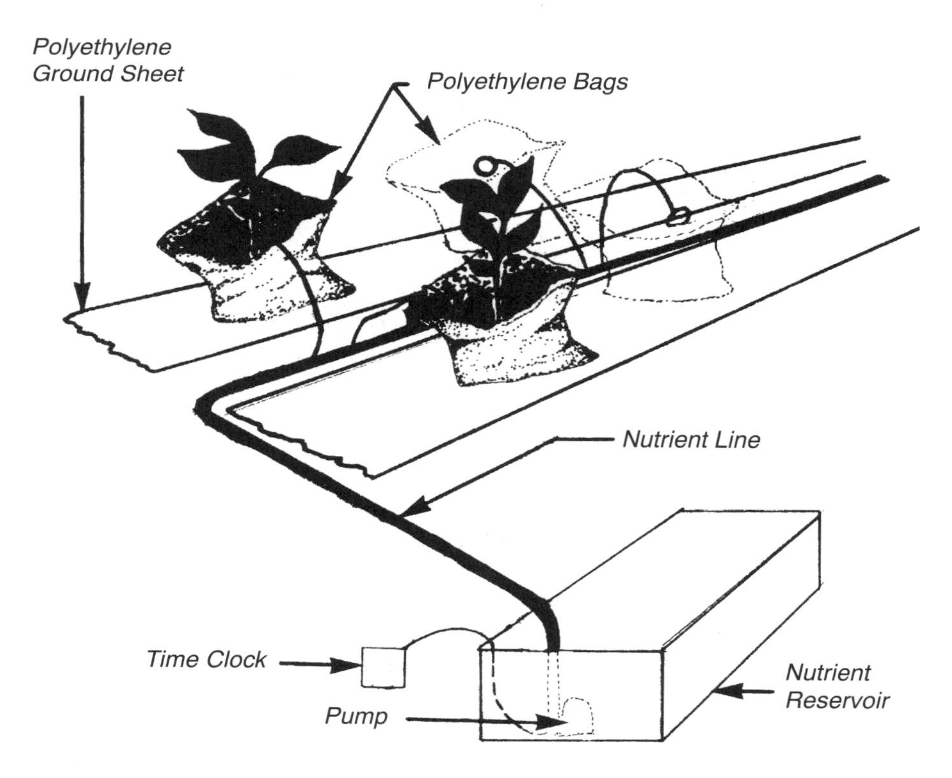

Figure 9.20. A home unit using a sawdust-bag culture system.

9.9 Advantages and Disadvantages of Sawdust Culture

Here are the advantages:

1. Since it, like sand culture, is an open system there is less chance of spread of disease such as *Fusarium* and *Verticillium* wilts, especially in tomatoes.
2. No problem of plugging of drainage pipes with roots.
3. Good lateral movement of nutrient solution throughout the root zone.
4. Good root aeration.
5. A new nutrient solution is added each irrigation cycle.
6. The system is simple, easy to maintain and repair.
7. The high water retention of the sawdust reduces any risk of rapid water stress should a pump fail.
8. It is adaptable to fertilizer injectors and therefore less space for tanks is required.
9. Sawdust has been relatively inexpensive and readily available in ares of extensive forest industries.

Here are the disadvantages:

1 It is only applicable to areas having a major forest industry; therefore, it is not feasible to use in arid desert countries.
2. It must be steam or chemically sterilized.
3. The availability of good sawdust, even in areas with an extensive forest industry, is decreasing.
4. Initially, there can be problems of sodium chloride toxicity to plants if the medium is not well leached before planting.
5. Over the cropping season, salt accumulation can occur in the medium to levels toxic to the plants. This toxicity can be reduced by leaching with pure water.
6. Plugging of trickle-feed lines may occur if proper filters are not used, or if cleaning of these filters is neglected.
7. If the sawdust used is very coarse, coning of water may occur, causing roots to grow downward rather than laterally.
8. As sawdust is organic in nature, it decomposes with time. Between crops it must be rototilled and a proportion of new sawdust added to make up for that decomposed and that lost on plant roots during the pulling of the plants at the end of each crop.

9.10 Bark Culture

In some areas where sawdust may not be readily available, but bark is, the bark can be used in a similar manner to that of sawdust as a hydroponic medium.

A 5-acre (2-hectare) greenhouse operation, Gourmet Garden Produce, at Sonora, California, grew European cucumbers in bark culture. This was a cogeneration project. The greenhouse was located on the property of a large sawmill (Fig. 9.21). Steam from the dry kilns of the sawmill were directed through large underground pipes to the greenhouse where it was converted to hot-water heating.

Figure 9.21. A 5-acre greenhouse cogeneration project. Sawmill operation is to the right. (Courtesy of Gourmet Garden Produce, Sonora, CA.)

Cucumbers were seeded in rockwool blocks placed in a plastic gutter NFT system (Fig. 9.22) in a separate seedling house. The greenhouses were divided into four sections each of 1¼ acre (0.5 hectare) to allow a sequence of planting dates. Each section was planted about 2 months apart. In this way continuous production was achieved. Seedlings were transplanted 2 weeks after seeding.

Prior to setting out the bags of bark, the greenhouse floor was covered with either white-on-black polyethylene or a weed mat to prevent contact with the underlying soil. Douglas fir bark, with a small percentage of sawdust to improve capillary movement of the solution, was placed in

5-gallon (20-liter) black plastic bags about the size of kitchen garbage bags (Fig. 9.23). The same bag of medium was re-used for 2½ years without sterilization, without the occurrence of *Pythium* fungus infection. The bark medium had naturally occuring *Trichoderma* fungi which are known to either produce antibiotics toxic to pathogens such as *Pythium, Fusarium,* and *Helminthosporium* (cause root diseases) or are otherwise antagonistic to their growth and activity (Agrios 1978).

Bags are spaced adjacent to each other in a single row with a drip-irrigation line to each bag (Fig. 9.24). Rows are approximately 4 feet (1.2 m) apart. Initially, one cucumber is planted in each bag and later one-third of them are removed. Within 1 week after transplanting, the cucumbers are well established and growing vigorously (Fig. 9.25). Note that in Figure 9.25 a fog cooling system was used in the greenhouse to lower temperatures below ambient during hot days. Alternate plants are strung apart in a V-cordon fashion; however, the one-third plants that are to be removed are strung inwards together with the second plant of each group of three (Figs. 9.26, 9.27). A very dense canopy forms as the plants reach the support wires (Fig. 9.28). The tops of the plants to be removed (every third one) are cut off and the fruit allowed to mature on the main stem before their removal (Figs. 9.29, 9.30). This thinning of the crop brings its density to 8,000 plants per 1.25 acres (0.5 hectare). This is still relatively high at 6.8 square feet (0.63 square meter) per plant.

Through drip irrigation and an injector system, nutrients are applied to the plants individually (Figs. 9.26, 9.29, 9.30, 9.31).

The cucumbers are wrapped with a shrink seal by machine and shipped 16 fruit to a case (Fig. 9.32). Smaller crooked fruit is shrink wrapped in bunches (Fig. 9.33) and sold by the pound. This crooked fruit must be removed from the plants and normally is discarded. Prices in the past were as high as $25 to $30 per box during the winter months for the number one fruit. Crooks were sold at about $1.50 per pound.

While the operation appeared to be financially successfully, the management in the sawmill operation recently changed and support for the greenhouse project was lost, resulting in a corporate decision to no longer operate it. The greenhouses were sold in 1992 and relocated to a site in Arizona.

Figure 9.22. Propagation of cucumber seedlings in rockwool blocks placed in an NFT gutter system. Inlet header and drip tubes run across the channels. (Courtesy of Gourmet Garden Produce, Sonora, CA.)

Figure 9.23. Weed mat on floor prevents weeds and pests emerging from soil below. Douglas fir bark and sawdust medium is contained in 5-gallon plastic bags. (Courtesy of Gourmet Garden Produce, Sonora, CA.)

Figure 9.24. Wide spacing of rows is required for cucumbers. (Courtesy of Gourmet Garden Produce, Sonora, CA.)

Figure 9.25. One week after transplanting the cucumber seedlings into the bags, plants grow vigorously. Note the use of a fog-cooling system. (Courtesy of Gourmet Garden Produce, Sonora, CA.)

Figure 9.26. V-cordon method of stringing cucumbers. (Courtesy of Gourmet Garden Produce, Sonora, CA.)

Figure 9.27. Alternate stringing with one-third of the plants to be removed later. (Courtesy of Gourmet Garden Produce, Sonora, CA.)

Figure 9.28. Dense canopy of plants as they reach the support wire above. Tops of one-third of the plants are cut off. (Courtesy of Gourmet Garden Produce, Sonora, CA.)

Figure 9.29. One-third of the plants have been removed and alternate stringing of V-cordon method of training is now evident. (Courtesy of Gourmet Garden Produce, Sonora, CA.)

Figure 9.30. Removal of one-third of the plants and all of the bottom dead leaves at the base of existing plants improves ventilation and light to the crop. Note the drip lines to the bags.(Courtesy of Gourmet Garden Produce.)

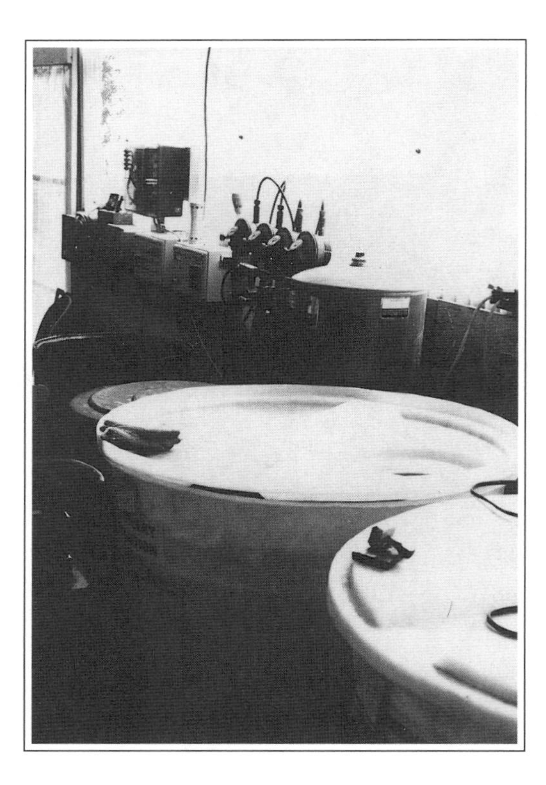

Figure 9.31. Anderson injector in background and stock solution tanks in the fore-ground. (Courtesy of Gourmet Garden Produce, Sonora, CA.)

Figure 9.32. Wrapping of cucumbers with a shrink-sealer machine. (Courtesy of Gourmet Garden Produce, Sonora, CA.)

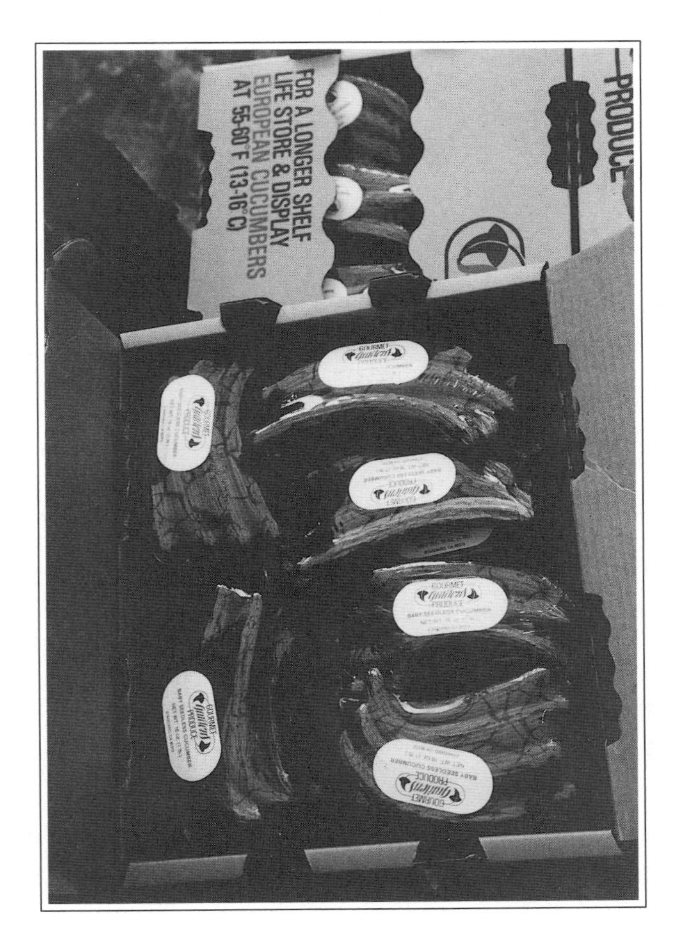

Figure 9.33. Crooked fruit wrapped in bunches with stretch film. (Courtesy of Gourmet Garden Produce, Sonora, CA.)

References

Agrios, George, N. 1978. *Plant Pathology*. 2nd Ed., Academic Press, New York, NY.

Maas, E. F. and R. M. Adamson. 1971. *Soilless culture of commercial greenhouse tomatoes.* Can. Dept. Agric. Publ. 1460.

Mason, E. B. B. and R. M. Adamson. 1973 T*rickle watering and liquid feeding system for greenhouse crops.* Can. Dept. Agric. Publ. 1510.

Chapter 10

Rockwool Culture

10.1 Introduction

Over the past 15 years rockwool culture has become one of the principal techniques for the growing of vine crops, especially tomatoes and cucumbers. Rockwool is widely used in the growing of greenhouse peppers in North America and Europe. In a 1986 study by C. J. Graves of the Glasshouse Crops Research Institute in Littlehampton, England, the total area (hectares) of tomato production in Great Britain using rockwool in 1978 was less than 1 hectare (2 acres). This increased to 77.5 (197 acres), 126 (320 acres), and 148 hectares (376 acres) from 1984 through 1986, respectively. Similarly, the area of cucumbers under rockwool increased from less than one hectare (2 acres) to 68 hectares (173 acres) over the same period from 1978 to 1986. In 1991, Desmond Day states that the estimated rockwool production in the United Kingdom for cucumbers is 230 hectares (575 acres) and for tomatoes is 160 hectares (400 acres).

Rockwool culture is the most extensively used form of hydroponics in the world, with more than 2,000 hectares (5,000 acres) of greenhouse crops being grown by this system in the Netherlands. The technology originated in Denmark in 1969 with the growing of tomatoes and cucumbers.

10.2 Rockwool Composition

Rockwool is an inert fibrous material produced from a mixture of volcanic rock, limestone, and coke, melted at 1500° to 2000° C. It is extruded as fine threads and pressed into loosely woven sheets. Surface tension is reduced by the addition of a phenol resin during cooling. While the composition of rockwool varies slightly from one manufacturer to another, it basically consists of silica dioxide (45 percent), aluminum oxide (15 percent), calcium oxide (15 percent), magnesium oxide (10 percent), iron oxide (10 percent), and other oxides (5 percent). Rockwool is slightly alkaline, but inert and biologically nondegradable. It has good water-holding capacity, with about 95 percent pore spaces. All fertilizers must be added to the irrigation water for plant growth. Rockwool has about an 80 percent water-holding capacity. The *p*H of rockwool is between 7 and 8.5. Since it has no buffering capacity, the *p*H can easily be reduced to optimal levels of 6.0 to 6.5 for tomatoes and cucumbers by the use of a slightly acid nutrient solution.

Rockwool culture is an open, nonrecycling hydroponic system, generally, with nutrients fed to the base of each plant with a drip-irrigation spaghetti line and emitter. Approximately a 10–15 percent excess of solution is supplied during each watering to allow leaching of minerals from the rockwool slabs.

10.3 Rockwool Cubes and Blocks

Plants may be seeded into small rockwool cubes, granular rockwool, or a peat-lite plug mix placed in styrofoam trays of 240 cells (Fig. 10.1) or plastic cell-pacs and flats. This system is particularly suitable for tomatoes and peppers, having relatively small and less expensive seed than cucumbers, where automatic sowing equipment may be used to reduce labor costs. With cucumbers it is better to either seed them into rockwool cubes and later transplant to rockwool blocks, or to sow them directly into the blocks (Fig. 10.2–10.3). After sowing the seed, cover the holes in the cubes or blocks with coarse vermiculite. This helps retain moisture during germination and assists the plant in removing its seed coat.

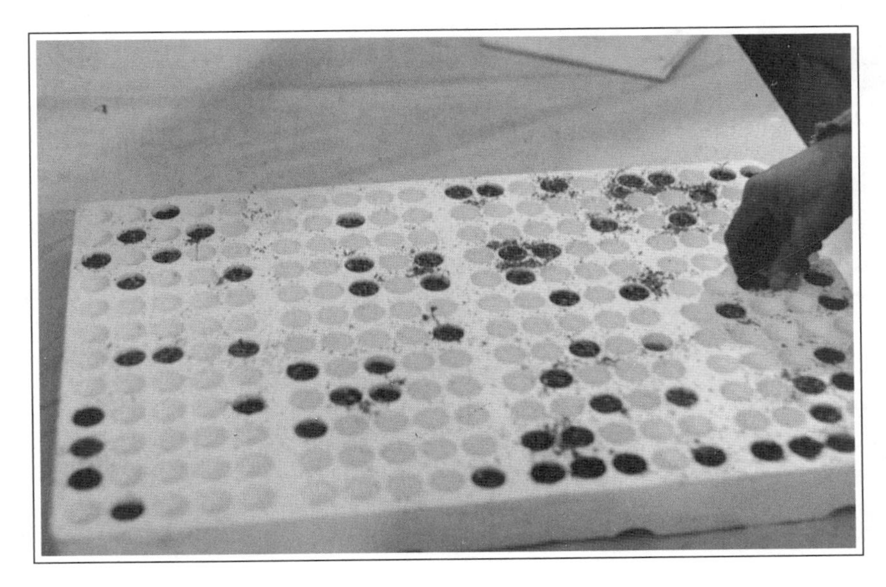

Figure 10.1. Tomato plants seeded in styrofoam trays, using a granular rockwool.

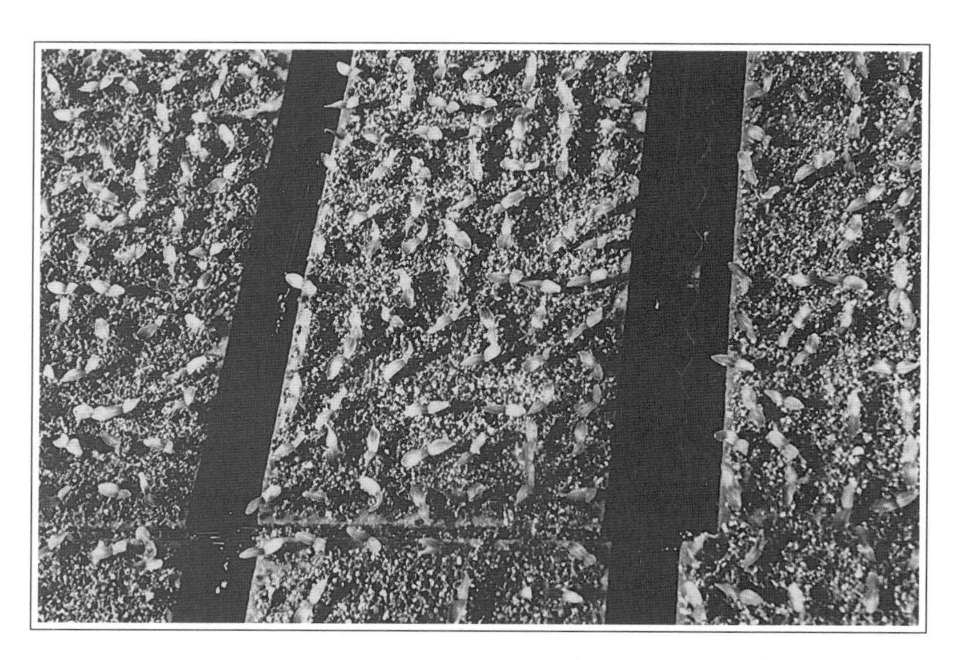

Figure 10.2. Cucumbers seeded in rockwool cubes. Note the cubes were covered with vermiculite to retain moisture while the seeds germinated. (Courtesy of Environmental Farms, Dundee, FL.)

Figure 10.3. Cucumbers seeded in rockwool blocks. (Courtesy of Environmental Farms, Dundee, FL.)

The trays are watered with raw water until germination takes place, then with a dilute nutrient solution thereafter until first true leaves begin to unfold. Then these plugs are transplanted into larger rockwool blocks having large holes (Figs. 10.4–10.6). Rockwool blocks are available in a number of sizes, 7.5 × 7.5 × 6.5 cm (3 × 3 × 2.5 in.), 7.5 × 7.5 × 10 cm (3 × 3 × 4 in.), 10 × 10 × 6.5 cm (4 × 4 × 2.5 in.), and 10 × 10 × 8 cm (4 × 4 × 3 in.) (length × width × height). The growing blocks are manufactured in strips, each block individually surrounded by a plastic wrapping. The choice of growing block is determined by the plant being grown and at which stage the grower wishes to transplant to the final rockwool slabs.

Figure 10.4. Tomato plants transplanted into rockwool blocks after first true leaves have developed.

The longer he wishes to hold the plants in a seedling area, the larger blocks he should use (Figs. 10.3–10.8). While the smaller blocks are good for tomatoes and peppers, the larger ones are more suitable for cucumbers which grow rapidly. By placing the blocks on wooden slats (Fig. 10.7) or on a wire-mesh bench (Fig. 10.8, 10.9), roots which grow out of the bottom are air pruned. This keeps the bulk of the roots inside the block and lessens transplant shock.

Figure 10.5. Cucumber seedling in rockwool cube is transplanted to rockwool block with a large hole. (Courtesy of Environmental Farms, Dundee, FL.)

Figure 10.6. Author transplanting cucumber seedlings in rockwool cubes to rockwool blocks. (Courtesy of Environmental Farms, Dundee, FL.)

Figure 10.7. Tomato seedlings growing in rockwool blocks under supplementary artificial lighting in a nursery area of the greenhouse. (Courtesy of Gipaanda Greenhouses Ltd., Surrey, B.C., Canada.)

Figure 10.8. Cucumber seedlings growing on wire-mesh trays to allow good root pruning. Seedlings are irrigated by hand with a nutrient solution. (Courtesy of Environmental Farms, Dundee, FL.)

Figure 10.9. Base of rockwool block shows good root pruning of cucumber plant. (Courtesy of Environmental Farms, Dundee, FL.)

10.4 Rockwool Slabs

Slabs are available as unsleeved or sleeved with a white polyethylene sheet. They come in a number of sizes, 90 × 30 × 5 cm (35.5 × 12 × 2 in.), 90 × 15 × 7.5 cm (35.5 × 6 × 3 in.), 90 × 20 × 7.5 cm (35.5 × 8 × 3 in.), 90 × 30 × 7.5 cm (35.5 × 12 × 3 in.) and 90 × 45 × 7.5 cm (35.5 × 18 × 3 in.) (length × width × thickness).

The slabs of 15–20 centimeters (6–8 inches) width are generally recommended for tomatoes and peppers, those of 20–30 centimeters (8–12 inches) for cucumbers and the 30 centimeters (12 in) width for melons. Wider slabs for tomatoes can lead to excessive vegetative growth if only two or three plants are grown in each. The trend now is to use the wide slabs and grow 4 to 5 plants in them. Slabs are also available in several densities. Higher-density slabs will maintain their structure when used for several crops over a period of two to three years and especially when steam sterilized between crops.

Wrapped slabs are ready to use for annual crops such as tomatoes, cucumbers and melons. They are designed for growers wishing to use a fresh growing medium each year to assure an ideal air/water ratio and sterile medium. However, due to their cost, it is more cost effective to use the same slabs for 2 to 3 years (crops) by steam sterilizing between crops or changing the crop. For example, if a greenhouse operation grows tomatoes cucumbers and peppers, the tomatoes can be grown in new slabs, which are then used subsequently for cucumbers and peppers. There will be some structural breakdown resulting in a 10 to 15 percent loss, but this may be reduced by using the higher-density slabs.

To sterilize the slabs, they must be stacked on pallets after removing the plastic wrapping. They should be stacked with each layer in opposite directions, with some air spaces between so that the steam will penetrate the entire pile. They are covered with canvas and steam is piped into the pile under the covering. The center slabs must reach a temperature of 60° to 80° Celsius (140°–180° F) throughout for 30 minutes to be assured of complete sterilization. The slabs are re-wrapped after sterilization.

10.5 Rockwool Layout

Rockwool culture was first designed as an open, non-recycling hydroponic system. More recently, the use of recycling systems with rockwool culture is emphasized as the awareness of the environment and conservation of water is becoming more important. Such recycling systems will be discussed later. Nutrients are fed to the base of each plant with a drip-irrigation line and emitter. Approximately 15–20 percent excess of solution is supplied during each watering to allow leaching of minerals from the rockwool slab.

A typical layout for an open-system rockwool culture is shown in Figure 10.10. Prior to placing out the growing slabs, the greenhouse floor area should be disinfected with a 2-percent formaldehyde solution. The soil surface (floor) should be leveled. Beds consist of two rockwool slabs set 60–75 cm (2–2.5 ft) apart for cucumbers. With tomatoes and peppers the slabs may be closer together, generally 40–45 cm (16–18 in.) depending upon the specific plant row spacing required. Soil or sand makes the best floor as it can be easily formed to obtain good drainage. A slight slope

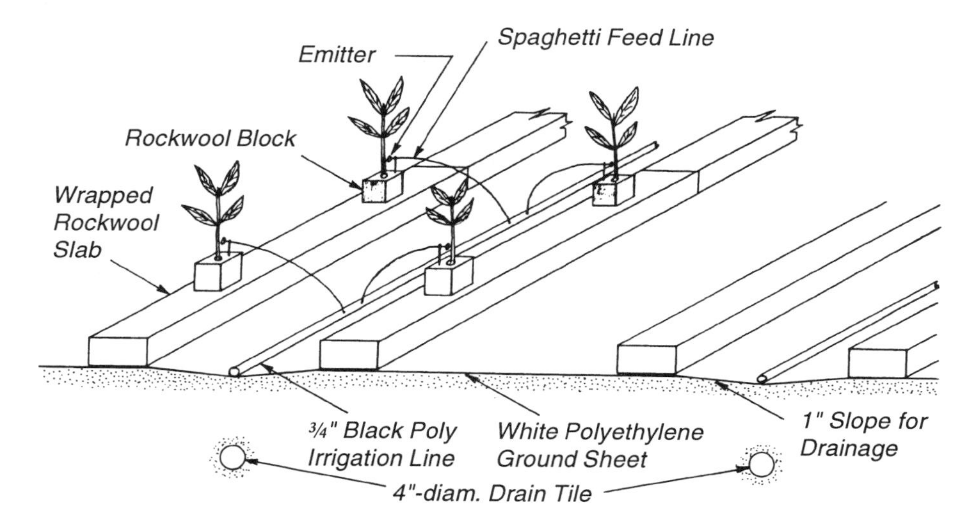

Figure 10.10. Sketch of an open system of rockwool culture.

toward the center between the two slabs will provide adequate drainage away from the slabs (Fig. 10.11). The entire greenhouse floor area is covered with a white-on-black polyethylene of 6-mil thickness to provide light reflection and good hygiene (Fig. 10.12). If the drainage conditions in the greenhouse are poor, it is necessary to lay a drainage pipe in the middle of the bed. The drainage pipe should be covered with pea gravel and/or coarse sand and the white ground sheet placed on top of it. If this method is used, the ground sheet should have small holes punched in it to allow percolation of the excess nutrient solution to the drain pipe. Alternatively, the unpunched polyethylene sheet will act as a drainage channel to conduct excess solution to a drain pipe at the far end of the greenhouse. But problems with algae may develop if stagnant water remains in the channel.

If the underlying greenhouse floor is backfilled with a sand or gravel base of high percolation, the rockwool slabs could be placed directly on the floor without a polyethylene ground sheet. Excess nutrients would

Figure 10.11. Sloping of floor to provide drainage for each two rows of slabs.

Figure 10.12. Laying of white polyethylene ground sheet.

drain away rapidly; however, weed growth would be promoted in these areas immediately adjacent to the growing slabs.

In more northerly latitudes where the ground temperature is cold, it may be advantageous to place the slabs on top of 1-inch (2.5-cm) thick styrofoam sheets.

10.6 Irrigation System

A drip-irrigation system with a fertilizer injector provides nutrients to each plant individually. A drip emitter of 0.5 gallon (2 liters) per hour feeds each plant via a drip or spaghetti line (Fig. 10.13). A PVC main line must be of sufficient diameter (minimum of 3 inches) to supply the area of the greenhouse section to be irrigated at a given time. From the main, sub-mains or headers, distribute solution to subsections. Risers (usually ¾-inch diameter) join the ½-inch black-poly lateral of each row to the header. If the emitters are punched directly into the lateral line at the plant spacing rather than putting them at the end of the drip line, less plugging will occur as they do not dry out between watering. Spitter-type emitters which give a small spray of nutrient solution near the base of the plant can be used also, but they would have to be attached to the ends of the drip lines.

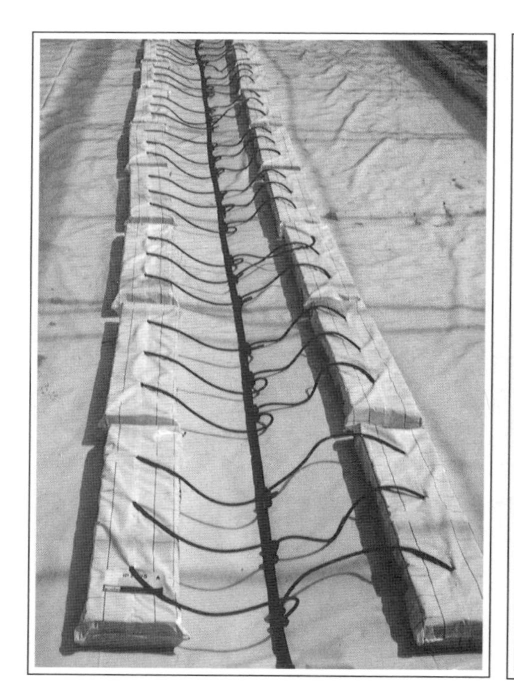

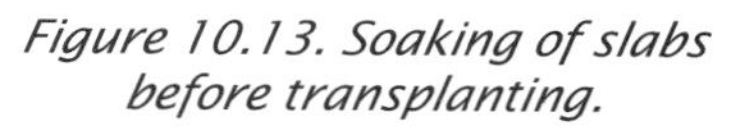

Figure 10.13. Soaking of slabs before transplanting.

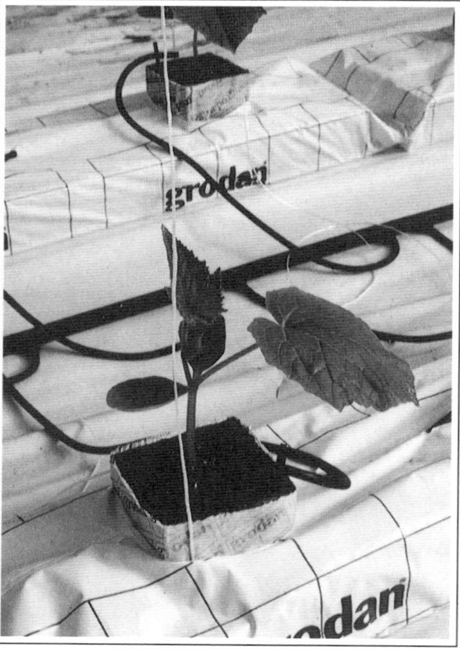

Figure 10.14. Drip line with emitter placed with stake on block for 3–4 days following transplanting, then moved to slab next to block for continued feeding.

A spitter has the advantage of giving wider application of solution and thus eliminating dry spots, but excessive amounts could easily be applied and moisture at the stem base will promote diseases. Drip lines are supported on the rockwool blocks or slabs with a small stake (Fig. 10.14).

The rockwool slabs must be soaked with nutrient solution prior to transplanting (Fig. 10.13). The slabs must be soaked with nutrient solution for 24 to 48 hours. This will adjust the *p*H and uniformly moisten the slabs. To do this, place the slabs in their final position with three drip lines entering a small slit on top of each slab at equal spacing. Operate the nutrient system until the slabs are full of solution. Do not cut drainage holes in the slabs until they have been soaked. Check that all the slabs are soaked. Usually, a few slabs are found that have holes in their wrappers and they do not soak properly. If this happens, the growth of the plants on such a dry slab will be greatly restricted. Following the first irrigation after transplanting, the slabs should again be checked for dry spots between the drip lines, as the slabs must be fully saturated to ensure sufficient solution reserves for the plant during the initial post-transplant period. Drainage holes must be cut on the slab sides at the bottom edge (Fig. 10.15). They should be in the shape of an inverted "T" or an angled straight cut approximately 4–5 centimeters (2 inches) in height. Two or three holes should be made in each slab on the inside face. Make the cuts between plant locations, not directly under them, or the solution will run immediately out of the holes and not be forced to flow laterally.

Figure 10.15. Cutting drainage slits on the inside bottom face of the slabs. (Courtesy of Environmental Farms, Dundee, FL.)

A special tool is available to cut the holes in the top of the slabs (Fig. 10.16). In transplanting, simply set the rockwool blocks growing the seed-

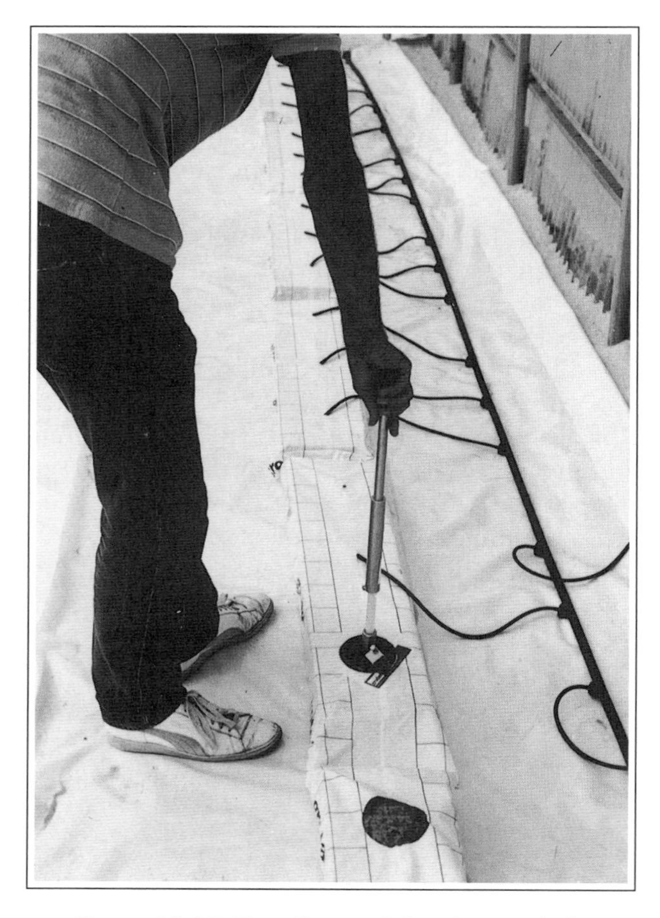

Figure 10.16. Use of a special tool to cut holes for plants in tops of the slabs. (Courtesy of Environmental Farms, Dundee, FL.)

lings on top of the slabs through the holes cut in the plastic sleeve (Figs. 10.14, 10.17, 10.18). Place the drip line at the edge of the block with the stake. With cucumbers it is very important not to allow the solution to fall onto the crown of the plant as this will promote diseases. Little transplant shock is encountered by the plants. The roots will grow from the blocks into the slabs within several days. Once the plants have rooted into the slabs (usually within a week), relocate the drip line to the slab next to the base of the rockwool block. By tilting the stake toward the block, the solution will drip off the end rather than run back (Fig. 10.17). Be careful that the end of the drip line is above the hole in the slab.

Prior to transplanting, all of the support strings should be strung to the overhead wires. When transplanting, place the bottom end of the support string underneath the rockwool block of the seedling (Fig. 10.18). This will secure the end of the string.

Figure 10.17. Drip line located on the edge of the rockwool block for the first week after transplanting.

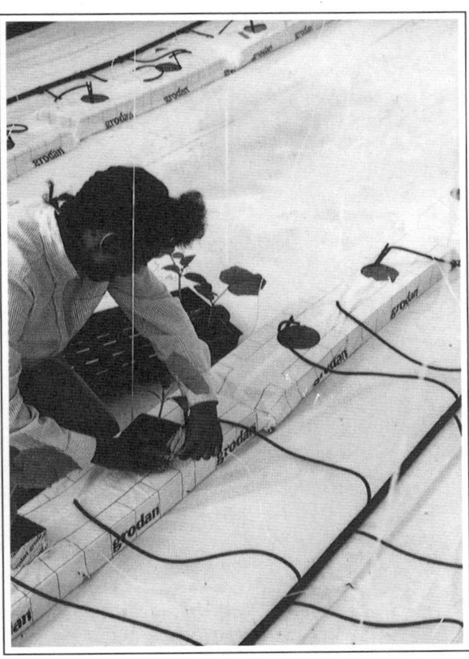

Figure 10.18. After cutting holes in slab with a special tool, cucumber plants are set onto slabs.

(Courtesy of Environmental Farms, Dundee, FL.)

More frequent irrigation is necessary immediately after transplanting until the plants become established. Later a frequency of 5 to 10 times per day should be adequate, with rates varying according to plant stage of growth and environmental conditions. Daily waterings up to 20 times may be needed for summer crops.

Large greenhouse operations generally use a computerized system to control environmental conditions and nutrition. In this case a slab, representative of all slabs growing the crop, is selected and placed in a "start tray" (Fig. 10.19). The start tray monitors the amount of solution present in a slab growing healthy plants. The bottom of the wrapper is removed from the slab placed in the start tray so that excess solution drains easily. A V-shaped groove in the bottom of the stainless-steel start tray conducts the solution to one end where an electrode is positioned. As long as there

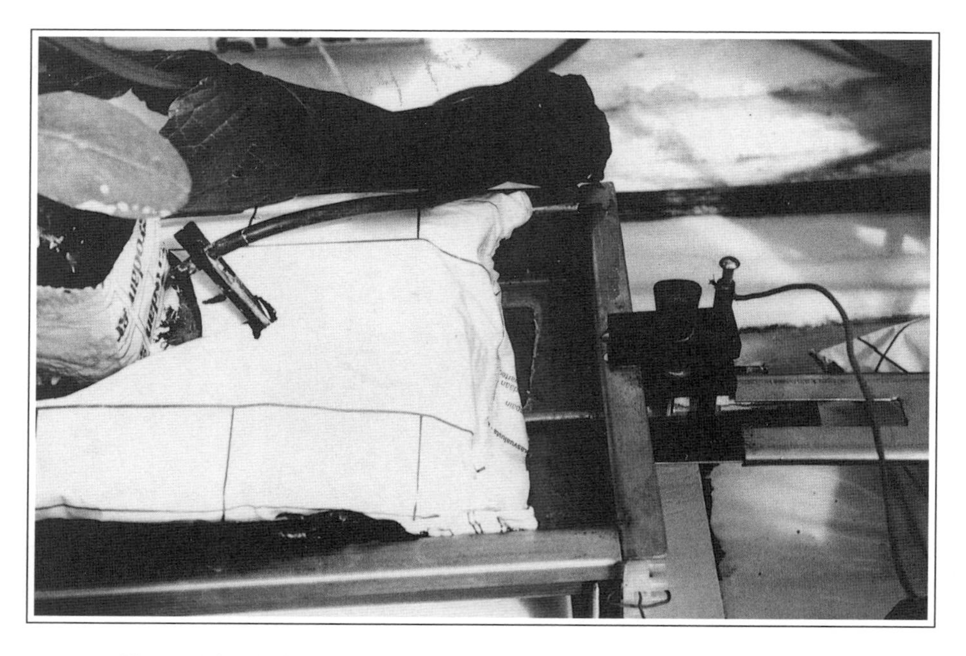

Figure 10.19. Start tray monitors the amount of solution present in the slab. Note the probe on the right and the position of the drip line in the slab next to the base of the rockwool block. (Courtesy of Environmental Farms, Dundee, FL.)

is adequate solution present, the solution is in contact with the electrode and the circuit complete. The signal then prevents an irrigation cycle from being initiated. As soon as the circuit is broken, as the solution falls with the drying of the slab due to uptake by the plants, an irrigation cycle begins and will continue for the preset time interval set in the irrigation controller or computer. Smith (1987) states that watering should be done before the slabs have lost more than 5–10 percent of their water-holding capacity. The moisture level in the slabs may be raised or lowered by raising or lowering the probe in the start tray. That is, if a higher moisture level is to be maintained between irrigation cycles, the probe may be raised and as a result, more frequent irrigation cycles will occur.

The duration of an irrigation cycle should be long enough to get a 15–20 percent runoff from the slabs. The drainage from the slabs may be determined by setting up a special collection tray under one of the slabs having healthy plants (Fig. 10.20). A pipe outlet at the lower end of the collection tray directs the waste solution to a container below, buried in the floor. A second container collects the solution from one emitter. This gives the amount of water which actually came out of each emitter, assuming they are all equal. If pressure-compensating emitters are used, the flow through each should be the same. To get the volume of solution that entered the slab, multiply the volume collected from the one emitter by the number of emitters in each slab. Divide the waste volume from the collection tray by this inlet volume from the emitters. It is easiest to use milliliters as a volume measurement. A graduated cylinder will suffice to measure the liquids.

Figure 10.20. Collection tray to monitor the amount of solution runoff from the slab. (Courtesy of Environmental Farms, Dundee, FL.)

Monitoring of salt levels in the rockwool slabs should be carried out at least every other day. Greenhouses having computers may have a number of electrical-conductivity (EC) and *p*H sensors in slabs at several locations in the house giving a continuous monitoring. Otherwise, a sample must be taken from the slab with a small syringe (Fig. 10.21) and tested for its *p*H and EC (Fig. 10.22). Values should be close to those of the nutrient solution provided to the plants. If significantly high levels of conductivity or nonoptimal *p*H levels are detected, the slabs must be irrigated more often until their solution concentration approaches that of the input solution. "Clear" raw water should not be used as it often contains sodium, calcium, and magnesium. These ions will accumulate in the slabs, and other nutrients such as potassium, nitrates, and phosphates will be depleted, causing an imbalance in the slab nutrition.

Figure 10.21. Use of syringe to sample EC and pH of solution in slab. (Courtesy of Environmental Farms, Dundee, FL.)

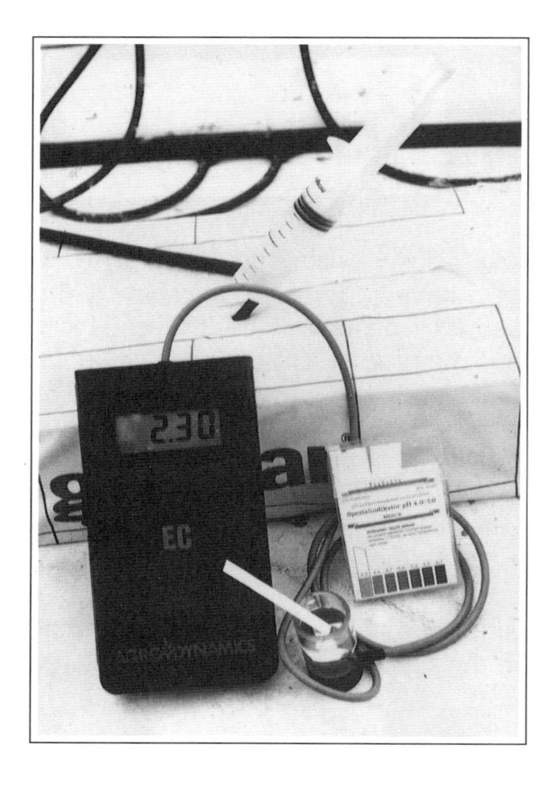

Figure 10.22. Testing the solution EC and pH with an EC meter and pH paper. (Courtesy of Environmental Farms, Dundee, FL.)

10.7 Cucumbers in Rockwool

Cucumbers were grown at Environmental Farms, Dundee, Florida. An existing greenhouse of 1½ acres (0.6 hectare) that had been previously a citrus nursery was converted to rockwool culture of cucumbers. The floor was filled with 6 inches (15 cm) of sand after drainage tiles were installed. The floor was contoured for the rockwool slabs and a white polyethylene liner placed on it (Figs. 10.11, 10.12). A drip-irrigation system was installed and the cucumbers seeded in rockwool cubes and transplanted to blocks (Figs. 10.13, 10.2, 10.3, 10.5, 10.6). Fourteen days from seeding they were transplanted to the slabs (Figs. 10.18, 10.23). They became

Figure 10.23. Fourteen days after seeding, cucumbers have been transplanted to the slabs. (Courtesy of Environmental Farms, Dundee, FL.)

established in the slabs within 2 to 3 days and grew close to 6 inches (15 cm) per day (Fig. 10.24). Eighteen days from transplanting (32 days from seed), they reached 5 feet (1.5 m) in height (Fig. 10.25). Many of the small stem fruit had to be removed (Fig. 10.26) leaving only 5 fruit on the stem once it reached the support wires at 6½ feet (2 m). The tendrils must also be removed as they tangle around the leaves and fruit, causing distortion in shape. At 40 days from seeding harvesting began (Fig. 10.27). The fruit was harvested in the late morning after the plants had dried out. Plastic tote bins were used to harvest the fruit and they were transported to the packing area by a tractor and trailer (Fig. 10.28). The cucumbers were

Figure 10.24. Cucumbers 28 days after seeding (15 days after transplanting) are close to 3 feet tall, growing about 6 inches per day. (Courtesy of Environmental Farms, Dundee, FL.)

Figure 10.25. Cucumbers 31 days after seeding (18 days after transplanting) are close to 5 feet tall.

Figure 10.26. Small fruit on stem, many of which are removed. Note the stringing of the plant.

(Courtesy of Environmental Farms, Dundee, FL.)

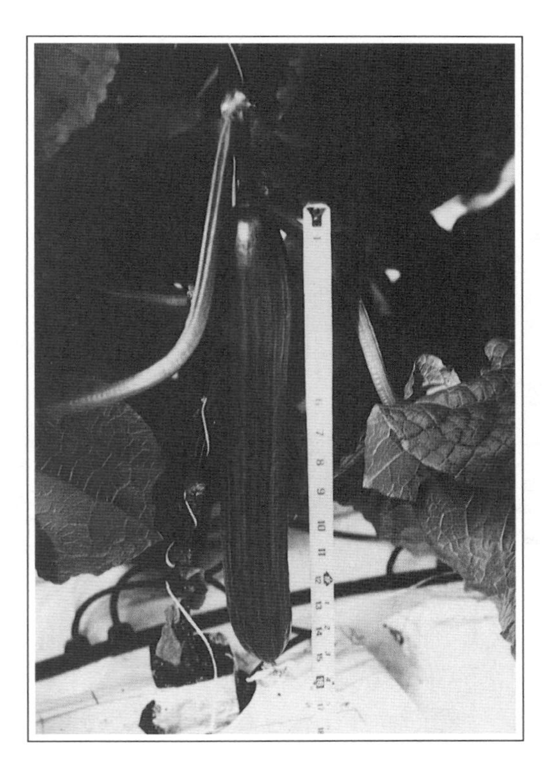

Figure 10.27. Beginning of harvest, 40 days after seeding. The fruit is straight and long — about 15 inches or 38 cm in length. (Courtesy of Environmental Farms, Dundee, FL.)

Figure 10.28. Fruit is harvested in plastic tote bins and transported to the packing area with a tractor and wagon. (Courtesy of Environmental Farms, Dundee, FL.)

shrink-wrapped with an L-bar sealer and oven (Fig. 10.29). They were packed 12 fruit to a case, palletized and placed in cold storage at 50 to 55° F (10 to 13° C) (Fig. 10.30). Three grades, regular, large and jumbo, were determined by the length of the fruit. Most of the product was shipped to the northeastern U.S. and Canada.

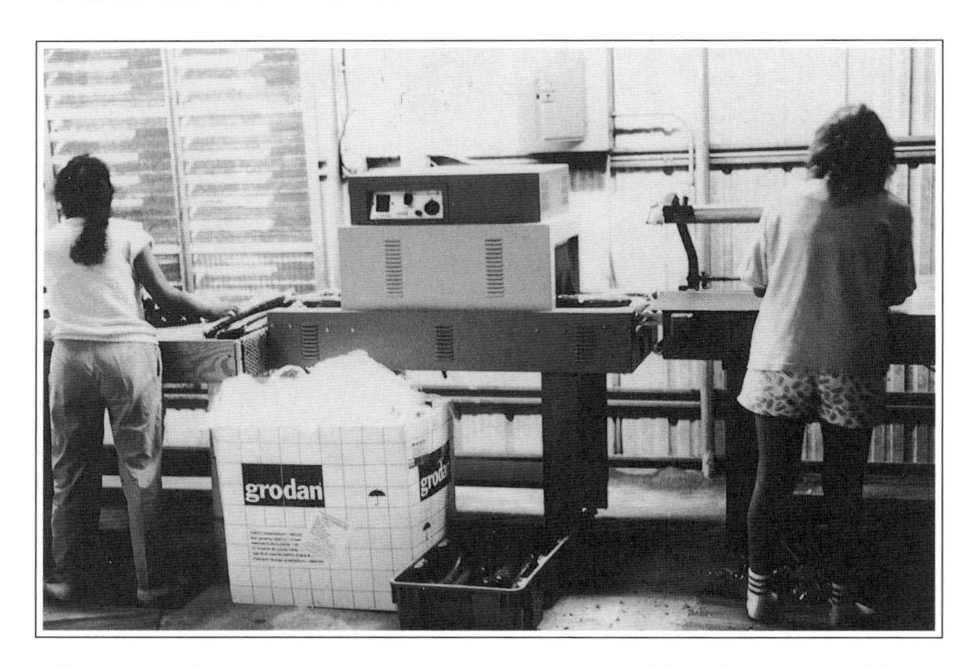

Figure 10.29. Cucumbers may be shrink-wrapped with an L-bar sealer (right) and an oven (center). This equipment is suitable for small operations of approximately one acre in size. Larger cucumber crops require equipment capable of greater volumes such as that shown earlier in Figure 9.32. (Courtesy of Environmental Farms, Dundee, FL.)

As with any agricultural crop, pest and disease problems arose. The most troublesome pests were aphids, whiteflies and thrips. With the humid conditions of Florida, gummy stem blight and mildew diseases were present. Powdery mildew was overcome by use of a resistant variety, Marillo. Gummy stem blight was prevented by painting a fungicide on the base of the plant stem (Fig. 10.31). Viruses became the most difficult problem. Aphids and whiteflies were the vectors. Control depended largely on control of the insect vectors. The spread of viruses during pruning, training and harvesting by the workers is prevented by dipping tools in a milk bath carried by the workers. This dip is necessary between touching each plant.

A fogger was used to apply pesticides (Fig. 10.32). It dispersed a suspension of fog through the crop canopy as it was pushed on a cart down every row (Fig. 10.33).

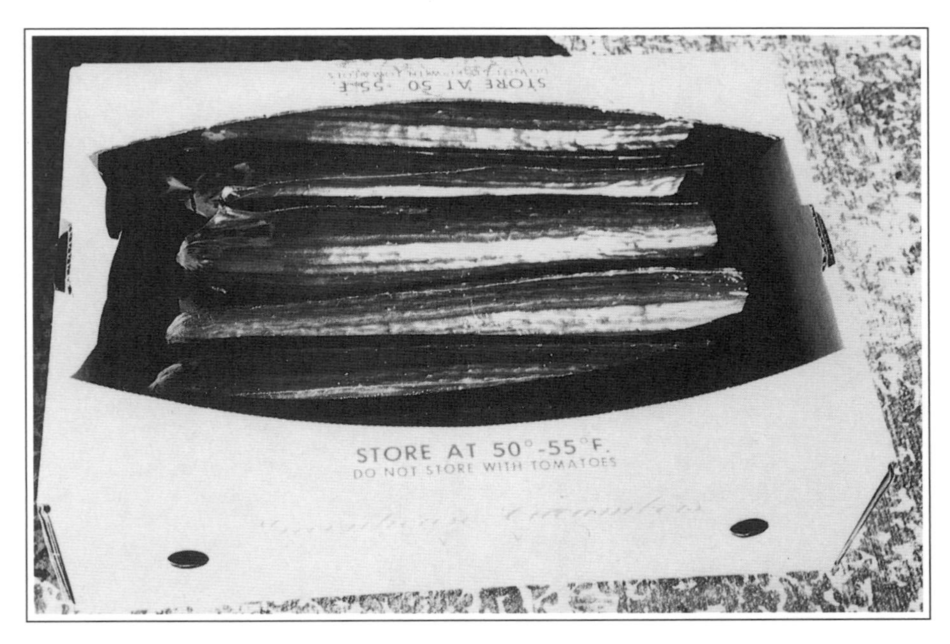

Figure 10.30. Cucumbers packed 12 fruit to a case. (Courtesy of Environmental Farms, Dundee, FL.)

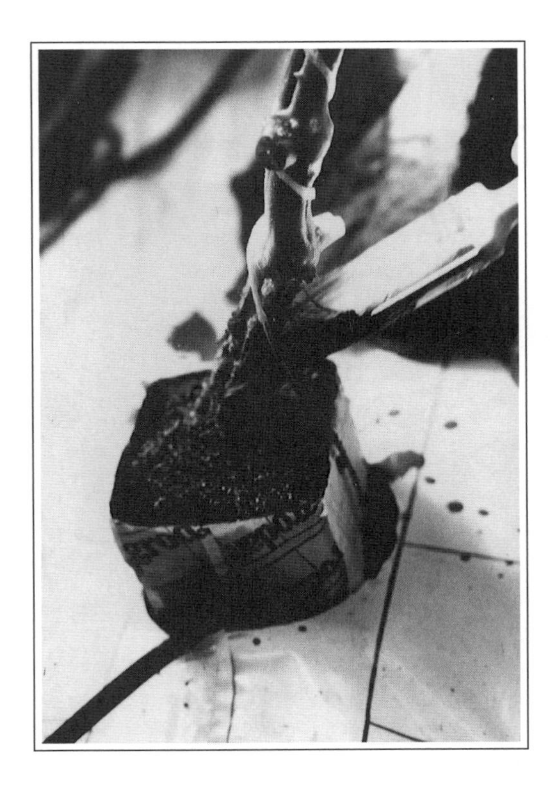

Figure 10.31. Application of a fungicide to the base of the stem prevented gummy stem blight. (Courtesy of Environmental Farms, Dundee, FL.)

Figure 10.32. A fogger can apply pesticides effectively to high-density crops. (Courtesy of Environmental Farms, Dundee, FL.)

Figure 10.33. It is important to use protective clothing when applying pesticides. Note that the fogger is pushed down the rows on a cart to make the task easier for the applicator. (Courtesy of Environmental Farms, Dundee, FL.)

A modified renewal umbrella training with only one leader growing over the wire was adopted as the cucumbers in Florida are extremely vigorous with the amount of sunlight present. With the existence of high insect populations and disease inoculum due to the semi-tropical conditions of Florida, most cucumber growers change their crops every 3-4 months as diseases and insects eventually infect the crop severely enough to substantially reduce yields.

Houweling Nurseries Ltd., of Delta, B.C. grows two crops of European cucumbers annually. The first crop is seeded December 1st and transplanted in late December with first harvest by early February. Production continues until June. A second crop is seeded in June in a separate seedling range. Three-week-old plants are transplanted and production commences within three weeks and continues until mid-November.

"Mustang," "Bronco," "1289," "Ventura" and "Exacta" varieties of cucumbers are grown. Cucumbers are planted at a density of 1.2–1.4 plants per square meter (10.76 sq ft) which is 5,000 plants per acre (12,500 plants per hectare). While the industry average annual production is 110 cukes per square meter, this operation has produced over 140 cukes/sq meter. This is equivalent to 10 to 13 cukes per square foot or 73 to 93 fruit per plant.

10.8 Tomatoes in Rockwool

Gipaanda Greenhouses Ltd., of Surrey, B.C., in 1993 grew 5.25 acres (2.1 hectares) of tomatoes in rockwool. They started the seedlings in flats of 1-inch rockwool cubes and after 12 to 14 days transplanted them to 3-inch (7.5-cm) rockwool blocks. It is now becoming common practice in the B.C. greenhouse industry to seed tomatoes in 2-inch (5-cm) thick styrofoam trays having 1-inch (2.5-cm) holes with rockwool cubes already in place. The tray lends itself to automatic seeding and dislodging by machine. The seedlings are transplanted to 4 inches × 4 inches × 3 inches (10 cm × 10 cm × 7.5 cm) rockwool blocks. The larger blocks are used so that the plants will not dry out as quickly when placed on top of the rockwool slabs before positioning them inside the slab wrapper.

To overcome *Fusarium* crown and root-rot disease, a resistant rootstock variety is used in the absence of a commercially resistant variety. Both varieties are germinated at 79° F (26° C). After germination, the two seedlings are grown in separate sections of the greenhouse under different temperature regimes. The rootstock is grown at 64° F (18° C) and the scion variety (Belmondo) at 72° F (22° C). To obtain a thicker-stemmed rootstock, the rootstock variety is sown 2 days earlier than the scion variety (Belmondo). The rootstock variety is grown for 14 days and the scions for 12 days before transplanting to the rockwool blocks.

One cube of each variety is transplanted into the same hole of the rockwool cube. The temperature has slowly been changed to 72° F (22° C). Ten days after transplanting, the plants are grafted. An incision is made in both plants and fitted together, followed by a clear poly-tape wrapper. This is an approach graft. After 7 days the top of the rootstock and the bottom of the scion are cut off. A plastic strip may be staked to the top of the rockwool block to prevent rooting of the stem of the scion into the block later when plants mature and their stems lowered (Fig. 10.34). As only 150 to 200 plants can be grafted per hour, with a greenhouse of this size requiring 40,000 plants, it can be costly. *Fusarium* disease was reduced in the past by interplanting lettuce. The lettuce roots produced an inhibitor to the disease, but the lettuce introduced aphids. Presently, new varieties are being developed that will be resistant to *Fusarium* and will eliminate the need to graft.

Figure 10.34. A strip of plastic is staked to the top of the rockwool block to prevent the scion of the tomato rooting into the block when the stem comes in contact with it. (Courtesy of Otsuki Greenhouses Ltd., Surrey, B.C., Canada.)

After the graft has taken and the stock and scion pruned (last week of November), the plants are set on top of the rockwool slabs in the greenhouse, but not placed inside. This spaces the plants so that they may continue growing without light restriction, allowing slower, thicker top growth and more root development. Each plant is fed with a drip line.

After one month, the transplants are set into the slabs. First harvest occurs in mid-February. Last harvest is mid-November, allowing a 10-day crop turnaround.

To save on the number of slabs required, a unique training system was adopted. Two plants, instead of the conventional three, were set onto each slab. Slabs and plants were spaced to get centers of 75 cm (30 in.) within rows by 45–50 cm (18–20 in.) between rows. This is equivalent to 0.6 square meters (6 square feet) per plant. This is reduced to the normal final spacing of tomatoes of 0.3 square meters (3 square feet) per plant by mid-April when light has improved. Plant density is increased by letting each plant form two shoots. By the end of February, one-sixth of the plants were allowed to form a second shoot by letting one healthy sucker grow (Fig. 10.35). This procedure is continued over a period of two months, so that by the end of April every plant has two stems and therefore the plant density is doubled.

Figure 10.35. A tomato plant with two shoots. (Courtesy of Gipaanda Greenhouses Ltd., Surrey, B.C., Canada.)

Plants are supported by polyethylene twine about 20 feet (6 m) long for each plant. The string is wound around bobbins having two hooks which hang from the overhead wire support. The plant vines reach 25 to 30 feet (7.6 to 9 m) by the end of the season. Plants are lowered by untying the string from the bobbin. Plant stems are looped around the ends of each bed by use of a 3-inch diameter plastic drain-tile pipe to prevent breakage (Fig. 10.36). The stems are laid along the top of the slabs. Wire hoops about 3/16 inch in diameter placed across the slabs support the plant stems (Fig. 10.37).

Figure 10.36. Plant stems are bent around a 3-inch diameter plastic drain-tile pipe at the ends of the rows when lowering the plants.

Figure 10.37. Wire hoops placed across the rockwool slabs keep the plant stems raised to improve ventilation and thus reduce fungus diseases.

(Courtesy of Gipaanda Greenhouses Ltd., Surrey, B.C., Canada.)

Some growers place stakes across the top of the slabs to support the stems (Fig. 10.38). This provides better aeration of the stems and reduces diseases (Fig. 10.39). It also keeps the scions above the rockwool blocks so that they will not root into the blocks. A strip of plastic is placed on top of the blocks as described earlier to help prevent rooting of the scion (Fig. 10.36).

Figure 10.38. Stakes may be placed across the rockwool slabs to raise the plant stems above the floor. (Courtesy of Houweling Nurseries Ltd., Delta, B.C., Canada.)

Figure 10.39. Wire hoops raise the plant stems well above the floor, creating good air exchange at the base of the plant. (Courtesy of Gipaanda Greenhouses Ltd., Surrey, B.C., Canada.)

Gipaanda Greenhouses Ltd. has found that the management of irrigation is important in rockwool slab oxygenation (Ryall 1993). The last irrigation is done 2½ to 3 hours before sunset to reduce the solution retention in the rockwool slabs to 65% of their capacity. This drying of the slabs to create a water deficit overnight allows better oxygenation of the roots. Active root growth is encouraged by the overnight water deficit in the slabs. The first irrigation occurs about 2 hours after sunrise. Subsequent irrigations are initiated by the computer according to light accumulation. On a bright sunny day it may irrigate up to 20 times. Overall, the slabs are irrigated to give a 40 percent leachate during irrigation cycles. During heavy demand for water during midday, the overdrain may be increased to 60 percent.

Tomatoes are harvested in tote bins and transported down the aisles on wheeled carts that run on the heating pipes (Fig. 10.40). All the product is graded, packed and marketed through the B.C. Greenhouse Vegetable Growers Cooperative.

Figure 10.40. Harvesting tomatoes and transporting them down the row, to the central passageway, in tote bins on a cart which runs along the heating pipes. (Courtesy of Houweling Nurseries Ltd., Delta, B.C., Canada.)

Houweling Nurseries Ltd., Delta, B.C., operates 30 acres (12 hectares) of greenhouses. They grow tomatoes, cucumbers and peppers separately in three sections. This greenhouse is unique in its use of sawdust as fuel to fire its boilers for generation of heat (Fig. 10.41). An auxillary natural gas furnace is used primarily to generate carbon dioxide to enrich the greenhouse atmosphere. Cooling is by natural ventilation through ridge vents assisted by overhead sprinklers to cool the glass surface. Water used to sprinkle the roof is collected in a reservoir and recycled.

Figure 10.41. Houweling Nurseries, a 30-acre greenhouse, uses sawdust to fuel their hot-water boilers for heating. (Courtesy of Houweling Nurseries Ltd., Delta, B.C., Canada.)

An open rockwool-culture system similar to those described above is used to grow tomatoes, cucumbers and peppers. Feeding is by a drip-irrigation system with an emitter at each plant. Electrical conductivity (EC) and *p*H levels are controlled by a computer, depending upon various light, heat and humidity conditions (Fig. 10.42).

Similar to Gipaanda Greenhouses Ltd., they had used the "Belmondo" variety of tomato grafted to a numbered resistant rootstock. Now they are growing "Apollo" and "Trust" which do not have to be grafted, as they are resistant to root rot. Planting density is approximately 10,000 plants per acre (25,000 plants per hectare) (4.35 square feet per plant) with an average yearly production of 48 kg/square meter (10 lb/square foot).

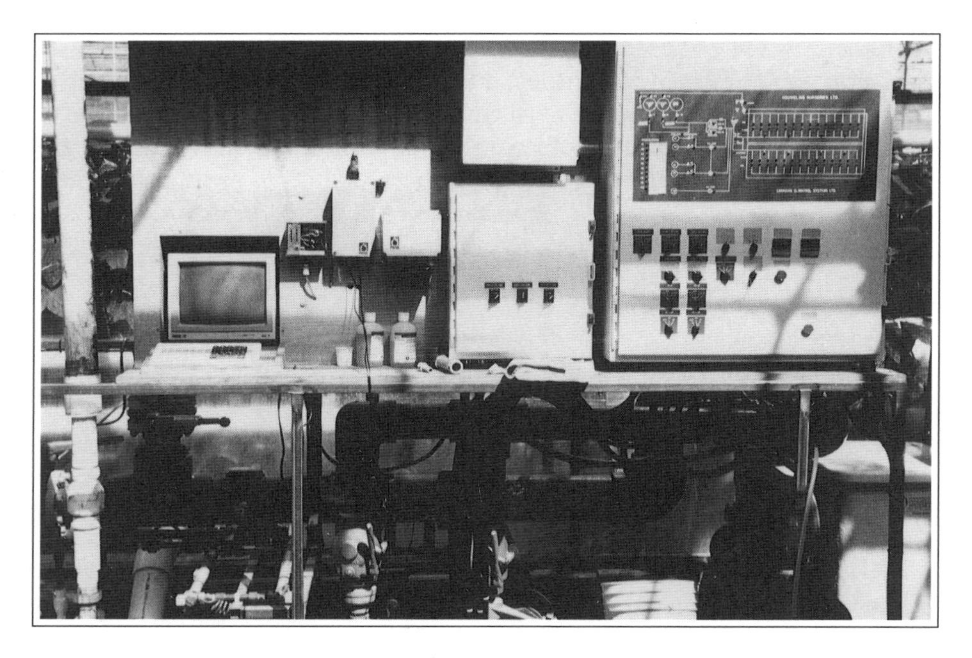

Figure 10.42. Feeding of the plants in a large greenhouse is controlled by a computer. (Courtesy of Houweling Nurseries Ltd., Delta, B.C., Canada.)

In a new 12-acre (5-hectare) range constructed in 1994, Houweling Nurseries is growing tomatoes in the wide rockwool slabs. The slabs are placed in an enameled-aluminum channel so that in the future the nutrient solution may be recycled (Fig. 10.43). They have a machine which forms the channel in continuous lengths, similar to aluminum gutters for houses (Fig. 10.44). The channel drains to a central 6-inch (15-cm) diameter PVC catchment pipe (Fig. 10.45). Five plants are grown per slab in alternating holes. The plants are trained in a V-cordon system with every other plant leaning in the same direction toward the aisle (Fig. 10.46). Using this method of training and five plants per slab, the cost of slabs can be reduced.

10.9 Peppers in Rockwool

Peppers are propagated similar to tomatoes. Seed is sown into rockwool cubes and seedlings transplanted to rockwool blocks. Peppers require a higher germination temperature than tomatoes, generally in the range from 21°–28° C (70°–82° F). As higher temperatures are more costly to heat, a suitable temperature for most conditions is about 24° C (75° F). Emergence occurs in 7 to 10 days.

Transplanting to rockwool blocks may be done after 3 weeks (Fig. 10.47) to become established in the blocks within 1 to 2 weeks (Fig. 10.48). Propagation temperatures may range from 18°–23° C (65°–73° F) with higher

Figure 10.43. Slabs with tomato plants placed in an aluminum channel so that the solution may be recycled. (Courtesy of Houweling Nurseries Ltd., Delta, B.C., Canada.)

day temperatures up to 25° C (77° F) according to sunlight conditions. Peppers are very temperature-dependent, requiring a high temperature regime to produce vigorous growth capable of supporting heavy crop production. As root temperatures are very important, it is best to propagate plants on slatted benches to provide good air movement, especially in more northerly climates where ambient temperatures are low during the propagation season. Piped hot-water heating under the benches allowing heat to rise up through the plants is ideal.

Plant density for greenhouse production is between 8,000 and 10,000 plants per acre (20,000 to 25,000 plants per hectare). This is similar to that of tomatoes. Three plants are located per slab with slabs spaced apart within the rows to give 50-cm (19-inch) centers for 8,000 plants per acre or 36-cm (14-inch) centers for 10,000 plants per acre. Rows of slabs are at 40–46 cm (16–18 inches) centers. Once good rooting into the slabs has become established and shoot growth has reached 40 cm (16 inches), the temperature regime should be reduced to 18°–19° C (65° F) at night and 22°–23° C (72° F) by day. A vigorous crop will tolerate day temperatures as high as 30° C (86° F) in full sunlight, but temperatures of 35° C (95° F) or above will be harmful.

In British Columbia, the growers seed peppers October 1st and trans-

Figure 10.44. Machine that forms continuous aluminum channel. (Courtesy of Houweling Nurseries Ltd., Delta, B.C., Canada.)

Figure 10.45. Central catchment pipe collects the solution from the channels. (Courtesy of Houweling Nurseries Ltd., Delta, B.C., Canada.)

Figure 10.46. V-cordon training of tomato plants. Note single rows of slabs are used to form double rows of plants. (Courtesy of Houweling Nurseries Ltd., Delta, B.C., Canada.)

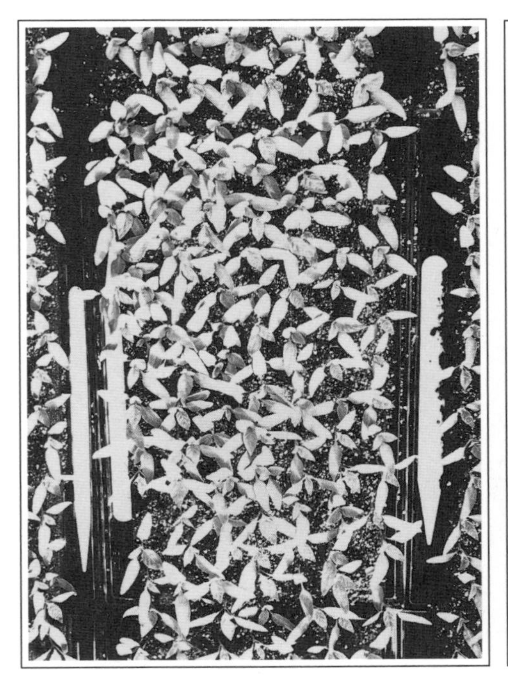

Figure 10.47. Peppers are seeded into rockwool cubes and seedlings are transplanted to rockwool blocks after 3 weeks. (Courtesy of Environ-mental Farms, Dundee, FL.)

Figure 10.48. Peppers transplanted to rockwool blocks become established within several weeks. These plants are about 5 weeks from seeding. (Courtesy of Environmental Farms, Dundee, FL.)

plant to the rockwool slabs by early December, with first harvest by early March (Figs. 10.49, 10.50). Harvesting continues until mid-November.

A number of varieties of each color (red, yellow, orange and purple) are grown. Houweling Nurseries Ltd. achieve annual yields of 21 kg/square meter (4.3 lb/sq ft) or about 190,000 lb per acre.

Peppers are trained as two stems per plant in a V-cordon arrangement supported by strings attached to overhead wires (Figs. 10.51, 10.52). Houweling Nurseries uses four wires at 10 feet (3 m) height. Plants are worked on by use of carts that travel on the heating pipes as described earlier.

Houweling Nurseries Ltd. ships all of its product to the Western Greenhouse Co-operative Association which grades, packs and markets the produce.

Between crops the rockwool slabs are not sterilized. Instead, the company shifts them to different crops. They use new slabs each year on tomatoes. The old slabs from the previous crop of tomatoes are used for the cucumbers and likewise the old slabs from the cucumbers are used for

Figure 10.49. Pepper growers in British Columbia begin harvesting in early March. (Courtesy of Houweling Nurseries Ltd., Delta, B.C., Canada.)

Figure 10.50. Peppers in rockwool culture at first harvest. Note the training of the plant into two stems. (Courtesy of Houweling Nurseries Ltd., Delta, B.C., Can.)

Figure 10.51. Peppers are trained as two stems in a V-cordon method. (Courtesy of Gipaanda Greenhouses Ltd., Surrey, B.C., Canada.)

Figure 10.52. Peppers are supported by plastic twine attached to four wires (two support each row of plants) located about 10 feet in height. (Courtesy of Houweling Nurseries Ltd., Delta, B.C., Can.)

the peppers. In this way they get three crops from the slabs over three years. There is some structural breakdown of the slabs during this period of use, so some of the slabs that have become compressed are doubled up, one on top of the other.

10.10 Recirculating Rockwool Systems

With increasing pressure of public awareness of the environment and water as a natural resource, it has become important to conserve water and minimize any runoff of fertilizer-enriched solution. In countries such as Holland with a high population and a large greenhouse industry using large volumes of water, the recirculation of nutrient solutions in hydroponic systems is imperative. One of the earliest forms of recirculating rockwool culture is the use of plastic semi-rigid channels in which the rockwool slabs are contained. There are a number of manufacturers of these channels. Some may be used as NFT systems as well as with rockwool. These channels come in rolls that when unwound form a channel by staking the sides upright with special clamps. The ends are folded up and stapled. A special drain-pipe adapter is fitted onto the return end of the channel. PVC piping conducts the solution back to a nutrient cistern. Each plant is irrigated with a drip system as discussed above. The only difference is that a nutrient tank of regular-strength nutrient solution, rather than an injector, is the source of nutrients. If the slabs are not wrapped, they will have to be soaked prior to covering them with white polyethylene. It is important that the channel is about ½ inch (1–2 cm) wider than the slabs to allow flow of the drain water past the slabs.

Another recirculating system of rockwool culture (Ryall 1993) also uses a prefabricated channel wider than the rockwool slab. A styrofoam slab 1–1.5 inches (2.5–4 cm) thick is placed under the rockwool slabs to elevate them above the drain water. This eliminates the need to accurately profile the soil underneath, as a drainage channel is formed next to the slabs (Fig. 10.53). The styrofoam also insulates the rockwool slabs from temperature fluctuations. Black polyethylene lines the channel and covers the styrofoam. If a prefabricated plastic channel is used, only the styrofoam itself needs to be wrapped with black polyethylene to protect it from moisture. The top of the slabs and channel are covered with white-on-black polyethylene. A slope of ¼ to ½ inch (1–2 cm) is adequate for drainage. One or two holes in the slab wrapping will provide good drainage. Runoff from the slabs is vertically away from the base of the slabs into the side drainage trough. This gives good aeration of the slabs. The solution is recirculated from the drain trough through a 1½-inch diameter PVC drain pipe connected to an underground 4-inch diameter collection pipe to the nutrient tank. Solution is applied to each plant individually through a drip-irrigation system.

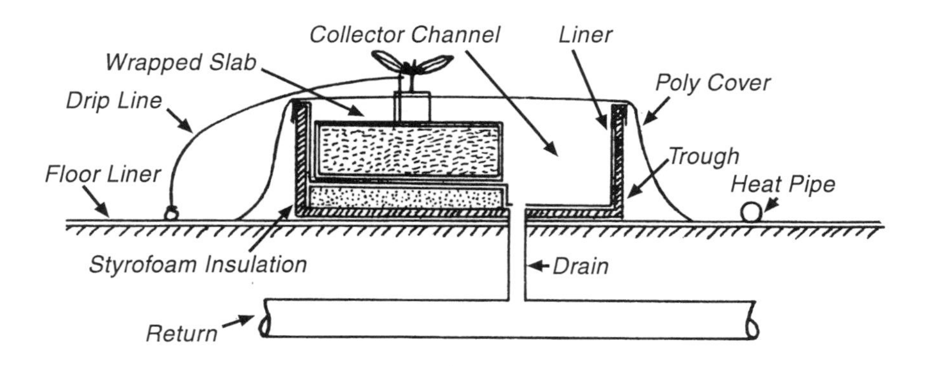

Figure 10.53. A proposed recirculating system of rockwool culture.

Ryall (1993) predicts that in the future, rockwool systems will use a single row of slabs with tomato plants trained in a V-cordon fashion. Four plants rather than two will be placed on each slab with alternating plants trained in opposite directions using the V-cordon system. This will greatly reduce the cost of rockwool slabs and also the handling costs during sterilization. A 2-mil-thick poly wrapper can be placed around every 2 slabs so that wicking cannot occur between slabs. Alternatively, white-on-black polyethylene may be lapped over the top of the slabs and taped with a clear packaging tape. Folds are raised between slabs to prevent wicking of the solution (Fig. 10.54). Although the volume of media per plant is reduced, previous methods used with NFT systems, as discussed in Chapter 6, indicate that roots do not require a full slab. In addition, twice as many drippers would be applied to each slab, increasing the uniformity of moisture.

The return solution is collected in a small cistern and then pumped up into an above-ground nutrient tank. Large particles are removed from the solution by a separator, followed by a filtration system (100-mesh strainer). The solution is then sterilized before re-entering the irrigation system. Sterilization may be by one or several methods: heat treatment, ozonation, ultraviolet radiation. The entire filtration, sterilization and pumping systems must be doubled up in case of failure.

The nutrient tank volume must be adequate for at least one day's supply for the plants (approximately 1 liter/sq ft/day). The irrigation system would consist of an injector system operated by a computer monitoring the *p*H and EC, pump, sand filters and emitters at the plants. A problem with emitters is their plugging within the cropping season. Locating the

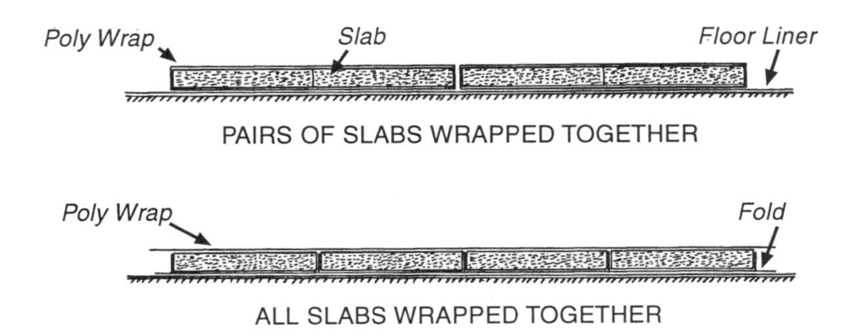

Figure 10.54. Wicking of solution between rockwool slabs is prevented by raising folds of the polyethylene wrapping between the slabs.

emitters on the poly lateral line instead of at the ends of the drip lines will help. Alternatively, spaghetti lines without emitters as used in the potted-plant industry could be used. Some growers free plugged emitters by forcing air through them which vibrates the precipitates from them. This is time-consuming and plants may undergo stress in the interim period before plugged emitters are discovered.

Nutrient analyses of both feed and drain solutions provide data on which to base adjustments of the solution. This must be done weekly. Returning solution has to be diluted with raw water to make up the loss in volume by the uptake of the plants. This decreases the EC level and changes the *p*H. The computer monitoring the *p*H and EC adds stock solutions and acid to bring these factors back to a preset level. However, as discussed in Chapter 3, electrical conductivity measures total dissolved salts, not the levels of individual ions.

Nutrient-solution analysis is imperative to enable the grower to make changes in the composition of the nutrient solution to keep it within optimum levels.

10.11 Advantages and Disadvantages of Rockwool Culture

Advantages:

1. As an open system, there is less chance for disease to spread throughout the crop.
2. Uniform application of nutrients to the plants; each plant is fed individually.
3. Since rockwool is light in weight, it can easily be handled.
4. It is simple to provide bottom heat.
5. It is easily steam-sterilized if the grower wishes to use it several

times. Structurally, it will not break down for up to three to four years.
6. Rapid crop turn-around is possible at a minimal labor cost.
7. It provides good root aeration.
8. There is less risk of crop failure due to mechanical breakdown in the system than with NFT.
9. Little growth setback occurs during transplanting.
10. With the use of the plug trays, the sowing of seeds can be automated.
11. It requires less capital cost for equipment and installation than many NFT systems.

Disadvantages:
1. Rockwool is relatively expensive in countries not manufacturing it locally.
2. Accumulation of carbonates and sodium may occur in the rockwool slabs in regions having high salts in the raw water. In such areas a reverse-osmosis water purifier would be required.

Overall, rockwool culture offers many positive factors in its rapidly accepted use in the growing of not only vine crops, but also cut flowers such as roses and potted flowering plants. It is also widely used as a propagation medium for vegetables such as lettuce, spinach and other low-profile crops.

References

Bijl, Jaap. 1990. Growing commercial vegetables in rockwool. *Proc. 11th Ann. Conf. on Hydroponics*, Hydroponic Society of America, Vancouver, B.C., March 30–April 1, 1990, pp. 18–24.

Hochmuth, George J. 1992. Production and economics of rockwool tomatoes in Florida. *Proc. 13th Ann. Conf. on Hydroponics*, Hydroponic Society of America, Orlando, Florida, April 9–12, 1992, pp. 40–46.

Ryall, David. 1993. Growing greenhouse vegetables in a recirculating rockwool system. *Proc. 14th Ann. Conf. on Hydroponics*, Hydroponic Society of America, Portland, Oregon, April 8–11, 1993, pp.33–39.

Smith, Dennis L. 1986. *Peppers & Aubergines. Grower Guide No. 3.* 92 pp. London: Grower Books.

———. 1987. *Rockwool in Horticulture.* 153 pp. London: Grower Books.

Chapter 11

Other Soilless Cultures

11.1 Introduction

Many other methods of soilless culture are being used successfully. Some of the media used are peat, vermiculite, perlite, pumice, and plastic styrofoam. Often mixtures of these media are used in various proportions. Growing trials with various mixtures determine which proportions are most suitable to the plants in question. For example, flowering potted plants such as chrysanthemums, poinsettias and Easter lilies and tropical foliage plants can be grown well in mixtures of peat-sand-pumice in a 2:1:2 ratio.

11.2 Media

11.2.1 Peat

Peat consists of partially decomposed aquatic, marsh, bog, or swamp vegetation. The composition of different peat deposits varies widely, depending upon the vegetation from which it originated, state of decomposition, mineral content and degree of acidity (Lucas et al. 1971; Patek 1965).

Of the three types of peat—moss peat (peat moss), reed sedge, and peat humus—peat moss is the least decomposed and is derived from sphagnum, hypnum, or other mosses. It has a high moisture-holding capacity (10 times its dry weight), is high in acidity (*p*H of 3.8 to 4.5) and contains a small amount of nitrogen (about 1.0 percent) but little or no phosphorus or potassium. Peat from hypnum and other kinds of mosses breaks down rapidly, as compared with sphagnum, and is not as desirable. Peat from sedges, reeds, and other swamp plants also decomposes rapidly.

Sphagnum moss is the dehydrated young residue or living portions of acid-bog plants in the genus *Sphagnum,* such as *S. papillosum,*

S. capillacium, and *S. palustre.* It is relatively sterile, light in weight, and has a very high water-holding capacity. It is generally shredded before being used as a growing medium.

11.2.2 Vermiculite

Vermiculite is a micaceous mineral which is expanded when heated in furnaces at temperatures near 2000° F (1093° C). The water turns to steam, popping the layers apart, forming small, porous, sponge-like kernels. Heating to this temperature gives complete sterilization. Chemically, it is a hydrated magnesium-aluminum-iron silicate. When expanded, it is very light in weight (6 to 10 pounds per cubic foot) (96 to 160 kilograms per cubic meter), neutral in reaction with good buffering properties, and insoluble in water. It is able to absorb large quantities of water—3 to 4 gallons per cubic foot (0.4 to 0.5 mililiters per cubic centimeter). It has a relatively high cation exchange capacity and thus can hold nutrients in reserve and later release them. It contains some magnesium and potassium which is available to plants.

Horticultural vermiculite is graded in four sizes: No. 1 has particles from 5 to 8 millimeters in diameter; No. 2, the regular horticultural grade, from 2 to 3 millimeters; No. 3, from 1 to 2 millimeters, and No. 4, which is most useful as a seed-germinating medium, from 0. 75 to 1 millimeter. Expanded vermiculite should not be pressed or compacted when wet, as this will destroy its desirable porous structure.

11.2.3 Perlite

Perlite is a siliceous material of volcanic origin, mined from lava flows. The crude ore is crushed and screened, then heated in furnaces to about 1400° F (760° C), at which temperature the small amount of moisture in the particles changes to steam, expanding the particles to small, sponge-like kernels which are very light, weighing only 5 to 8 pounds per cubic foot (80 to 128 kilograms per cubic meter). The high processing temperature gives a sterile product. A particle size of 1/16 to 1/8 inch (1.6 to 3.1 millimeters) in diameter is usually used in horticultural applications. Perlite will hold 3 to 4 times its weight of water. It is essentially neutral with a *p*H of 6.0 to 8.0 but with no buffering capacity; unlike vermiculite, it has no cation exchange capacity and contains no mineral nutrients. It is most useful in increasing aeration in a mixture since it has a very rigid structure. While it does not decay, the particle size can become smaller by fracturing as it is handled. A fine grade is useful primarily for seed germination, while a coarser type or horticultural grade is best suited for mixing with peat, in equal parts, for propagation or with mixtures of peat and sand for growing plants.

11.2.4 Pumice

Pumice, like perlite, is a siliceous material of volcanic origin. It, however, is the crude ore after crushing and screening without any heating process. It has essentially the same properties as perlite, but is heavier and does not absorb water as readily since it has not been hydrated. It is used in mixtures of peat and sand for the growing of potted plants.

11.2.5 Soilless Mixtures

Most mixtures contain some combination of sand, peat, perlite, pumice and vermiculite. The specific proportions of each used depends upon the plants grown. Some useful mixtures are:

1. peat:perlite:sand	2:2:1 for potted plants
2. peat:perlite	1:1 for propagation of cuttings
3. peat:sand	1:1 for propagation of cuttings and for potted plants
4. peat:sand	1:3 for bedding plants and nursery container-grown stocks
5. peat:vermiculite	1:1 for propagation of cuttings
6. peat:sand	3:1 light weight, excellent aeration, for pots and beds, good for azaleas, gardenias and camellias which like acid conditions
7. vermiculite:perlite	1:1 light weight, good for propagation of cuttings
8. peat:pumice:sand	2:2:1 for potted plants

In general, pumice, which costs less, may be substituted for perlite in most mixes.

The most common mixtures are the University of California mixes of peat and fine sand, and the Cornell "Peat-Lite" mixes. The U.C. mixes were derived from the California Agricultural Experiment Station in Berkeley. The U.C. mixes vary from fine sand only to peat moss only, but the mixes that are used more commonly contain from 25 to 75 percent fine sand and 75 to 25 percent peat moss. These mixes are used for growing potted plants and container-grown nursery stock. The Peat-Lite mixes were devised by Cornell University, New York, from equal proportions of peat and vermiculite. They have been used primarily for seed germination, growing of transplants and for container growing of spring bedding plants and annuals. Some growers have used them to grow tomatoes commercially in beds similar to that of sawdust culture.

All the required minerals must be added to these mixes, and some or all of them are added at the time of mixing. The Cornell Peat-Lite mixes are considerably lighter in weight than the U.C. mixes, as either perlite or vermiculite weighs about one-tenth as much as fine sand. The Peat-Lite mixes are made from equal parts of sphagnum peat moss and either horticultural perlite or number 2 vermiculite.

The U.C. Mix. The basic fertilizer additions recommended for a U.C. mix of 50 percent fine sand and 50 percent peat moss are as follows (Matkin and Chandler, 1957).

To each cubic yard (0.7645 cubic meter) of the mix, add

- 2½ lb (1.136 kg) hoof-and-horn or blood meal (13 percent nitrogen)
- 4 oz (113.4 g) potassium nitrate
- 4 oz (113.4 g) potassium sulfate
- 2½ lb (1.136 kg) single superphosphate
- 7½ lb (3.41 kg) dolomite lime
- 2½ lb (1.136 kg) calcium carbonate lime

The fine sand, peat moss, and fertilizer must be mixed together thoroughly. The peat moss should be moistened before mixing. As the crop grows, additional nitrogen and potassium fertilizer must be provided.

The Cornell "Peat-Lite" Mixes. Instructions for three Peat-Lite mixes are as follows (Boodley and Sheldrake 1964; Sheldrake and Boodley 1965).

1. Peat-Lite Mix A (to make 1 cubic yard or 0.7645 cubic meter):
   - 11 bu (88 U.S. gal) (333 liters) sphagnum peat moss
   - 11 bu (88 U.S. gal) (333 liters) horticultural vermiculite No. 2 grade
   - 5 lb (2.27 kg) ground limestone
   - 1 lb (0.4545 kg) superphosphate (20%)
   - 2 to 12 lb (0.909 to 5.4545 kg) 5-10-5 fertilizer
2. Peat-Lite Mix B:
   Same as A, except that horticultural perlite is substituted for the vermiculite.
3. Peat-Lite Mix C (for germinating seeds):
   - 1 bu (8 U.S. gal) (30.3 liters) sphagnum peat moss
   - 1 bu (8 U.S. gal) (30.3 liters) horticultural vermiculite No.4
   - 1½ oz (42.5 g) ammonium nitrate
   - 1½ oz (42.5 g) superphosphate (20%)
   - 7½ oz (212.6 g) ground limestone, dolomitic

The materials should be mixed thoroughly, with special attention given to wetting the peat moss during mixing. Adding a nonionic wetting agent, such as Aqua-Gro (1 ounce per 6 U.S. gallons water) (28.35 grams per 22.7 liters water) to the initial wetting usually will aid in wetting the peat moss.

Fertilizer, Sphagnum Peat Moss, and Vermiculite Mixture. The Vineland Research Station in Ontario, Canada (Sangster 1974), uses a slight modification in fertilizer ingredients for addition to a mixture of equal volumes of sphagnum peat moss and vermiculite (50:50 peat-vermiculite) per cubic yard (0.7645 cubic meter) as shown below.

1¾ 6-cu ft (0.17-cu. meter) compressed bales	peat moss
2 6-cu ft (0.17-cu. meter)	bags vermiculite (No.2)
12 lb (5.45 kg)	ground limestone (dolomitic)
5 lb (2.273 kg)	calcium sulfate (gypsum)
1.5 lb (0.682 kg)	calcium nitrate
2.5 lb (1.136 kg)	20% superphosphate
8-10 lb (3.64-4.54 kg)	Osmocote 18-6-12 (9 month)
6 oz (170 g)	fritted trace elements (FTE 503)
1 oz (28.35 g)	iron (chelated such as Sequestrene 330)
0.5 lb (227 g)	magnesium sulfate

Osmocote 18-6-12 provides a continuous supply of nitrogen, phosphorus, and potassium throughout the growing season. FTE 503 slowly releases iron, manganese, copper, zinc, boron and molybdenum.

Mixing the fertilizer ingredients with the peat-moss mixture can be done in several ways. Small volumes can be mixed with a shovel on a concrete floor. When mixing on a floor, first disinfect the floor area with a solution of 5 parts water to 1 part Javex (5.25 percent sodium hypochlorite). Spread the fertilizer evenly over the medium and turn the mix back and forth from one pile to another several times with a shovel. A large garbage can is useful for mixing a two-bushel batch. Scoop the mix into the garbage can, pour the mix back onto the floor, and repeat this procedure several times.

A large concrete mixer works well for mixing large amounts. Often commercial growers acquire a "ready-mix" unit from an old concrete truck. This unit can be mounted on a concrete slab and a motor attached to operate it. A series of conveyors can feed in ingredients and also pile the finished product in the potting area of the greenhouse.

If plastic-lined beds are used for the plants, the medium can be mixed directly in the beds. It can be mixed with a padded hoe, taking care not to rupture the plastic liner. For large greenhouses a ready-mix unit should be used, with conveyors conducting the finished product directly to the beds.

Dry peat is usually hard to wet. Adding two ounces (56.7 g) of a non-ionic wetting agent such as Aqua-Gro in ten gallons (37.85 liters) of water will help wet the peat in one cubic yard (0.7645 cu. meter) of mix. Micronutrients should be dissolved in water, which is then sprinkled over the medium or added directly to the mixer while mixing. For a two-bushel (16 U.S. gallons) (60.6 liter) batch, the nutrients can be dissolved in one gallon (3.785 liter) of warm water and then sprinkled over the medium prior to mixing.

11.2.6 Synthetic Foams (Plastoponics)

Attempts are now being made in different parts of the world to develop a completely reproducible synthetic compost in which peat is replaced totally or in part by a synthetic foam which may be urea-formaldehyde, polyurethane or polystyrene.

Foams can be produced with varying proportions of open cells. The thickness of the cell walls and the size of the pores can be varied. This will affect the density of the foam and its water-retention capacity. They are excellent aerators of soil and yet they can entrap large volumes of water per unit volume. For example, 1 pound (454 grams) of urea-formaldehyde foam will hold up to 12 gallons (45 liters) of water. The foams are very light and when used for a pot-plant compost, they must be weighted with dense inert particles such as sand.

Various foam-sand mixtures have been tested successfully in the growing of orchids, carnations, bulbs, numerous houseplants and tomatoes (Cook 1971). They have also worked well as rooting blocks for the propagation of cuttings.

11.3 Growing Herbs in a Peat-Lite Mix

California Watercress, Inc. of Fillmore, California grows a number of herbs in a peat medium containing 60 percent peat, 15 percent sand, 15 percent fir bark and 10 percent perlite. Dolomite lime is added to stabilize the *p*H between 6.0 and 6.5. A wetting agent and a 9-month formulation of slow-release fertilizer is incorporated into the medium. Mint, chives, thyme, basil and oregano have been grown successfully in the medium.

About one acre (0.4 hectare) of greenhouses were constructed to grow these herbs hydroponically year-round, but emphasis is on winter production in the greenhouses. Beds 8 ft by 155 ft (2.4 m by 47 m) were constructed of 8 inch × 8 inch × 16 inch (20 cm × 20 cm × 40.5 cm) cement blocks and 4 ft by 4 ft (1.2 m by 1.2 m) pallets (Fig. 11.1). Sides were formed using 1 inch × 8 inch (2.5 cm × 20 cm) lumber nailed to the edges of the pallets. This gave a bed depth of 5 inches (13 cm). The beds were lined with chicken wire and 6-mil black polyethylene (Fig. 11.2). The chicken wire prevented the black poly from sagging between the spaces of the pallet tops. Small slits were made in the bottom of the poly liner at centers of 16–18 inches (40–46 cm) to allow adequate drainage. The polyethylene liner was stapled along the top edges, sides and bottom of the beds (Fig. 11.3). Support braces to secure the sides of the beds were nailed every 20 feet (6 m) along the bed length before putting the medium in the beds (Fig. 11.4). The medium was moistened prior to placing it in the beds.

Figure 11.1. Beds constructed of cement blocks and pallets. (Courtesy of California Watercress, Inc., Fillmore, CA.)

Figure 11.2. Beds are lined with chicken wire to support the polyethylene liner. (Courtesy of California Watercress, Inc., Fillmore, CA.)

Figure 11.3. Black polyethylene liner is stapled to the bed. (Courtesy of California Watercress, Inc., Fillmore, CA.)

Figure 11.4—Placing of a Peat-Lite medium in the completed bed. Note the beds under construction to the left. (Courtesy of California Watercress, Inc., Fillmore, CA.)

The irrigation system consisted of a central injector and with 2-inch diameter PVC mains and 1-inch diameter PVC headers supplying nutrients via "T-tape" drip hoses at 12-inch (30.5-cm) centers (Figs. 11.5, 11.6).

Figure 11.5. Irrigation system showing 2-inch main overhead tee'd to 1-inch headers to each bed with eight "T-tape" drip lines running along the length of the beds. Note the solenoid valve on the header line and the second line that is the header for the latter half of each bed. (Courtesy of California Watercress, Inc., Fillmore, CA.)

Two headers, one at the front of the beds and the second in the middle connected to the drip lines so that each set ran only half the length of the beds (about 75 feet) (23 m) (Fig. 11.6). The drip tape was pre-punched at 12-inch (30.5-cm) centers along its length. The irrigation system was

Figure 11.6. Irrigation header with eight T-tape drip lines running along the length of the bed. Chives were planted in the middle bed. (Courtesy of California Watercress, Inc., Fillmore, CA.)

operated by two timer controllers. Irrigation cycles were generally every 2–3 days for 1 to 1½ hours depending upon the stage of plant growth and weather conditions. The duration of each cycle was long enough to produce sufficient leaching to flush any salt buildup.

Seedlings or rooted vegetative cuttings grown in small pots or trays were transplanted to the beds (Fig. 11.7). In some cases, as for chives and mint, field plants were divided and roots washed before transplanting to the beds.

It is expected that since these herbs are perennials, the plants could remain in the beds for several years under continual cropping (Figs. 11.8–11.11). When they need to be replaced, the medium should be steam sterilized, some new medium added and the beds tilled.

The greatest problem with this medium is the root buildup. After more than two years of cropping the same plants, a highly compacted root system formed throughout the medium. This reduced oxygenation with resultant root die-back. The crop should be changed each year and the medium sterilized to prevent root rots.

Figure 11.7. Transplanting herb seedlings into a bed of Peat-Lite medium. (Courtesy of California Watercress, Inc., Fillmore, CA.)

Figure 11.8. Chives 15 days after previous cut in left bed and 12 days after cutting of bed on right. (Courtesy of California Watercress, Inc., Fillmore, CA.)

Figure 11.9. Mint 35 days after transplanting, ready for first harvest. (Courtesy of California Watercress, Inc., Fillmore, CA.)

Figure 11.10. English thyme growing in a Peat-Lite medium. Note the T-tape drip-irrigation line. (Courtesy of California Watercress, Inc., Fillmore, CA.)

Figure 11.11. Oregano ready to harvest. The pipe in the middle of the bed is the header from the overhead line which feeds the lower half of the bed. (Courtesy of California Watercress, Inc., Fillmore, CA.)

11.4 Herbs in Foam

The author has tested a number of types of Hypol polymer foams developed by W.R. Grace & Co. These hydrophilic foams treated with different surfactants and having differing densities were tested with basil and mint. One bed 6 ft by 12 ft (1.8 m by 3.6 m) out of 40 beds using an NFT system as described in Chapter 6, Section 6.16, was set up with the foam polymer test. This provided information on the use of foam as a substrate in commercial greenhouse production.

Five-week-old basil seedlings were transplanted at 6-inch (15-cm) centers on the foam trials and on the NFT capillary matting beds. While the foam had high capillary action, its 2-inch (5-cm) thickness, did not allow adequate moisture to reach some of the transplants along the edges of the bed. As a result, these peripheral plants were set back in rooting (Figs. 11.12, 11.13). By watering overhead once per hour, rooting into the foam was improved.

Once the plants became established, they yielded greater than the control plants in the capillary matting (Fig. 11.14). The foam kept the crowns of the basil above the moisture at the bottom of the beds, whereas the control plants suffered from rotting at their crowns due to the presence of the solution in proximity with their bases (Fig. 11.15). Basil does not like excessive moisture at the base of the plant, therefore, in the longer term, the foam produced healthier plants than the capillary matting.

Similar trials were carried out with mint after the basil crop was removed. Two types of foam, 1-inch (2.5-cm) thick having different densities and surfactants, were compared with a control using the capillary-matting NFT system. Rooted cuttings were placed on the foam and capillary matting of the different treatments (Fig. 11.16). After approximately two months, the time of first harvest, the mint grew over the entire surface of the beds forming a mat of plants, stolons and roots. A subsequent harvest took place 45 days later (Fig. 11.17). During that period of cropping, there was no significant difference among treatments; all produced high yields. However, less leaf die-back and fungal infection occurred in the foam trials compared with the control using the capillary matting. This continued vigorous growth in the foam was attributed to the presence of less moisture at the base of the plants in the foam compared to the capillary matting. Root growth was exceptionally healthy in the foam treatments (Fig. 11.18).

From these trials and discussions with growers, it is evident that there is a potential commercial application of foam polymers in the growing of hydroponic vegetables. It may be an alternative medium to rockwool. Trials in the future need to include the use of foams in slabs similar to rockwool in the growing of vine crops such as tomatoes, cucumbers and peppers.

Figure 11.12. Basil growing in foam medium of an NFT system, 25 days after transplanting. Note the setback of plants on the left side due to lack of lateral movement of the solution in the foam. (Courtesy of California Watercress, Inc., Fillmore, CA.)

Figure 11.13. Basil growing in NFT capillary mat system, 25 days after transplanting. Compare this with previous photo and note that initially the basil in the capillary matting grew faster and more uniformly than in the foam. (Courtesy of California Watercress, Inc., Fillmore, CA.)

Figure 11.14. Basil in foam NFT system, 19 days after first harvest. Plants are now more uniform in growth than previously. (Courtesy of California Watercress, Inc., Fillmore, CA.)

Figure 11.15. Basil in capillary-mat NFT system, 19 days after first harvest. Note presence of crown rot. (Courtesy of California Watercress, Inc., Fillmore, CA.)

Figure 11.16. Rooted cuttings of mint placed on 1-inch-thick foam NFT system. Note the location of the drip-irrigation lines and the division of the beds into two. (Courtesy of California Watercress, Inc., Fillmore, CA.)

Figure 11.17. Mint in foam NFT system, 45 days after previous cut, ready to harvest. (Courtesy of California Watercress, Inc., Fillmore, CA.)

Figure 11.18. Healthy root growth of mint in foam NFT system. (Courtesy of California Watercress, Inc., Fillmore, CA.)

With its ability to be sterilized easily, and durability to last up to 5 years or more, it is expected to become an important hydroponic medium of the future.

11.5 Perlite Culture

Perlite culture is an alternative to rockwool culture. In Great Britain it is considered the third most important hydroponic system behind rockwool and NFT. Day (1991) states that it is mainly used in Scotland and the North of England in the growing of 16 hectares (40 acres) of greenhouse tomatoes. There are a number of smaller growers using perlite in the United States. Two of the larger well-respected growers are Ken and Richard Gerhart. European cucumbers are grown by Gerhart Greenhouses of Daggett, California, operating on one-half acre (0.2 hectare), and Gerhart Inc. of North Ridgeville, Ohio, grows cucumbers and tomatoes on 3 acres (1.2 hectares). Gerhart Greenhouses in California is located adjacent to a Southern California Edison power-generating plant (Fig. 11.19). The greenhouse utilizes deionized water obtained from the cooling towers of the natural gas electrical-generating plant. This clean raw water is a great advantage to a greenhouse located in an area having very hard raw water. Without this cogeneration association, the greenhouse would require a reverse-osmosis treatment unit to deionize the water. In a third operation

Figure 11.19. Gerhart Greenhouses cogeneration project. Greenhouses are located adjacent to a California Edison power-generating plant. (Courtesy of Gerhart Greenhouses, Daggett, CA.)

using perlite bags, American Foodsource Corp. of La Junta, Colorado, was growing almost 3 acres of cucumbers until 1992. This operation in 1993 was closed due to management problems.

The perlite-bag system is set up in the same way as rockwool culture, with a few differences. Using an injector with drip irrigation, each bag has two emitters, one at the base of each plant (Fig. 11.20).

The Gerhart brothers designed the perlite system to be low cost and easy to install. They found that bags 7½ inches (19 cm) in diameter by 43 inches (109 cm) long were required for 3 plants per bag, and 6¼ inches (16 cm) in diameter by 43 inches (109 cm) were sufficient for 2 plants per bag. These sizes of bags provided about 7 liters of medium per plant for tomatoes, and 10 liters per plant for cucumbers which stored adequate solution for the plants. Bags made from 6-mil thick white opaque polyethylene "layflat" 12 inches × 43 inches (30.5 cm × 109 cm) sealed on one end are manufactured by Cleveland Plastics. A hopper with a chute is used to fill the bags with perlite, then the other end is heat sealed. The average cost of the filled grow bag is $0.65 per plant.

As root-rot diseases are a serious threat to seedlings less than four weeks old, a system of propagation that would prevent such diseases was important to successful growing. All containers, if not new, must be

Figure 11.20. Perlite-bag hydroponic system. Convection tube to the left of the cucumber plants conducts heat to the base of the plants. Note the use of a drip-irrigation system similar to that of rockwool culture. (Courtesy of Gerhart Greenhouses, Daggett, CA.)

adequately sterilized. Seed can be sown in rockwool cubes and transplanted several days after germination to a 4-inch-square (10-cm-square) band pot containing #522 mix from Paygro Company (Fig. 11.21).

The medium consists of perlite, hardwood bark (composted) and peat. Micro-organisms and chemicals, with properties similar to fungicides contained in the composted bark, suppress *Pythium* and *Phytophthora* root-rots. When these pots with their seedlings are transplanted to the perlite bags, beneficial micro-organisms follow the roots into the grow bags and establish long-term protection of the plants' roots. The Gerharts have grown up to four consecutive crops in the same perlite bags without sterilization between cropping cycles (Fig. 11.22). There was no significant occurrence of root diseases, and production quality and yield remained high (Fig. 11.23). Potassium silicate is added to the nutrient solution at a rate of 100 ppm to also protect cucumbers against *Pythium*. Between crops, formaldehyde is sprayed through the fogging system after the last harvest, before removing the plants, to kill any insects.

Two crops are grown per year. The fall crop is seeded at the end of October, transplanted early November with first harvest in mid-December and final harvest in April. The summer crop is seeded in May and transplanted by early June with production from mid-June through September.

Figure 11.21. Cucumber seedlings are grown in a bark medium in 4-inch-square plastic pots which are later set on top of the perlite bags (two per bag) during transplanting. Note hole in bag under the pot. (Courtesy of Gerhart Greenhouses, Daggett, CA.)

Figure 11.22. Cucumbers growing in perlite bags. Note the poly-convection heating tube between the rows of plants. (Courtesy of Gerhart Greenhouses, Daggett, CA.)

Figure 11.23. Three to four productive crops of cucumbers are grown in the perlite bags without sterilizing the bags between crops. (Courtesy of Gerhart Greenhouses, Daggett, CA.)

Calcium carbonate is added to the deionized water from the cooling towers, as it is free of all minerals. The calcium carbonate gives it some buffering capacity to stabilize the *p*H. Oxygen is added to the water in the reservoir with an air pump and air stones placed in the storage tank. The level of oxygen in the water is maintained at 8 ppm at the emitters. The storage tank is necessary since the water coming from the cooling towers is less than 10 gallons (38 l) per minute, but it operates 24 hours per day, so adequate water can be collected over this period for the greenhouse needs.

The EC and *p*H of the nutrient solution entering and leaving the perlite bags are recorded and adjustments made to the formulation and watering cycles as required to maintain an optimum feeding program. The number of cycles, duration of each cycle, formulation and leachate rate are all monitored. The leachate may vary up to 20 percent depending upon the status of the grow bag and crop performance. About 1 to 1½ inches (2.5 to 4 cm) of solution is maintained in the bag as a reservoir by cutting a slit in the side of the bag facing the irrigation line at a height of 1 to 1½ inches (2.5 to 4 cm) up from the bottom. After cutting a hole in the top of the bag at the location of each plant, the transplants, in their pots, are placed in these positions, the roots are allowed to grow into the bags several days before the emitters are positioned beside the pots in the bag (Fig. 11.20).

11.6 Ring Culture

Ring culture was originally developed in England for the growing of tomatoes. The plants are set into a round ring, 8 to 10 inches (20 to 25 cm) in diameter, of plastic film or tar paper filled with a sterile medium such as the Cornell Peat-Lite or U.C. mixes. The rings have no top or bottom, and are spaced out on the bed of lightweight aggregate 4 to 6 inches (10 to 15 cm) deep. The beds containing the aggregate are lined with a plastic sheeting to prevent roots from penetrating into the infested soil beneath.

The basic difference between "ring" and "bed" culture is that in the bed method, the plants are simply set in a long, narrow trough filled with a growing medium. Both methods work satisfactorily, but in very early plantings during the cold and cloudy periods of February and March, the growing medium in the rings warms up much more readily and the plants begin growth faster due to the higher temperature around the roots. Also, the taller medium column will provide for better root aeration.

Ring Culture Layout. Beds are prepared similarly to that of sand and sawdust cultures. The width of the beds should be 24 to 30 inches (61 to 76 cm) with 24- to 30-inch (61- to 76-cm) walkways. Tomatoes should be set in two rows about 6 inches (15 cm) from the edge of the bed to give 12 to 18 inches (30.5 to 46 cm) between rows. The plants should be 12 to 14 inches (30.5 to 36 cm) apart in each row. The sides of the beds are constructed of 2 × 6-inch (5 × 15 cm) or rough 1 × 6-inch (2.5 × 15 cm) cedar or redwood lumber on edge, as shown in Figure 11.24. The beds are lined with 6-mil polyethylene. While no drainage pipes are needed, holes should be cut in the plastic on the side wall about 1 to 2 inches (2.5 to 5 cm) from the bottom. One-inch (2.5-cm) holes every 10 feet (3 meters) on each side should be adequate. Plastic rings 9 inches (23 cm) in diameter and 9 inches (23 cm) high are placed on the aggregate-filled bed (Fig. 11.24).

The beds may be set on concrete floors, or the entire greenhouse floor may be covered with polyethylene prior to constructing the beds. Holes for drainage can be cut next to the beds.

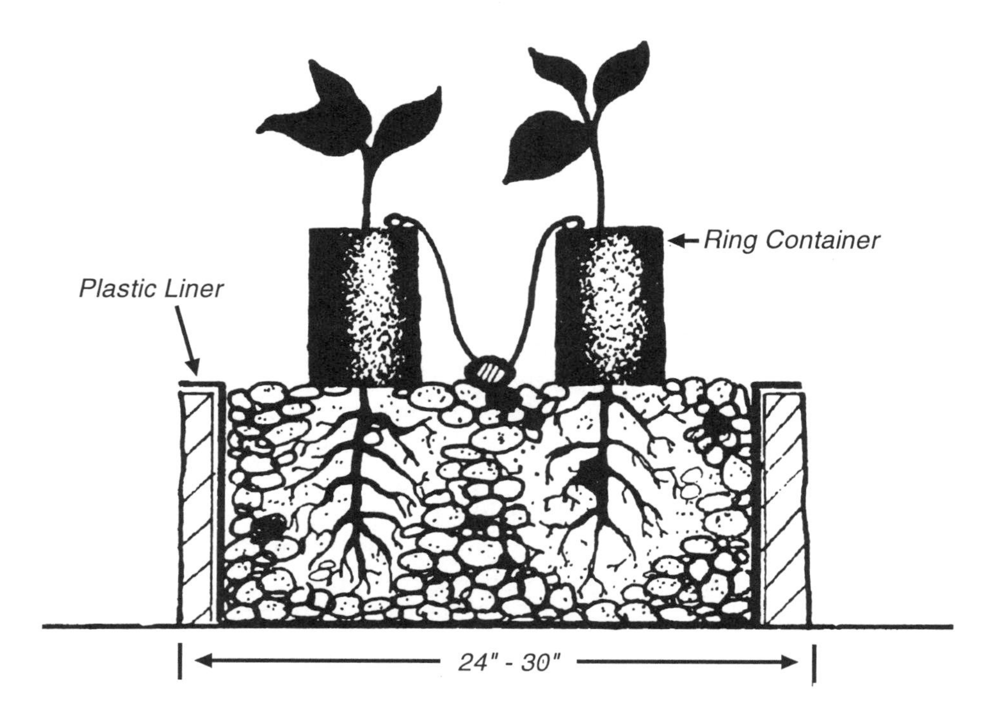

Figure 11.24. Cross section of ring culture and underneath aggregate bed.

Rings can be prepared by cutting 9-inch (23-cm) lengths from layflat white-on-black polyethylene tubing of 6-mil thickness. If tubing of 14-inch (36-cm) layflat width is used, a 9-inch (23-cm) diameter ring will be produced when opened. Also, rolled roofing paper (about 40-pound grade and usually 36-inch (91-cm) width) can be cut into 9-inch (23-cm) strips and formed around a cylinder such as a tin can or plastic pipe and the edges stapled to form the ring.

Rings are filled with a sterile medium such as the Cornell Peat-Lite mix. If polyethylene tubing is used for rings, a funnel device is needed, such as a length of PVC pipe (8-inch or 20-cm diameter), and a small shovel to handle the medium. If tarpaper rings are used, no funnel is needed. The rings are filled about two-thirds full if potted transplants are to be set. They can be filled to within 1 inch (2.5 cm) of the top after transplanting.

Watering and Feeding. Watering and feeding can be applied by an automatic trickle-feeding system using spaghetti tubes, such as outlined for sawdust culture.

11.7 Peat Modules

On Guernsey Island in the United Kingdom, a large number of commercial greenhouse growers use peat modules. The cropping module measures 39 inches (1 meter) long and 7 to 9 inches (18 to 23 cm) wide. It was basically designed for the commercial tomato grower, allowing three plants per module. These modules are now being used also by amateur gardeners. They can be used outside during the summer growing season for flowers, bedding plants and vegetables. The modules contain a sterile peat medium.

11.8 Column Culture

The growing of plants in vertical columns has been developed in Europe, particularly in Italy and Spain. This system originated from the use of barrels or metal drums (Fig. 11.25) stacked vertically and filled with gravel or a peat mixture. Holes were punched in the sides around the containers in order to place the plants into the medium. Later, asbestos-cement pipes with spirally-positioned holes were used.

Watering and feeding is supplied by a trickle-irrigation system mounted at the top of each column. If gravel is used as a medium, the nutrient solution can be recycled by placing the column over a collecting trough which conducts the solution back to a centrally located reservoir (Fig. 11.26).

Figure 11.25. Column culture of strawberries in the Canary Islands using metal drums.

Figure 11.26. Column culture in Costa Rica using asbestos-cement pipes and a cyclic gravel-culture system.

In Italy the system has been refined to a column built of smaller modular pipes placed one on top of the other (Fig. 11.27). Each module has several cup-shaped protuberances in which the plants are placed, rather than merely holes located around the periphery. A peat mixture is used for the medium. It is an open system which allows any excess nutrient solution to drain from the bottom end of the column. This system is particularly useful for the growing of strawberries.

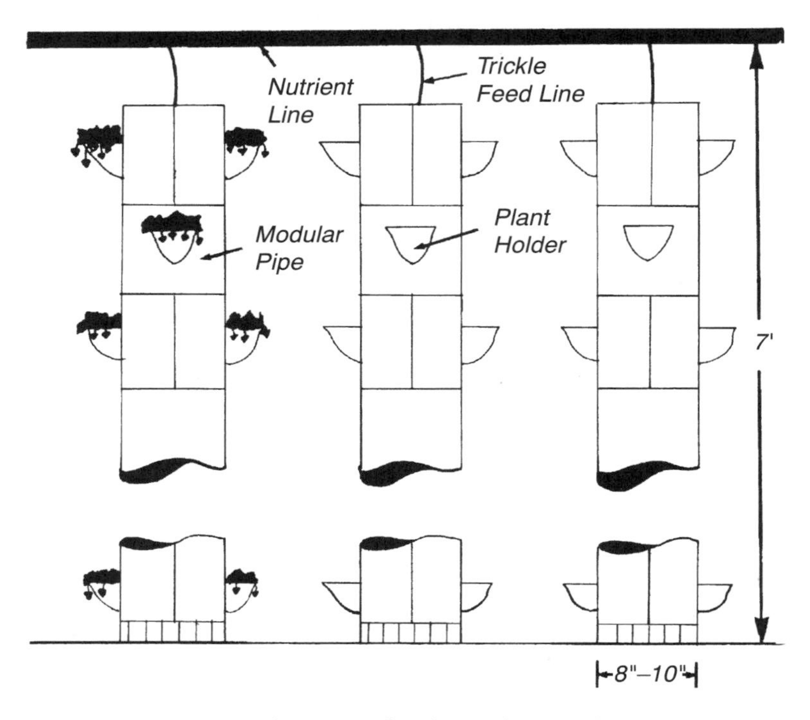

Figure 11.27. Schematic of Italian column-culture system.

11.9 Sack Culture

Sack culture is a simplification of column culture. The system is basically the same except that polyethylene "sacks" are used instead of rigid drums or pipes. Black layflat 0.15-millimeter thick polyethylene of about 6-inch (150-millimeter) diameter and 6 feet (2.0 meters) in length is filled with a peat-vermiculite mixture. The bottom end is tied to prevent the medium from falling through, and the top end is tied to constrain the medium into a sausage-like form. The top end is tied by a wire or rope to the greenhouse and the sack hangs down, giving a column effect.

Watering and feeding is automated by use of a trickle spaghetti-tube system to each sack from a central nutrient reservoir or fertilizer injector (Fig. 11.28). Small holes 2.5 to 5 centimeters (1 to 2 inches) in diameter are

Figure 11.28. Main irrigation pipe and lateral lines running along the greenhouse superstructure above the "sacks."
(Courtesy of M. Tropea, University of Catania, Italy.)

cut around the sack's periphery, into which plants are placed (Fig. 11.29). The nutrient solution is applied at the top end of the sack and percolated down through the entire sack.

The plant-holding containers, or sacks, supported by the greenhouse superstructure are spaced 80 centimeters (32 inches) apart within the rows, and the rows are spaced at 1.2 meters (about 4 feet), as shown in Figure 11.30.

Watering and feeding cycles are generally from 2 to 5 minutes, giving a volume of 1 to 2 liters of nutrient solution per sack per irrigation cycle. Nutrients are not recycled but allowed to percolate from the top to the bottom of the sack and out the drainage holes. Once a month the system is flushed with pure water to remove any salt buildup. At the end of each growing period the entire sack and substrate is disposed of and new ones are made up with sterile medium (Fig. 11.31).

This system is particularly useful for lettuce and strawberries which normally require a lot of greenhouse floor area with little utilization of the vertical space (Fig. 11.32). However, tests with tomatoes (Fig. 11.33), peppers (Fig. 11.34), eggplant (Fig. 11.35), cucumbers and other vegetables have also been very successful (Tropea 1976).

In 1976 an industrial installation was operating in Campania, Italy, growing about 8 hectares (about 20 acres) of sack culture in polyethylene greenhouses. They were growing mainly strawberries, but had experimented successfully with many other crops. They developed an "electronic brain" which is programmed to automatically feed the entire 20-acre complex, which is divided into 32 sections, each of them covering an area of 2,500 square meters (about 0.6 acre) (Fig. 11.36). The electronic brain controls the timing of irrigation cycles according to the environmental conditions and plant stage of development.

The cost of the entire system, including the metal-framed greenhouses, heating, irrigation system and electronic brain has been moderate, and operating costs are substantially reduced by the efficient use of the vertical space within the greenhouse and by reduction of labor.

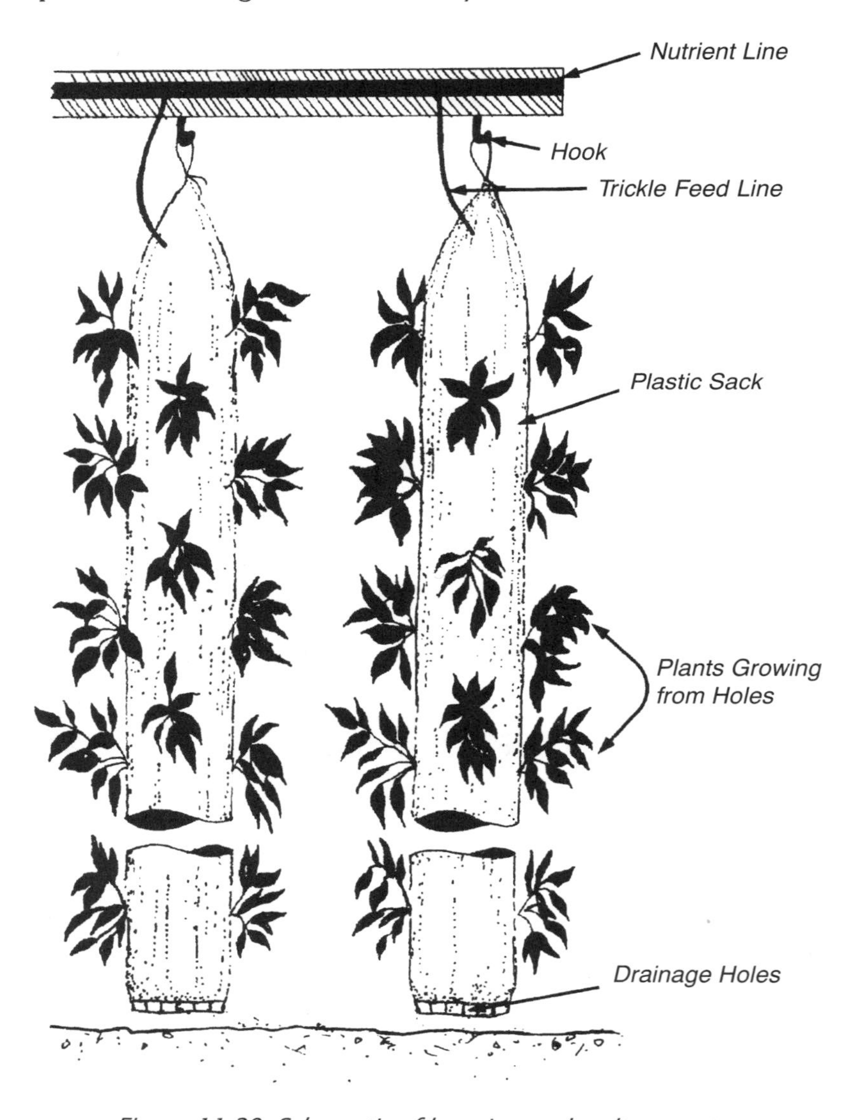

Figure 11.29. Schematic of hanging sack-culture system.

Figure 11.30. Rows of "sacks" supported by the greenhouse superstructure. (Courtesy of M. Tropea, University of Catania, Italy.)

Figure 11.31. Mixing medium and filling "sacks" for next crop. (Courtesy of M. Tropea, University of Catania, Italy.)

Figure 11.32. Strawberries growing in vertical sacks. Figure 11.33. Tomatoes.
Figure 11.34. Peppers. Figure 11.35. Eggplants.
(Courtesy of M. Tropea, University of Catania, Italy.)

Figure 11.36. An "electronic brain" controls the irrigation cycles. (Courtesy of M. Tropea, University of Catania, Italy.)

A sack-culture system has been developed for the growing of foliage houseplants, annuals, flowering plants, vegetables, small fruits and herbs. These "Gro-Tubes" have specifically been designed for the hobbyist and interior landscaping trades. The plastic tubes are approximately 2½ inches (6 cm) in diameter by 4 feet (1.2 m) in length filled with a medium of peat, sand, and vermiculite. Osmocote, fritted trace elements and lime fertilizers added to the medium provide the basic plant nutrients in some makes of plastic tubes, while others may have less fertilizers incorporated into the mix and therefore require a liquid fertilizer on a regular basis several weeks after planting. A water reservoir secured on the top of the sack and a collecting cup at the bottom provide ease of watering and prevention of loss of drainage water. The water percolates from the reservoir bottle through a wick which is incorporated in the medium the entire length of the sack.

Such sacks are ideal for the growing of herbs, lettuce, strawberries, and houseplants such as coleus, peperomia, begonias, ferns, wandering Jew, pothos, petunias, marigolds and impatiens. Most houseplants are set in the bags as unrooted cuttings through slits made in the sides of the bags. The bags will contain 30 to 50 coleus cuttings, 40 to 60 wandering Jew cuttings, or 12 to 20 fibrous begonias. They root and grow laterally and vertically from the suspended bag. Many vegetables and flowering plants can be placed in the bags as seedling transplants. As the plants grow to maturity, they fill the sacks making a very showy display of massive vertical plants as illustrated in Figures 11.37, 11.38.

Figure 11.37. Coleus growing in sacks showing recently planted cuttings on the right in the foreground.

Figure 11.38. Fully grown house plants in Gro-Tube. Note water reservoir at the top.

(Courtesy of Poly-Tech Growing Systems Ltd., left.; Seimer, Inc., right.)

11.10 Small-Scale Units

Single home units may be constructed very similar to those outlined in chapter 9 on sawdust culture. Either growing trays or plastic bags may be used (Fig. 11.39). On a small scale sterilization should not be attempted. Merely replace the medium with each new crop.

The simplest and most inexpensive method of growing herbs, lettuce and bedding plants hydroponically is the use of a plastic flat, "com-pack" filler tray and one-gallon (3.75-liter) nutrient bottle reservoir. A 10½-inch by 21-inch (27-cm by 53-cm) plastic flat with no drainage holes is used with a 48-compartment "com-pack" tray (Chapter 12). One of the eight sections (6 compartments) is removed at one corner and the nutrient tank is placed in that location. The nutrient tank can be a rigid polyethylene bottle with a large plastic cap through which a small hole is drilled and a split cork ring glued. The bottle is filled with nutrient solution and inverted into the flat. The hole in the cap allows solution to flow out of the bottle until the level in the flat reaches that of the hole; then, with the air supply cut off, the solution stops draining from the reservoir bottle.

Figure 11.39. Small "home units" about 2 feet long by 10 inches wide (60 cm by 25 cm) using a perlite-vermiculite medium. Ideal for use on windowsills or balconies of homes and apartments.

The compartments of the "com-pack" are filled with coarse vermiculite. Seeds are planted, covered with a shallow layer of additional vermiculite, then moistened with water alone. A black plastic cover may be placed on top of the tray for several days until germination takes place. This prevents desiccation of the seedlings during germination. Once the seeds have germinated, the cover is removed and the growing tray exposed to full light.

Such a nursery tray is meant to grow small plants such as herbs or bedding plants which will later be transplanted. However, the tray can be used to grow rapid crops of leaf lettuce. The lettuce must be harvested within several weeks of seeding when it reaches three to four inches (8 to 10 cm) in height (Fig. 11.40).

Figure 11.40. Hydroponic nursery tray for bedding plants, herbs (sweet basil) and lettuce. One-gallon nutrient reservoir.

11.11 Sterilization of the Medium

All the media mentioned in this chapter must be sterilized either chemically or by steam, as outlined in Chapter 8.

11.12 Advantages and Disadvantages of Peat Mixtures

Advantages:

1. Like sand and sawdust cultures, they are open systems, therefore there is less spread of diseases such as *Fusarium* and *Verticillium* wilts, especially in tomatoes.
2. No problem of plugging drainage pipes with roots.
3. Good lateral movement of nutrient solution throughout the root zone.
4. Good root aeration.
5. A new nutrient solution is added each irrigation cycle.
6. The system is simple, easy to maintain and repair.
7. The high water-holding capacity of the medium reduces risk of water stress should a pump fail.
8. It is adaptable to fertilizer injectors and therefore less space is required for storage tanks.
9. Peat, perlite and vermiculite are generally readily available in most regions of the world.

10. Sack culture enables a greenhouse operator to efficiently utilize vertical space for crops such as lettuce and strawberries which normally use a large amount of floor area. Therefore, a much greater number of plants can be grown in a given area of greenhouse.
11. Sack and column cultures keep plant parts and fruit off the underlying medium, thus reducing disease problems of fruit and vegetation.

Disadvantages:

1. The medium must be sterilized between crops by steam or chemically, which requires more time than in gravel culture. However, sterilization is very thorough.
2. Peat, pumice and vermiculite are more costly than sawdust in areas having a large forest industry.
3. Over the cropping season, salt accumulation can build up in the medium to toxic levels. Proper and regular leaching with pure water can overcome this problem.
4. Plugging of trickle feed lines may occur if proper filters are not used or if cleaning of these filters is neglected.
5. Since peat is organic in nature, it decomposes over time with continual cropping. Between crops it should be roto-tilled and additional peat must be added.
6. Perlite, pumice and vermiculite break down with continued use, resulting in compaction of the medium. For this reason, the peat mixtures are generally replaced between crops, resulting in replacement costs (of both medium and labor) each year.
7. If compaction occurs during the cropping period root, aeration will be greatly restricted, resulting in poor crop yields. Both the original mixture ratios and handling are important to prevent compaction.

In summary, peat mixtures are used extensively in container-grown plants. In beds other than ring culture, another medium such as sand or sawdust would be more suitable. In sack culture, peat or sawdust mixtures are most suitable because of their light weight. Of these cultures, perlite bags and foam slabs offer most potential for future commercial production of vine crops.

References

Broodley, J.W. and R. Sheldrake, Jr. 1964. *Cornell "Peat-Lite" mixes for container growing.* Dept. Flor. and Orn. Hort., Cornell Univ. Mimeo Rpt.

Cook, C.D. 1971. Plastoponics in ornamental horticulture. *Gardeners Chronicle/HTJ,* Oct. 7, 1971, Oct. 15, 1971.

Day, Desmond. 1991. *Growing in perlite.* Grower Digest No. 12. 35 pp. London: Grower Books.

Gerhart, K.A. and R.C. Gerhart. 1992. Commercial vegetable production in a perlite system. *Proc. 13th Ann. Conf. on Hydroponics.* Hydroponic Society of America, Orlando, FL, April 9–12, 1992, pp. 35–39.

Graves, C.J. 1986. Growing plants without soil. *The Plantsman*, May 1986, pp. 43–57.

Hartmann, H.T. and D.E. Kester. 1975. *Plant propagation principles and practices*, 3rd ed. Englewood Cliffs, N.J.: Prentice-Hall.

Irish peat modules for amateurs. *Gardeners Chronicle/HTJ*, Aug.10, 1973, p. 17.

Lucas, R.E., P.E. Riecke and R.S. Farnham. 1971. *Peats for soil improvement and soil mixes.* Mich. Coop. Ext. Ser. Bull. No. E-516.

Matkin, O.A. and P.A. Chandler. 1957. *The U.C. type soil mixes.* Section 5 in Calif. Agr. Exp. Sta. Man. 23.

Patek, J.M. 1965. Peat moss. *Amer. Hort. Mag.* 44: 132–41.

Sangster, D.M. 1974. *Soilless culture of tomatoes with slow-release fertilizers.* Ontario Min. of Agric. and Food Factsheet 73–057.

Sheldrake, R., Jr. and J.W. Broodley. 1965. *Commercial production of vegetable and flower plants.* Cornell Ext. Bull. 1056.

Sheldrake, R., Jr. and S. Dallyn. 1969. *Production of greenhouse tomatoes in ring culture or in trough culture.* Cornell Veg. Crops, Mimeo No. 149.

Tropea, M. 1976. The controlled nutrition of plants, II—a new system of "vertical" hydroponics. *Proc. 4th Int. Congress on Soilless Culture,* Las Palmas, pp. 75–83.

Chapter 12

Plant Culture

12.1 Introduction

While this book is an outline and review of soilless culture methods, it is appropriate to discuss briefly some of the methods and products available for the growing of seedlings to be transplanted into a soilless culture system. In order to successfully grow plants you must start out with vigorous and healthy disease-free seedlings. Tomatoes are used as an example. We shall then proceed to other aspects of successful plant cultures.

12.2 Seeding of Tomatoes

Three basic systems may be used for growing tomato transplants. One approach is to broadcast seed or sow it in rows in flats of a soilless mix. Either the U.C. or Cornell Peat-Lite mixes may be used (see Chapter 11). When the seedlings reach the cotyledon stage (Fig. 12.1), they are transplanted at wider spacing in other flats or in individual plant-growing containers.

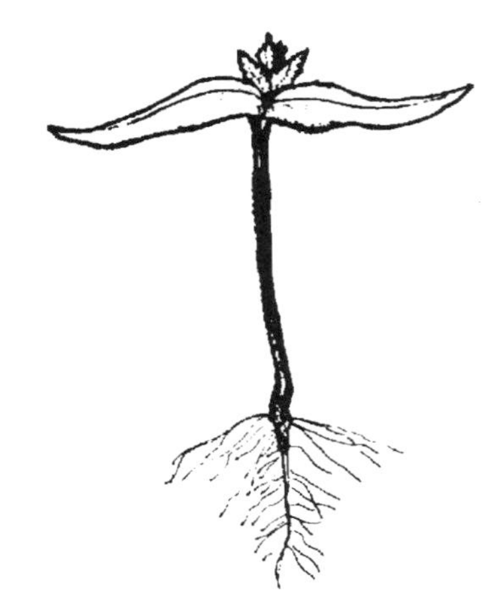

Figure 12.1. Seedling in cotyledon and early first true-leaf stage.

Another system is to seed directly into the growing flats or containers. Many types of containers are on the market. Plastic "com-packs" and "multi-pots" are shaped like ice-cube trays with from one large compartment to 12 compartments per pack (5¼ inches by 5¼ inches by 2⁵⁄₁₆ inches deep) (13 cm by 13 cm by 6 cm) (Fig. 12.2). These come in 8 packs per 10½-inch-by-21-inch (27-cm by 53-cm) sheet, which fits into a corresponding plastic flat. A similar product (Jiffy strips) is made of peat rather than plastic. Also, individual plastic and peat pots of various sizes are available. In all of these containers a soilless mix must be used. Most vegetable plants can be grown satisfactorily in 2¼- or 3-inch (6- or 8-cm) square or round pots. In the peat pots and strips, transplanting is done by placing the pot and medium into the beds. The plant roots will grow through the peat-container walls.

The third system is to seed directly into Jiffy peat pellets, Oasis "Horticubes," rockwool cubes, Kys cubes, or BR-8 blocks. The latter two are about 1½-inch (4 cm) paper-fiber cubes. Seeds are placed in small precut seeding holes and filled with vermiculite or a peat mix. Peat pellets are compressed peat discs contained by a nylon mesh. These discs are about 1½ inches (4 cm) in diameter and ¼ inch (0.6 cm) thick when shipped dry. After soaking in water for 5 to 10 minutes they swell to about 1½ inches (4 cm) in height (Fig. 12.2). Seeds are placed in the top and covered with their peat medium. They contain sufficient nutrients to carry most plants for 3 to 4 weeks. Plant roots grow out through the nylon mesh. During transplanting, the peat pellet (Jiffy-7) or paper cube (Kys cube, BR-8) with the plant is placed in the growing bed.

Oasis cubes, rockwool cubes and rockwool blocks (Fig. 12.2) are commonly used with NFT, rockwool and sawdust cultures. Use of rockwool blocks originated in Denmark and Sweden. They are the most widely used form of growing blocks for tomatoes, cucumbers and peppers.

Rockwool is composed of coke and limestone molten at 1600° C (2912° F) and spun into fibers. The fibers are woven into slabs, stabilized with phenolic resin and provided with a wetting agent. The most widely used rockwool block is manufactured under the trade mark "Grodan." Other manufacturers of rockwool products now exist in North America, Europe and Japan.

Chemically, rockwool blocks are relatively inert as the elements they contain are practically unavailable to plants. While they are slightly alkaline in *p*H, they are quickly neutralized after a short time of passing nutrient solutions of lower *p*H through them.

Rockwool blocks have good physical properties of a low volume weight, a large pore volume and a great water-retention capacity. Both rockwool cubes and Oasis blocks are often used with NFT systems. Both of these growing blocks are sterile, so that disinfecting measures are unnecessary.

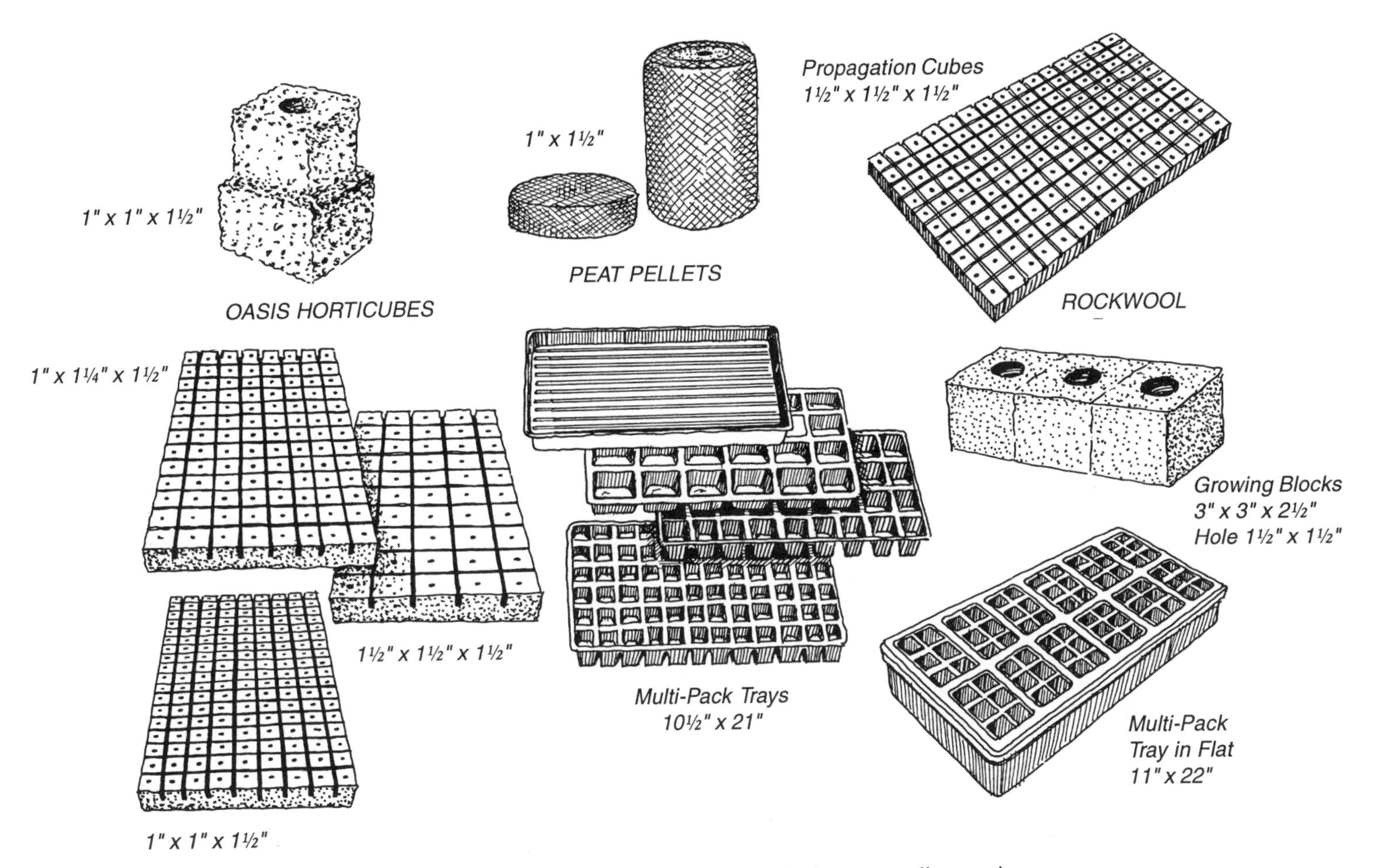

Figure 12.2. Seedling propagation cubes, blocks, peat-pellets and trays.

Rockwool blocks are available as propagation blocks and growing blocks. The propagation blocks are provided with a hole to facilitate placing cuttings into the blocks. The growing blocks are large enough to allow growing on by use of a "pot-in-pot" system. Seeds can be sown in propagation blocks which in turn can be placed into growing blocks after several weeks of growth. In this way, plants can continue growing in the same medium with a minimum of transplanting shock by means of the "pot-in-pot" system. Growing blocks are 10 centimeters by 10 centimeters (4 inches by 4 inches) surrounded by black polyethylene wrapping on the four sides. Rockwool blocks can be placed directly into NFT troughs. Rockwool blocks are completely dry, and therefore like peat pellets, must be properly moistened before seeding.

Oasis "Horticubes" are widely used in North America specifically for hydroponic and NFT applications. These cubes have good drainage, making them ideal for seed germination. The "Horticube" medium is available in 1 inch by 1 inch by 1½ inches (2.5 cm by 2.5 cm by 3.8 cm) blocks for maximum planting density. Each medium tray is composed of 162 cubes joined at their bases to allow easy separation for transplanting. Their main advantages are: they are sterile, easy to handle, and have a balanced *p*H.

They, like peat pellets, must be saturated with water prior to seeding. This can be accomplished in several ways: misting for one hour, soaking through hand submersion, or by overhead watering until the cubes are completely saturated. Seeds can then be placed in the pre-punched hole in each block. When roots appear on the outside of the cubes and true leaves develop, the plants are ready for transplanting. Since the plants are transplanted intact with their blocks, a minimum of transplant setback occurs.

Usually it is wise to sow at least two tomato seeds per pot or pellet, and as soon as the plants reach their cotyledon stage simply cut the less vigorous ones out at "soil" level with a scissors, leaving the desired number (one per pot or pellet for tomatoes). If this is not done at an early stage, the seedlings will compete for light and nutrients, resulting in spindly, thin-stemmed transplants. When automatic seeders are used, it may not be possible to sow more than one seed per cube. For European cucumbers, sow only one seed per pot, pellet or cube, as germination is very high (usually greater than 95%) and seeds are very costly.

A seedling house temperature range of 65° F (18° C) night minimum to 80° F (27° C) day maximum will provide good germination and growth of tomatoes. Optimum germination temperatures will vary with different plant species.

Plant tomato seed ¼ to ⅜ inch (0.6 to 1 cm) deep. Most large-seeded vegetables should be covered about ¼ inch (0.6 cm) deep. A good rule-of-thumb is to cover the seed about twice its greatest diameter. After watering the flats or containers, cover them with a sheet of clear plastic in order to prevent their medium from drying out during germination and emergence. This will save the labor of daily watering to maintain moisture. Clear plastic should not be used during hot weather unless shading from direct sunlight is provided. Otherwise, soil temperature under the plastic film may become too hot for proper germination. The plastic sheet must be removed at the first signs of seedling emergence.

Some growers use a low-pressure, intermittent mist system to provide uniform moisture levels for seed germination. The mist cycle should be 4 seconds every 6 minutes, from approximately 8:30 a.m. until 4:00 p.m., or it can be operated from an automatic moisture sensor. If mist is used, bottom heat should be supplied by steam pipes under the bench, or by imbedding electric heating cables under the flats and covering with 1 inch of gravel. As soon as seedlings emerge, flats should be removed from under the mist and grown under full light at the proper temperature.

Most commercial growers place the seeded cubes contained in trays in a special seedling section of the greenhouse. This area of the greenhouse has overhead sprinklers controlled intermittently from a time-clock or moisture sensor.

In northerly latitudes, during winter months when cloudy weather prevails, supplementary artificial lighting should be installed in the seedling house. Cool-white, high-output fluorescent tubes, metal-halide or mercury vapor lamps should be used in sufficient numbers to obtain at least 1,000 foot-candles of light intensity at plant surface. The lights should operate at least 14 hours per day for tomatoes. Lighting requirements, particularly day length (photoperiod), are very important for many plant species such as chrysanthemums and poinsettias; therefore, check out such factors before deciding on the type and amount of lighting needs.

Research by Dr. Gerald Wilcox of Purdue University (personal communication) has shown that wind action causes the development of a hormone within the plant that thickens its stem. Wind action can be duplicated in the greenhouse by use of blowers in the seedling area or by drawing a foam-rubber mat across the seedlings for one minute per day over a period of three weeks. The result is a shortening of the internodes which produces a shorter, stalkier plant that will be more productive at later stages.

12.3 Plant-Growing Temperature

Usually better-quality plants are grown when night temperature is 10° F (5.5° C) lower than day temperature. The best temperature range for warm-season crops is 60° F (16° C) night and 75° F (24° C) day. Cool-season crops will do better at 50° F (10° C) night and 60° F (16° C) day temperature. A 10-degree Fahrenheit (5.5-degree Celsius) lower day temperature is best during periods of cloudy weather. These ranges are only guidelines. Specific optimum minimum and maximum temperatures depend on species and even varietal differences. In tomatoes, if the day-night temperature differential is too great, long upright clusters form that will later kink and possibly break under fruit development.

If temperature is too low, plant growth will be slower and some purpling of leaves may occur, especially on tomatoes. If the temperature is too high, plant growth will be soft and "leggy," resulting in poor-quality plants.

The optimum temperature for the tomato plant varies with the stage of plant development (Table 12.1). Tomato plants grown under optimum temperature and light conditions develop large cotyledons and thick stems, with fewer leaves formed before the first flower cluster, greater number of flowers in the first, and often the second clusters, and higher early and total yields.

TABLE 12.1 Night and Day Temperatures from Seed Germination to Fruiting of Greenhouse Tomatoes

	Temperature	
Growth Stage	***Night***	***Day***
Seed germination	24°–26° C (75°–79° F)	24°–26° C (75°–79° F)
After emergence, until 1 week before transplanting	20°–22° C (68°–72° F)	20°–22° C (68°–72° F)
For 1 week	18°–19° C (64°–66° F)	18°–19° C (64°–66° F)
After transplanting, until start of harvesting	16°–18.5° C (61°–65° F)	21°–26° C (70°–79° F)
During harvesting	17.5°–18-5° C (63.5°–65° F)	21°–24° C (70°–75° F)

(To convert from temperatures in Fahrenheit to those in Celsius, use the following formula: (F − 32) × 5/9 = C ; where: F = degrees in Fahrenheit; C = degrees in Celsius.)

12.4 Watering

It is important that each watering soaks the medium thoroughly. Water should be applied uniformly over all plants, or uneven growth is likely to occur. Plants should be watered early enough so foliage will dry before dark. Remember that plants require less water during periods of dark, cloudy weather than during periods of bright, warm weather. As plants grow larger, they require more water. Normally, increasing day length and average temperature will increase water needs of plants.

Excess moisture may result in fast growth and poor fruit set in tomatoes, especially when combined with higher nitrogen levels and poor light conditions. The fertilizer program is most effective when moisture and temperature are controlled to fit available light conditions. Water requirements vary, depending on size of plant, day length, light intensity, and temperature.

Overhead irrigation systems for watering plants have been used by some growers. This type of watering saves considerable time, but it is difficult to obtain completely uniform distribution. Uniform watering is especially critical where plants are grown in small, individual containers or multi-packs.

Large greenhouse operations may use an overhead traveling-boom sprinkler irrigation system such as is used in the growing of forestry seedlings. In this way seedlings can be placed over the entire greenhouse area rather than leaving pathways for people to walk for manual watering. Also, moveable, rolling benches could be used. Inexpensive benching of concrete blocks, pipe, lumber and wire mesh would serve as a temporary seedling area. After the seedlings have been transplanted the temporary benching is removed and that section of the greenhouse is planted to a regular crop.

12.5 Light

During cloudy weather, tomato leaves become low in sugars. Leaves and stems become pale and thin and the fruit clusters may be smaller or fail to set. Excess nitrogen during such a period can be harmful. Supplementary artificial lighting has generally been found to be economically impractical except in the raising of the seedlings, as mentioned earlier.

During bright, sunny weather, sugar production in the leaves is high. The leaves are dark and thick, stems dark green and sturdy, fruit sets well with large clusters, and root systems vigorous. Nitrogen can be supplied at a heavier rate during this period.

When the weather is cloudy for more than 1 or 2 days, it may be necessary to:

1. Reduce day and night temperatures in the greenhouse 3° to 4° F.
2. Use as little water as possible but do not let the plants wilt.
3. Adjust the nutrient-solution formulation to increase the E.C.

These steps will help to keep plant growth balanced between leaf growth and fruit production. If they are not observed, the plants are likely to be very dark green and healthy, but are not likely to set many fruit.

During seedling stage, plants should be transplanted to wider spacing in flats (at least 2 inches (5 cm) apart) or individual containers at full cotyledon development or very early first-true-leaf stage (Fig. 12.1) to reduce mutual shading. In flats, plants should be set out in about 3 weeks to avoid spindly development. In individual small containers or growing blocks, plants can develop longer than 3 weeks, provided they are spaced out as they grow to avoid crowding. Excellent plants can be grown in 3- or 4-inch (7.6- or 10-cm) pots or blocks, but they require more nursery space than do small containers. Space sufficiently to prevent leaf overlapping during growth. Good containerized plants have pencil-size stems and are about as wide, from leaf tip to leaf tip, as they are tall (Fig. 12.3).

Figure 12.3. Vigorous, healthy tomato plants (5-6 weeks old) ready for transplanting. Plant on the left was grown in a 3-inch square peat pot, while the one on the right was grown in a Jiffy-7 peat pellet. Note roots emerging from the containers.

12.6 Carbon Dioxide Enrichment

In northern regions, carbon dioxide enrichment in greenhouses has improved productivity substantially. Commercial greenhouse operators claim a 20 to 30 percent increase in tomato yields; better fruit set in early clusters, especially when low light levels would normally reduce fruit setting; and larger fruits. In Ohio, cucumber yields have increased as much as 40 percent. Lettuce yields have increased 20 to 30 percent per crop, and faster growth rates have allowed production of an extra crop each year.

Carbon dioxide enrichment and supplementary artificial lighting is economically feasible to produce vegetable seedlings and bedding plants. These practices produce sturdier plants in much less time than conventional systems.

For maximum profits there is an optimum CO_2 concentration inside the greenhouse that depends upon the stage of growth and species of crop, as well as the location, time of year, and type of greenhouse. Generally levels at 2 to 5 times the normal atmospheric levels (1,000 to 1,500 ppm CO_2) may be taken as the optimum levels.

On cool days, when greenhouses are being heated and air vents are closed, plants will deplete the CO_2 in the greenhouse atmosphere. In one hour, plants in a closed greenhouse may lower the CO_2 concentration enough to significantly reduce growth rates. Air venting to maintain CO_2 concentration at outside levels would increase heating costs substantially. In these circumstances, special heaters using propane, natural gas, or fuel oil can supply CO_2 while simultaneously raising temperatures.

Carbon dioxide enrichment may be expected to increase fertilizer and water requirements, since plants will be growing more vigorously. Young tomato plants are especially responsive to CO_2 enrichment. Wittwer and Honma (1969) claim that growth rates can be increased by 50 percent and early flowering and fruiting accelerated by a week or 10 days. Effects are carried into the fruiting period. Not only is top growth and flower formation accelerated, but root growth is promoted even more. Carbon dioxide enrichment is of special significance in hydroponic culture since one of the sources of the gas, decaying organic matter in the soil, is not present.

12.7 Transplanting

Good quality transplants are essential to a good greenhouse crop. They must be set into the beds properly and cared for afterward to avoid any check in growth. For the least amount of transplanting shock, plants should be the right age and moderately hardened. Tomato or pepper plants should not have any fruit set.

Less transplant shock will be undergone by containerized plants than by bare-root plants because there is little or no disturbance to the root system. The transplanting shock of bare-root plants can be minimized, however, with proper planting and care.

When transplanting tomatoes to gravel medium, the crown area of the stem may be set below the bed surface up to 1 to 2 inches (2.5 to 5 cm) to provide better initial support for the plant. This also allows new roots to develop on the buried stem section. With leggy plants, 4 to 5 inches (10 to 12 cm) of stem may be buried on an angle. The above-ground portion will grow vertically from where it enters the medium. This, however, is not advisable for other media of higher water-retention, such as rockwool, as burial of the crown can result in rot diseases. Most other vegetables including cucumbers, peppers and lettuce must have their crowns positioned at the medium surface, not below it, or severe disease infection can occur in the crown area.

Irrigate the plants as soon as possible after they are set to avoid or minimize wilting. In gravel culture, the beds can be flooded while transplanting to maintain a high level of water in the coarse medium. Before planting, keep the containers or flats of plants moist. Properly set container plants will not wilt following transplanting, if their root balls are moist when planted and irrigation follows soon afterward. Bare-root plants may wilt after planting, even though irrigated immediately, but they will recover overnight.

12.8 Spacing

Most publications on greenhouse tomato production suggest space allowances of 3 to 4 square feet (0.28 to 0.37 square meters) per plant under soil culture. This is a population of 14,520 or 10,890 plants per acre (36,300 or 27,225 plants per hectare), respectively. Plantings as dense as 18,500 per acre (about 2 square feet per plant) or 46,250 per hectare (0.2 square meter per plant) have been used in hydroponic culture in California and Arizona with good yields and quality. Normal greenhouse spacing of tomatoes is about 3.5 square feet (0.3 square meter) per plant (12,000 plants per acre). Plants are spaced in double rows per bed. Plant rows should be 16 to 20 inches (40 to 50 cm) apart and plants 12 to 14 inches (30 to 36 cm) apart within each row. The precise spacing is a function of sunlight conditions. In more southerly areas having high light conditions, the closer spacing may be used, whereas, in the more northerly latitudes the wider spacing should be used. Plants may be placed in staggered positions in adjacent rows, in order to maximize the exposure of the leaves to sunlight, and to minimize physical interference of the leaves between adjacent plants.

Peppers are spaced similarly to tomatoes with planting densities of 8,000 to 10,000 plants per acre (20,000 to 25,000 plants per hectare). Rows are spaced 16 to 18 inches (40 to 46 cm) apart. Plant spacing within rows of 19 inches (48 cm) gives 8,000 plants per acre and at 14-inch (36-cm) centers provides 10,000 plants per acre.

European cucumbers require a minimum row width of 5 feet (1.5 meters). For the spring/summer crop, the plants require 7 to 9 square feet (0.65 to 0.84 square meter) for each plant, for the fall crop 9 to 10 square feet (0.84 to 0.93 square meter). This is a population of 3,500 to 5,000 plants per acre or 8,750 to 12,500 plants per hectare. Plants can be placed in single rows 14 to 16 inches (35.5 to 41 cm) apart within the row. The plants are then tied alternately to overhead wires spaced 2½ to 3 feet (0.76 to 0.9 meter) apart so that the plants are inclined away from the row on each side. They will then be growing in a V-arrangement down the row, allowing light to fall more uniformly on the plants.

12.9 Feeding and Watering

Once the plants have been transplanted, the feeding-irrigation system cycles must be set. The time between irrigation cycles depends on a number of factors, as discussed earlier, including the type of growing medium used. In sand and sawdust culture several irrigations per day will be sufficient while gravel culture requires watering every few hours during the daylight period. A typical watering program for a trickle gravel-culture system is given in Table 12.2.

TABLE 12.2 Annual Watering Scheme for Tomatoes, Cucumbers, Lettuce

Date	*Plant Stage*	*Cycle Frequency Per Day*	*Cycle Duration*
Feb. 1–15	Transplants	2 hr., 8:00 a.m.–5:00 p.m.	15 min.
March 1	Seedlings	3 hr., 7:00 a.m.–6:00 p.m.	15 min.
March 31	Mature Plants	2 hr., 7:00 a.m.–7:00 p.m.	15 min.
April 15	Mature Plants	2 hr., 7:00 a.m.–7:00 p.m.	15 min.
May 15–July 15	Mature Plants	1 hr., 7:00 a.m.–7:00 p.m.	15 min.
July 15	Transplants	1 hr., 7:00 a.m.–8:00 p.m.	15 min.
July 31–Sept. 1	Mature Plants	2 hr., 7:00 a.m.–7:00 p.m.	15 min.
Sept. 1–Oct. 1	Mature Plants	2 hr., 8:00 a.m.–6:00 p.m.	15 min.
Oct. 1–Jan. 1	Mature Plants	3 hr., 8:00 a.m.–5:00 p.m.	15 min.

Note: These timings are only approximate. Plants should be watched closely, especially during the long, hot summer days and after transplanting. If any wilting of the tops of the plants occurs, it is likely that they need to be irrigated more frequently. During winter months, particularly just after transplanting, care should be taken not to overwater. Water only several times a day.

After transplanting, irrigate once every 2 hours during the daylight hours for 15 minutes each time for the first two weeks until the plants take hold, that is, until the roots start growing out of the peat pellets and into the gravel. This condition will be expressed as vigorous leafy growth. After this initial hold period, regular irrigation cycles similar to those outlined in Table 12.2 should be followed. If the plants wilt at any time during the day, increase the frequency of irrigation cycles.

12.10 Plant Support

Plants such as tomatoes, peppers and cucumbers which are trained vertically must be strung and clamped. Use plastic twine and plastic clips. The string should be tied to support cables at gutter height of the greenhouse (8 to 10 feet) (2.5 to 3 meters) directly above the plants. Leave about 6 feet (2 meters) extra length of the string at the cable end so that the plants can be lowered once they have grown as high as the cable.

A more recent method of stringing is especially useful with one-year single crops of tomatoes. Extra string is wound on a thick wire hook having a flat spool shape (Fig. 12.4). Using this method, plants can easily be lowered with the additional string available. The string may be attached to the bottom of the plant by use of a plant clip, or simply by tying a loose loop around the stem below a healthy leaf. The string is then wound around the stem of the plant, always in the same direction (generally clockwise), and occasionally a plant clip may be placed under a leaf petiole to give additional support (Fig.12.5).

Figure 12.4. Wire spools with a hook on one end are used for extra support string during lowering of plants.

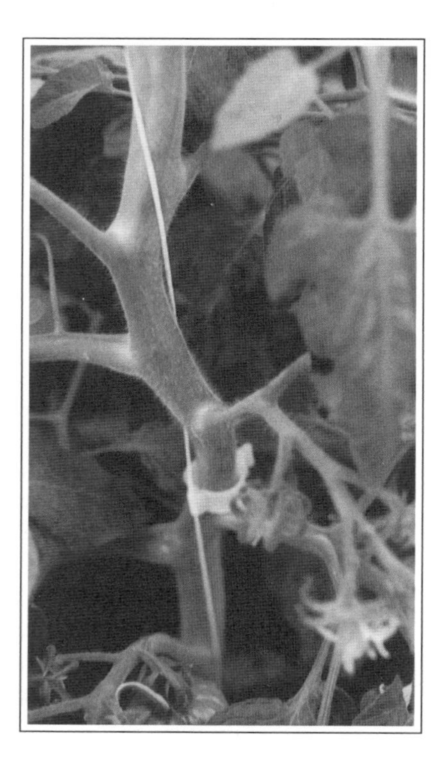

Figure 12.5. Support string wound around tomato plant. Several plastic plant clips are attached for additional support.

The plants must be clamped to the string before they fall over. Secure the first clamp under a pair of large leaves pulling the twine taut, but not so tight that when released it will pull the leaves of the plant up. Clamping should be done as shown in Figure 12.6.

The clamps should be placed directly under the leaf petioles, not above them, as such a position gives no support. Clamps should not be placed under flower clusters as the weight of maturing tomatoes later may break off the fruit cluster, or fruit may be punctured by the clamps. Clamps

Figure 12.6. The use of plastic plant clips to support plants vertically. In the photo on the right, receptive flowers at front have their petals curled back. Note the flowers which have closed have been pollinated.

should be secured along the plant every foot or so to give adequate support as fruit matures. The back (hinge) of the clamp must pinch the string, as shown in Figure 12.6.

12.11 Suckering and Trimming (Tomatoes, Cucumbers and Peppers)

Suckers are the small shoots which grow between the main stem and the leaf petioles. They must be removed before they grow too large as they drain the nutrients of the plant which could otherwise go into fruit development. They should be removed when they are about 1 to 2 inches (2.5 to 5 cm) long (Fig. 12.7). At this stage they can be easily snapped off by hand without creating a large scar of the axillary region (area between stem and petiole), as shown in Figure 12.8.

Removing suckers by hand rather than using a knife presents less hazard of spreading disease. Rubber or plastic gloves can be worn to protect your hands from the acidic sap of the plants. Larger suckers which may develop due to late suckering will have to be removed with pruning shears or a knife.

Often, with tomatoes, terminated plants (those no longer having a growing point) will be encountered. Select a vigorous sucker near the growing

Figure 12.7. Suckers or side shoots on tomato plants are removed while they are 1 to 2 inches long. Note that the small fruit on the cluster above has been pruned to select the most uniform four fruit.

point and allow it to continue growing, but remove other less vigorous suckers. Some plants may fork or split. Again, select the most vigorous branch and prune off the other growing point (plant apex).

As the plant matures and the fruit has been harvested from the lower trusses, the older leaves on the bottom of the plant will begin to senesce (yellow) and die. These leaves should be removed to allow better air ventilation and thus lower the relative humidity around the base of the plants. These leaves should be removed from the time the second truss (fruit cluster) has been completely harvested. After that, continue removing leaves which turn yellow, up to the truss presently bearing mature fruit. Simply snap off the leaves with your fingers to get a clean break to mini-

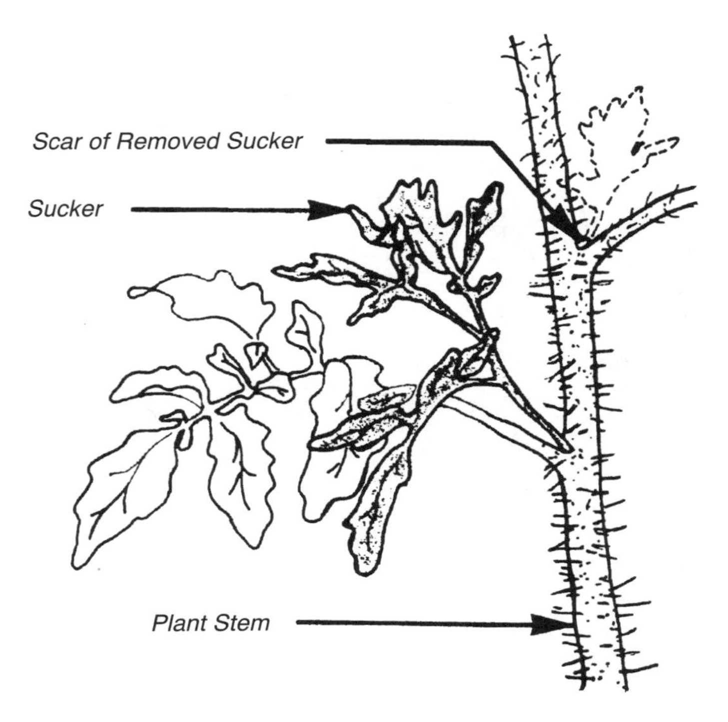

Figure 12.8. Removal of tomato suckers at an early stage.

mize scars. Remove all the dead leaves from the greenhouse and dispose of them. This leaf pruning can be repeated several times as the plants mature, but never prune out green leaves, as they nourish the maturing fruit. Generally, no more than 3 to 4 leaves should be removed at any time and not more frequently than once per week.

As the tomato plants reach the support cables above, loosen or unwind the support strings and lower the plants 1 to 2 feet (0.5 meter) each time. Since the lower leaves and fruit have been removed, the stems can be twisted or looped along the top of the bed or on support stakes or wires as described in Chapter 10. If care is not taken, stems may break. At all times, about 4 to 5 feet (1.2 to 1.5 meters) of foliage and fruit clusters should remain on the upper part of the plant (Fig. 12.9).

Flower trusses on tomatoes should be pruned to select the most uniform four- to five-set fruit on a truss (Fig. 12.7). Any misshapen flowers, double set fruit, and often the furthest flower out on the truss are removed. This gives uniform fruit development, shape, size, and color of tomatoes. Blossoms and small fruit should be removed as soon as two or three fruit set to pea-size.

With many of the new varieties, which set fruit heavily, as the fruit develop on the truss the weight of the fruit causes the truss to buckle or

Figure 12.9. Removal of lower leaves from tomatoes and lowering of the stems. Four to five feet of foliage and fruit clusters should remain on the upper part of the plants.

break resulting in lost production. Recently, a truss-support clip has been developed to prevent such breakage. This clip is an inexpensive plastic flexible support that easily snaps onto the truss stem (Fig. 12.10). It is attached early after fruit set (Fig. 12.11).

Long English (European) cucumbers are grown in greenhouses since they can be trained vertically and cannot withstand the temperature fluctuations they would encounter under outside conditions.

European cucumbers should be trained under two systems: the renewal umbrella system (Fig. 12.12), and the V-cordon system (Figs. 9.29, 12.13). A V-cordon system trains the plants to make best use of available light within the greenhouse. Two support wires are placed 6½ to 7 feet (about 2

Figure 12.10. Plastic truss support clip attached to fruit cluster.

Figure 12.11. Attach truss supports to fruit clusters early after fruit set.

(Courtesy of Houweling Nurseries Ltd., Delta, B.C., Canada.)

Figure 12.12. Renewal umbrella system of training European cucumbers.

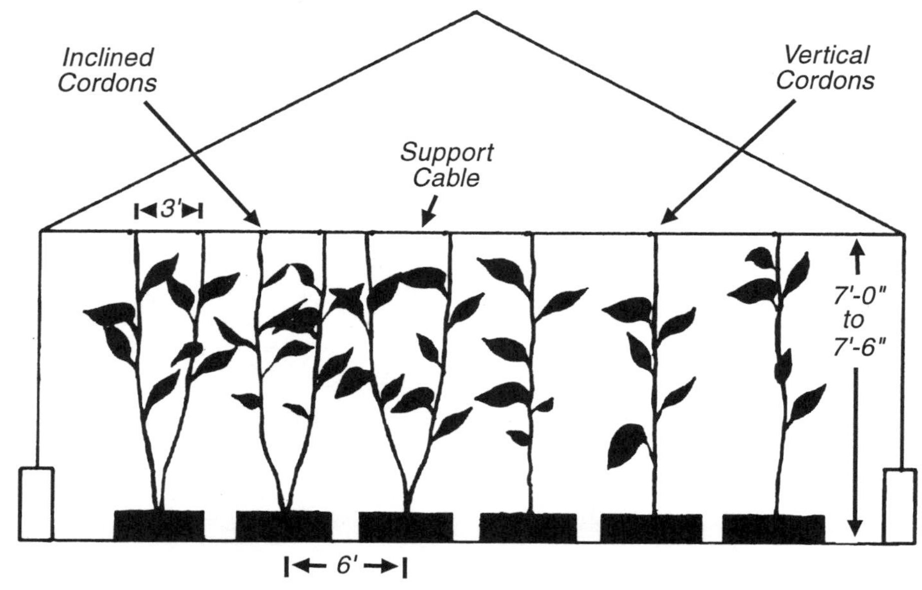

Figure 12.13. V-cordon system of training European cucumbers.

meters) above the floor, or at gutter height when using carts to work on the plants, over each row of plants. The wires are spaced 2½ to 3 feet (0.75 to 1 meter) apart. The support strings are then tied alternately to the two overhead wires so that the plants are inclined away from the row on each side (Figs. 9.29, 12.13). Suckering and trimming of cucumbers is necessary to obtain a balance between the vegetative vigor and fruit load of each plant. Suckers, and also all blossoms and fruit, should be removed up to the fifth true leaf.

If too many fruits are allowed to form at any one time, a large proportion will abort because the plant may not have sufficient food reserves to develop them. If a heavy load of fruit sets, often malformed or poorly colored fruits develop which are unmarketable. Remove them at an early stage. Multiple fruits in one axil should be thinned to one.

The renewal umbrella system of training plants is achieved by the following steps (Loughton 1975):

1. The main stem should be stopped at one leaf above the support wire. Pinch out the growing point at that level. Tie a small loop of string around the wire and below the top leaf so that the plant will not slide down the main string.
2. Do not allow fruit to develop on the main stem up to about 4 feet.
3. Remove all laterals in the leaf axils on the main stem, except 2 at the top.
4. Train the top 2 laterals over the wire to hang down on either side of the main stem. Allow these to grow to two-thirds of the way down the main stem.Training the main stem and laterals over the support

wire can damage the plant if care is not taken. To overcome this problem, a plastic stem-support which attaches to the stem of the plant and hangs from the support wire has been developed (Fig. 12.14).

5. Remove all secondary laterals, except 2 at the top.
6. While the fruit on the first laterals are maturing, allow the second laterals to grow out and downward.
7. When the fruits on the first laterals have been harvested, remove them completely, allowing the second laterals to develop.
8. Repeating steps 5, 6 and 7 will maintain fruit production.

Some growers carry a cucumber crop for 10 months using this renewal umbrella system and obtain over 100 cucumbers per plant.

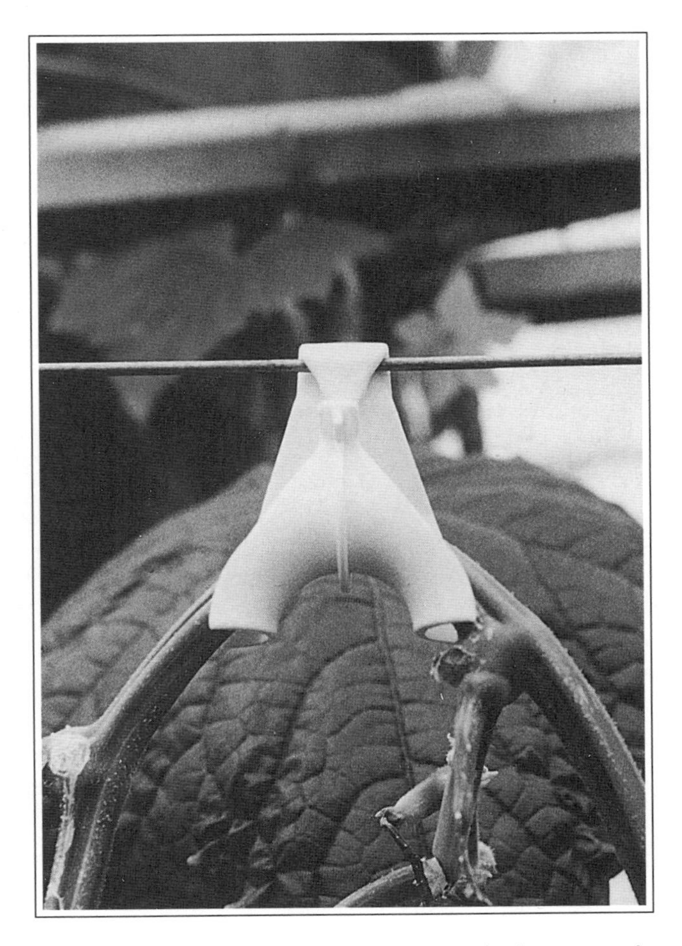

Figure 12.14. Plastic stem support attached to cucumber main stem as it reaches the overhead wire. (Courtesy of Houweling Nurseries Ltd., Delta, B.C., Canada.)

Peppers are trained as two stems per plant in a V-cordon system supported by strings attached to overhead wires (Figs. 10.51, 10.52).

While the pepper plant starts out as a single stem, it soon bifurcates to form two stems and then continues to do so producing additional stems. Flower buds form at these division points on the stem (Fig. 12.15). The flower in the first division of the stem is called a "crown" bud. Most greenhouse peppers are trimmed to two stems. Additional stems must be pruned to maintain a balance between vegetative growth and fruit development. To encourage vigorous initial vegetative growth capable of supporting later fruit production, flowers must be removed from the first and second stem layers (bifurcations) generally to a height of 16 inches (40 cm). After two stems have formed remove all sideshoots after the second leaf (Fig. 12.16). This suckering is necessary every several weeks. The support strings should be wound around the main stems in a clockwise direction every two weeks (Fig. 10.51). Plastic stem-support clips may also be used as was described for tomatoes. Peppers grow slower than tomatoes or cucumbers. Generally, one crop per year is grown.

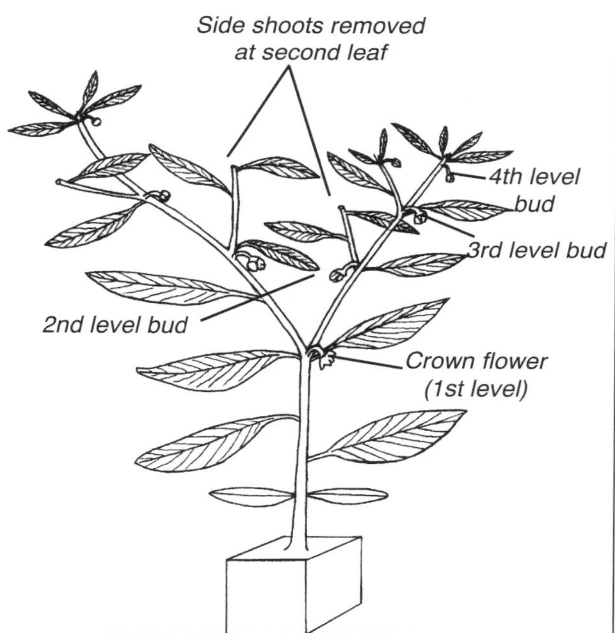

Figure 12.15. Position of flower buds and training of pepper plants in their early growth.

Figure 12.16. Training of peppers into two stems and removal of additional side shoots at the second leaf axis.

12.12 Pollination

Tomatoes are normally wind pollinated when grown outside. In greenhouses, however, air movement is insufficient for flowers to pollinate themselves. Vibration of the flower clusters is essential for good pollination in a greenhouse. This can be done by tapping the flowers with a stick, your fingers, or an electric vibrator such as an electric toothbrush which has had its bristles removed. The vibrator is momentarily held lightly against the flower-cluster branch. If environmental conditions are favorable and the flower is receptive, the fine yellow pollen can be seen flowing from the flower upon vibration.

Pollination must be done while the flowers are in a receptive state. This is indicated by their petals curling back (Figs. 12.6, 12.11). Plants should be pollinated at least every other day, since blossoms remain receptive for about two days. Pollinating should be done between 11:00 a.m. and 3:00 p.m. under sunny conditions for best results. Research has shown that relative humidity of 70 percent is optimum for pollination, fruit set, and fruit development. High humidity keeps pollen damp and sticky, except around midday, and lessens the chances of sufficient pollen transfer from anthers to stigma. Too-dry conditions (relative humidity less than 60 to 65 percent) cause desiccation of pollen.

Greenhouse temperatures should not fall below 60° F (15° C) at night nor exceed 85° F (29° C) during the day. At higher or lower temperatures, pollen germination and pollen tube growth are greatly reduced. Chemical growth regulators can be used to induce fruit development under lower than optimum temperatures, but these fruits are usually seedless. Open locules and thin outer walls may make the fruit soft and greatly reduce their quality.

Prolonged dark cloudy weather retards pollen development and germination, resulting in poor fruit set. Even if weather conditions are not ideal, pollination must be done at least every other day during the early afternoon to assist fruit set.

If pollination has been done correctly, small bead-like fruit will develop within a week or so (Figs. 12.7, 12.11). This is called fruit set. When young plants produce their first trusses, pollinate each day until fruit set is visible. It is important to get the fruit set on these first trusses as it throws the plant into a reproductive state which favors greater flower and fruit production as the plant ages. After the first few trusses have set, pollinating can be done every other day.

In the past, pollination of tomatoes was done by use of vibrators. Pollination of a 5¼ -acre (2.1-hectare) greenhouse tomato crop (see Gipaanda Greenhouses, Chapter 10) took two people full time. Research has shown that brushing the plants when moving down the aisles or while pollinating reduces yields up to 8 percent. With the use of bumble bees (*Bombus sp.*), now the accepted form of pollination of greenhouse tomatoes, a minimum of a 3% increase in yield is achieved according to Gipaanda Greenhouses Ltd. It is important to maintain the correct population levels of bumble bees as overpopulation may result in the bees overworking the tomato flowers. This is particularly true with peppers which require a lower population so they will not aggressively enter the blossoms and as a result cause a zipper-pattern scar on the fruit.

Bumble-bee hives, costing about $350 each, are available from a number of suppliers (see Appendix 5) (Fig. 12.17). The population of one hive is sufficient to work one-half acre (0.2 hectare) of tomatoes. A 66% sugar-water (by weight) solution with a preservative, provides food for the bees as the tomato flowers do not provide nectar. A supply reservoir sits above the hive (Fig. 12.18). The bees form a round hive inside the box container (Fig. 12.19). A plexiglass plastic top under the reservoir allows the grower to visually observe the progress of the bee population (Fig. 12.19). A slide stopper is positioned across the entrance to the hive to contain the bees when examining the inside of the hive.

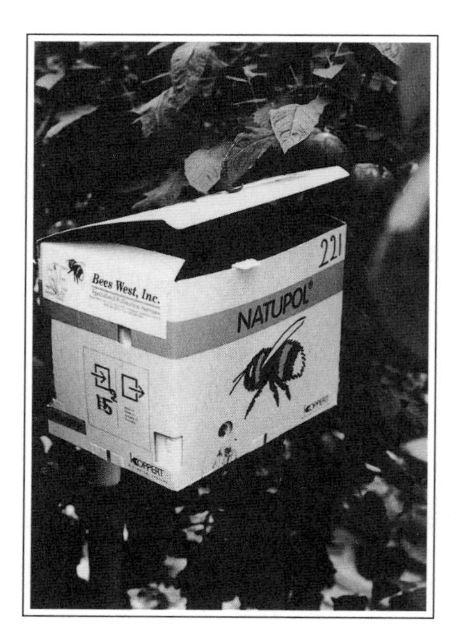

Figure 12.17. A commercial container of bumble bees for pollination of tomatoes.

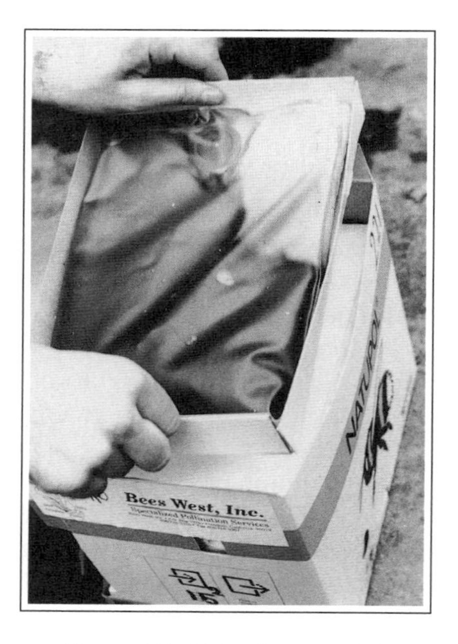

Figure 12.18. In the top of the container is a reservoir of sugar solution to feed the bumble bees.

Life expectancy of hives is generally guaranteed for 2 months, but on the average, they survive 3 months and up to 8 months. The activity of the hive and the percentage bruising of the corolla (all petals together) should

Figure 12.19. Bumble bees form a round hive inside the top of the container.

be monitored. A sampling of blossoms at post-receptive (folding-up of petals) stage should show 90–100% bruising. This indicates the number of flowers worked by the bees. Examination of the sugar-water consumption every 2 to 3 days should reveal that it is being depleted at a constant or increasing rate.

European cucumbers, unlike the regular North American seeded cucumbers, do not require pollinating to set fruit and the fruits are therefore seedless. Pollination may occur by the presence of bees and from male flowers on the same or neighboring plants in the greenhouse. Pollination causes seeds to form, and the fruits become clubbed at the end and develop a bitter taste. To prevent pollination, bees must not be allowed to enter the greenhouse, and male blossoms should be removed from the plants as soon as they develop. Now, all-female and gynodioecious (less than 10 percent male blooms) varieties have been developed so that male blossoms seldom—if at all—develop. In this way self-pollination is reduced, if not eliminated.

12.13 Physiological Disorders

Hydroponics has many advantages but it does not free the grower from the need for alertness in dealing with many physiological disorders common to all forms of food production. Physiological disorders are those fruit-quality defects caused by undesirable temperatures, faulty nutrition, or improper irrigation. Some varieties are more susceptible than others to some of these disorders. Because we are using tomatoes, peppers and cucumbers as examples, we will present as a reference source several disorders common to those fruits.

1. *Blossom-end rot (tomatoes, peppers).* This disorder appears as a brown, sunburned, leathery tissue at the blossom end of the fruit (Fig. 12.20). In the early stages, the affected area will have a green, water-soaked appearance. While the cause of blossom-end rot is a low supply of calcium in the fruit, the indirect cause is plant stress. This stress may be due to: (a) low soil moisture, (b) excess soluble salts in the growing medium, (c) high rates of transpiration, and (d) high soil moisture, which leads to poor root aeration. Prevention involves avoiding these conditions.
2. *Fruit cracking (tomatoes, peppers).* Symptoms are cracks radiating from the stem, almost always on maturing fruit, at any time during ripening. This is usually caused by a water deficit and excessively high fruit-temperatures followed by a sudden change in moisture supply to the plants. Prevention is by avoidance of high fruit-temperatures and maintenance of uniform soil (medium) moisture conditions.
3. *Blotchy ripening (tomatoes).* This is uneven coloring of the fruit wall in the form of irregular, light green to colorless areas with brown areas in the vascular tissue inside the fruit. It is associated with low light intensity, cool temperatures, high soil moisture, high nitrogen, and low potassium. It can be avoided under low light intensity conditions by less frequent applications of irrigation and fertilizer (especially nitrogen).
4. *Green shoulder, sunscald (tomatoes, peppers).* These disorders are associated with high temperature or high light intensity. Avoid removing leaves which offer protection to fruit clusters during the spring and summer months when sunlight is intense, and keep greenhouse temperatures down.
5. *Roughness and catfacing (tomatoes).* This is radial wrinkling of the fruit shoulders and walls and fruit-shape distortion due to protuberances and indentations (Fig. 12.21). This is caused by poor pollination and environmental factors such as low temperatures and high relative humidity which cause the flower parts to develop abnormally.
6. *Crooking (cucumbers).* This is excessive fruit curvature. It can be caused by a leaf or stem interfering with the growth of the young fruit or the sticking of a flower petal on the spines of a leaf stem or another young fruit. Adverse temperature, excessive soil moisture, and poor nutrition have also been suggested as causes. Remove severely curved fruit from the plant immediately.

Figure 12.20. Blossom-end-rot of tomatoes showing leathery, brown tissue at the blossom end of the fruit.

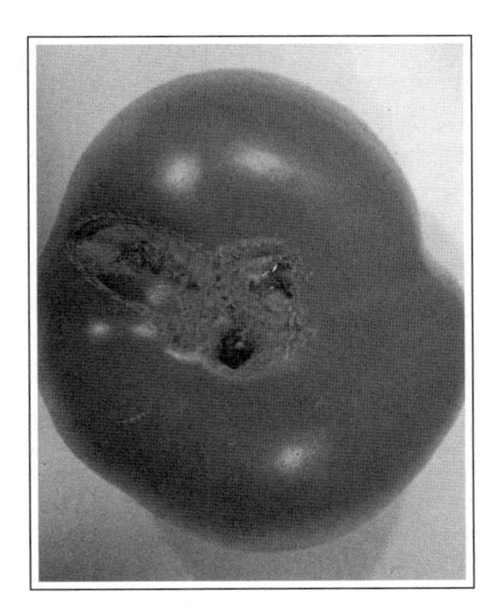

Figure 12.21. Catfacing of tomato.

12.14 Diseases and Insects

Since the control of diseases and insects is very specific, only a brief description of the most common problems are given here.

Some common tomato diseases:

1. *Leaf mold (Cladosporium).* This starts as a small gray spot on the lower side of the leaf and spreads until a definite pale area also develops on the upper surface (Fig. 12.22). Additional infection sites develop and the initially small spots expand. The basic control measure is sanitation within the greenhouse and careful ventilation and temperature control to prevent high humidity. Some fungicide sprays can be helpful.
2. *Wilt (Fusarium and Verticillium).* Initially plants wilt on hot days, then eventually they wilt all the time and the leaves become yellow. If the plant is cut just above the soil surface, a darkened ring is visible inside the outer green layer of cells. No spray or cultural treatment will control these diseases. They can only be controlled through sterilization of the growing medium.

Greenhouse growers of tomatoes in British Columbia lose about 10–15 percent of their plants annually to *Fusarium* wilt when growing in soilless culture. Experimental work supports the possibility of reducing infection through intercropping several lettuce plants in each row with the tomatoes. Apparently, lettuce gives off an exudate that slows the activity of *Fusarium* in the medium and thereby reduces infection of tomatoes. As discussed earlier (Chapter 10), grafting of resistant rootstocks and development of new resistant varieties, such as "Trust" and "Apollo" to root and crown rots, is preferred as lettuce promotes aphid infestation.

3. *Early blight* and *leaf spot (Alternaria* and *Septoria).* These cause discolored or dead spots on the leaves. Early blight has dark rings on a brown background. Leaf spot has small black dots on the affected area. Both organisms attack the oldest leaves first and cause severe defoliation of the lower part of the plant. Proper ventilation and removal of lower senescing leaves to improve air circulation and reduce relative humidity helps reduce the disease.

4. *Gray mold (Botrytis).* Under high humidity conditions, fungal spores infect wounds such as leaf scars, and a watery rot and fluffy gray growth forms over the affected area. The disease may develop along the stem for several inches and eventually girdle it, killing the plant. Proper ventilation to reduce the relative humidity will prevent the spread of the disease. Remove infected plants immediately. Affected areas at early stages of infection can be scraped and covered with a fungicide (Ferbam) paste or the plants can be sprayed with Ferbam.

5. *Virus (Tobacco mosaic virus, TMV).* Several viruses will attack tomatoes. TMV is the most common, causing distortion of the leaves and stunting of growth with resultant yield reductions (Fig. 12.23). Plant juice containing

Figure 12.22. Leaf mold of tomato plant. Lower side of leaf showing pale-colored areas and brown lesions.

Figure 12.23. TMV infected tomato leaf. Distorted leaf pattern and mosaic appearance.

the virus is spread by sucking insects or on the hands or tools of the people working in the crop. Sanitation, control of sucking insects and no smoking (TMV is present in tobacco) in greenhouses will help to avoid infection. Presently, an antibody type of spray has been used successfully. The plants are inoculated with an attenuated (mild) strain of TMV which offers cross-protection against infection by the severe strain. This use of TMV protection is no longer required as most commercial tomato varieties have TMV resistance bred into them.

Some common cucumber diseases:

1. *Powdery mildew.* Small snowy white spots initially appear on the upper surface of the leaves. They quickly spread in size and to other leaves (Fig. 12.24). Proper sanitation and adequate ventilation are the primary control measures. Chemical control is also possible.

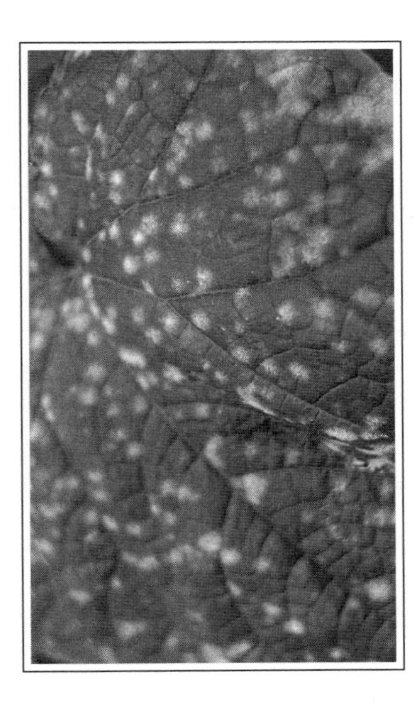

Figure 12.24. Powdery mildew of cucumber with white spots (lesions) on upper leaf surface.

2. *Cucumber mosaic virus (CMV).* Some of the same strains of virus found on tomatoes also infect cucumbers. The affected leaves are dwarfed or become long and narrow. No control measure other than prevention through sanitation is known.
3. *Wilt (Fusarium).* This is the same wilt disease as found commonly in tomatoes. Proper sterilization of the growing medium between crops and sanitation are the only control measures.
4. *Gray Mold (Botrytis).* Again, the symptoms and control measure are the same as for tomatoes.

5. *Gummy Stem Blight (Didymella bryoniae).* Tan-colored lesions appear on petiole stubs and base of main stem. Also infects flowers and developing fruit, causing shriveled fruit at flower end and internal browning of tissue. Avoid condensation on plants and guttation by adequate ventilation. Do not leave stubs when pruning, and disinfect pruning knives. Remove crop debris and bury away from greenhouse. Apply fungicide sprays of Rovral 50 WP or Benlate 50 WP and Manzate 200 at 4- or 7-day intervals, respectively, when infection appears or during high humidity, low light levels. Do not apply within 5 days of harvest. Follow rate and application directions of fungicide label.

Insects:

Biological control of insect pests is now widely accepted in the greenhouse industry. A number of companies (Appendix 2) sell these biological agents. Biological control refers to the use of living organisms to control other living pest organisms.

There are a number of advantages to using biological control agents. Pests cannot build up resistance against biological agents as easily as they can with pesticides. The cost of biological control is less than with the use of pesticides. There is no fear of phytotoxicity or persistence of chemicals potentially hazardous to human health. Generally, biological control does not completely eliminate the pest, as a certain level of pest population is necessary to sustain the predator population. For this reason, emphasis is on integrated control, using biological-control agents with cultural- and chemical-control measures. The objective is not complete pest elimination, but pest management whereby pest populations are maintained at a level below any significant plant damage.

With an integrated pest-management program, only certain pesticides must be used which do not harm the predatory agents. However, spot treatment of highly infested areas may be controlled by other more poisonous pesticides if pest populations cannot be controlled by the biological agents. Once such infestations are reduced, the use of these chemicals should be restricted. Such pesticides as insecticidal soaps, insect-growth regulators or hormones, yellow sticky traps and bacterial or fungi insecticides may be used with the biological agents without harming them. The grower must check with government agencies and the companies selling biological agents to determine which pesticides may be used without harming the specific predators in their integrated pest-management program. For example, if a program of mite control using the predatory mite *Phytoseiulus persimilis* has been introduced into the greenhouse, outbreaks of the two-spotted mite may be controlled using fenbutatin-oxide (Vendex), which is harmless to the predatory mite. Yellow sticky traps (Fig. 6.38) may be used to trap adult whiteflies. These traps are available commercially or may be made from pieces of stiff cardboard painted a bright yellow and coated with Vaseline, Vaseline/mineral oil mix, or other sticky material.

1. *Whitefly (Trialeurodes vaporariorum).* The whitefly life cycle is four to five weeks during which time it undergoes a number of molts in its nymph stage as shown in Figure 12.25. This is the most common pest in the greenhouse tomato crop. It is usually located on the undersides of the leaves. When at rest on a leaf, its triangular white body makes it easy to identify. The insect secretes a sticky substance on the leaves and fruit in

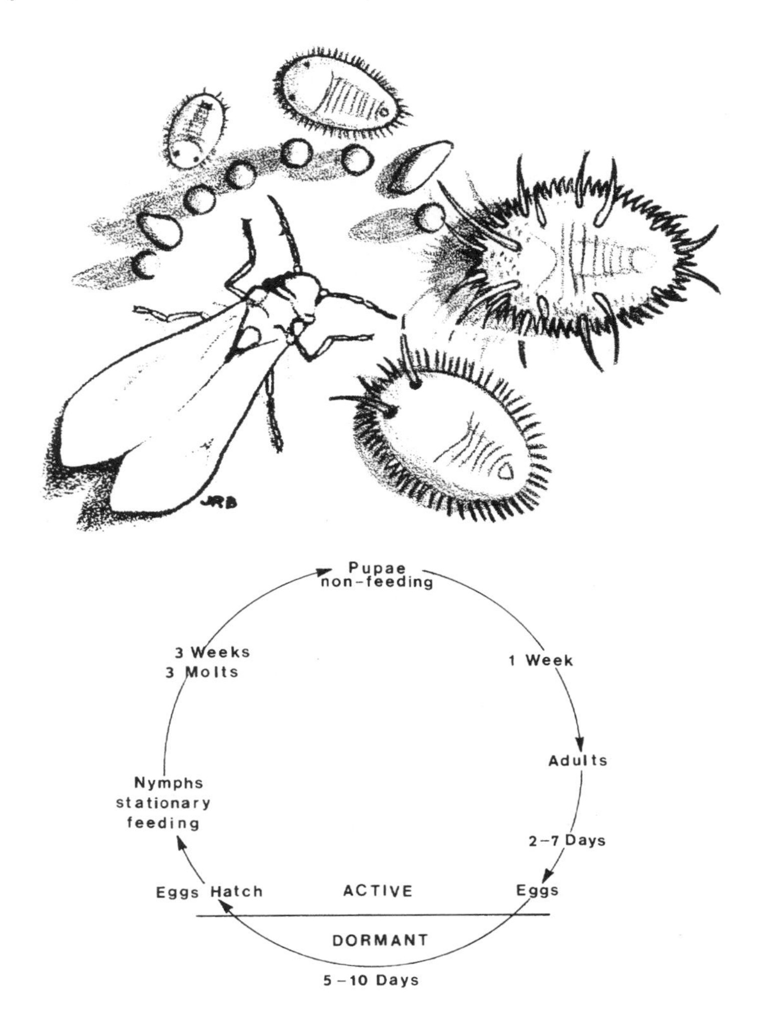

Figure 12.25. Life cycle of whitefly. (Insect drawings courtesy of J. R. Baker, North Carolina Agricultural Extension Service, Raleigh, NC.)

which a black fungus later grows, making it necessary to clean the fruit before marketing. Several pesticides such as parathion, pyrethrin and thiodan can be used for control. However, the insects quickly build up resistance, making it necessary to use different pesticides to obtain reasonable control. Whitefly is also a common pest on cucumbers and peppers.

Biological control may be achieved by use of the chalcidoid wasp *Encarsia formosa.* This parasite is available from biological-control-agent suppliers. Adult female wasps lay eggs in the whitefly larvae and also eat early larval stages of the whitefly. The parasitic grub feeds within the larva, which turns black within two weeks providing an easy method of determining the success of the *Encarsia.* Successful control is temperature- and humidity-dependent, as the wasp reproduces best at a mean temperature of 23° C (73° F) and a relative humidity not exceeding 70 percent. *Encarsia* are purchased as pupae stuck to paper strips (Fig. 12.26) which are hung on plants in the greenhouse. Each card contains the black pupae from which the parasite emerges. As soon as the first whiteflies are found, a program of repeated introductions of the wasp must be initiated.

To use *Encarsia formosa* successfully, follow these steps:

a. No residual pesticides should be used for a month prior to introduction.

Figure 12.26. Paper containing Encarsia *pupae attached to a leaf petiole of a tomato plant.*

b. Existing whitefly populations should be reduced, using insecticidal soap or insect hormones, to an average of less than one adult per upper leaf.
c. Adjust temperature and humidity to 23–27° C (73–81° F) and 50–70 percent respectively.
d. Introduce *Encarsia* at the rate of 10/square meter (1/square foot) of planted area or 1–5/infested plant.
e. Reintroduce *Encarsia* weekly for up to nine introductions.
f. Monitor plants weekly for the number of whitefly and black scales.

Several parasitic fungi, *Verticillium lecanii* and *Achersonia aleyrodis,* are being tested commercially for their effectiveness in controlling whitefly. They are safe to use with *Encarsia formosa.*

2. *Two-spotted spider mite (Tetranychus urticae).* Mites are closely related to spiders and ticks, having four pair of legs, unlike insects which have three pair of legs. Their life cycle goes through a number of nymph stages, as indicated in Figure 12.27. Their life cycle is from 10 to 14 days, depending upon temperature. At 26° C (80° F) the cycle is shortened to 10 days, whereas, at lower temperatures the cycle may be up to 2 months. Low relative humidity also favors their development. Misting plants frequently during dry periods will discourage spider mites, as they dislike high humidity.

This pest is particularly common on cucumbers. A webbing appearance on the undersides of leaves indicates an already heavy infestation. A magnifying glass is needed to observe the mite closely in order to see the dark-colored spot on each side of its body. The mite causes a yellowing of leaves starting as very small yellow pin-sized dots which coalesce to eventually form a very characteristic bronzed appearance (Fig. 12.28). Severe infestations will cause leaves to become entirely bleached as the mites suck the contents of the leaf cells, leaving a dead shell.

Mites live on plant refuse and on the greenhouse framework between crops. A thorough eradication and sterilization between crops is necessary. Clear the greenhouse of all plant material and fumigate or spray with appropriate chemicals. If populations are very large they will spread to tomatoes, lettuce, chard and almost all vegetables and flowers. Control by chemical miticides such as "Pentac," "Vendex," and "Mavrik" are effective only against adults.

A predatory mite, *Phytoseiulus persimilis,* is available commercially. *Phytoseiulus* differs from the two-spotted mite by its lack of spots, pear-shaped body, longer front legs, and especially its rapid movement when disturbed. It prefers temperatures in the range of 21–27° C (70–81° F). Adults develop from eggs in less than a week. This is twice as fast as their prey. However, since bright light and high temperatures are unfavorable

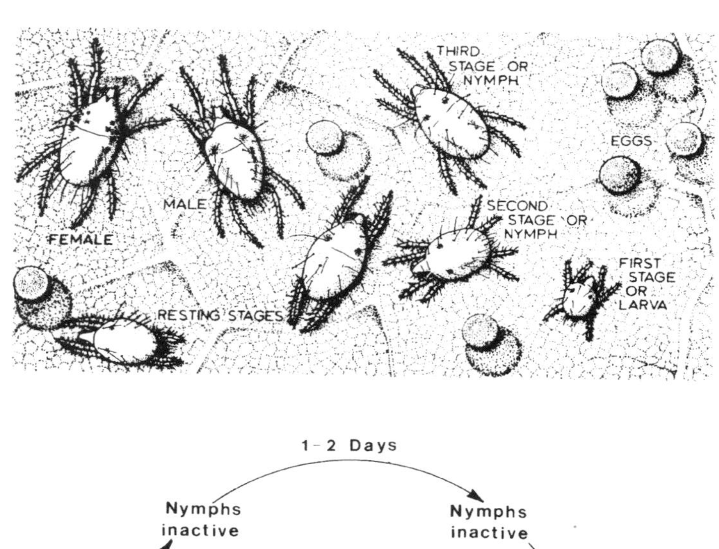

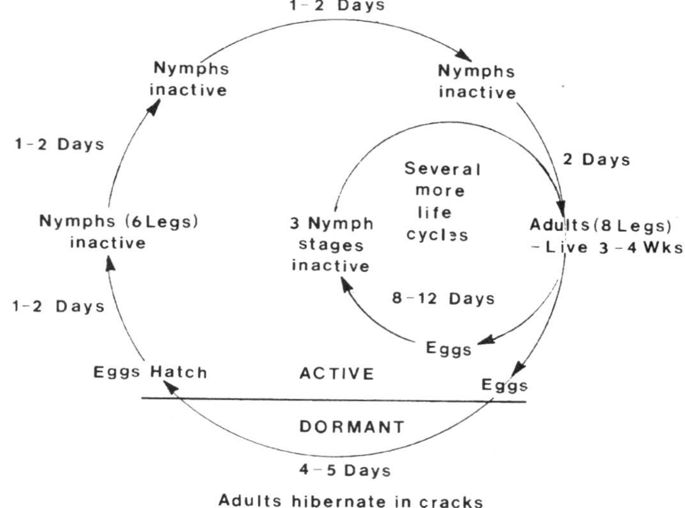

Figure 12.27. Life cycle of two-spotted spider mite. (Insect drawings courtesy of J. R. Baker, North Carolina Agricultural Extension Service, Raleigh, NC.)

to their development, such hot conditions should be avoided, since these conditions also favor spider mites.

Other predatory mites are available: *Metaseiulus occidentalis,* which performs well under cooler temperatures, and *Amblyseius californicus,* which prefers warmer temperatures.

These predators are applied mainly to cucumbers and peppers but may also be used on tomatoes, beans, gherkins, melons, grapes, strawberries, and various flower crops.

To use *Phytoseiulus persimilis* or others:

a. For one month prior to introduction, do not use residual pesticides.
b. Introduce the predator at the first sign of spider-mite damage. If

Figure 12.28. Spider-mite infestation of cucumber leaf. Severe infestation causing yellowing of leaves as pin-sized dots coalesce.

there is greater than 1 mite/leaf, reduce the population with insecticidal soap or fenbutatin-oxide (Vendex), until only 10 percent of the leaves are infested.

c. Maintain optimum temperatures for the predator and high relative humidity.
d. Generally, 8–10 predators/square meter (80–100/sq ft) should be introduced. Release in early morning onto the mid and upper foliage. Predator mites are available in shaker bottles.
e. Monitor spider-mite populations once per week.
f. Reintroduce the predator at monthly intervals. Populations of predators should be roughly maintained at one predator for every five spider mites. Good control should be achieved within four to six weeks.

3. *Aphids:* Green peach aphid *(Myzus persicae),* foxglove aphid *(Alacorthum solani),* and potato aphid *(Macrosiphum euphorbiae).*

The life cycle varies in length from 7–10 days to three weeks depending upon temperature and season, as shown in Figure 12.29.

Aphids are usually clustered in large colonies on new succulent growth, at the base of buds, and on the underside of leaves. The most common greenhouse species is the green peach aphid. The wingless forms are yellowish green in summer and pink to red in the fall and spring. The winged forms are brown. They have a pear-shaped body 1–6 millimeters in length with four wings, if winged. They excrete "honeydew" from their abdomen, which is a food for ants. The presence of large ant populations

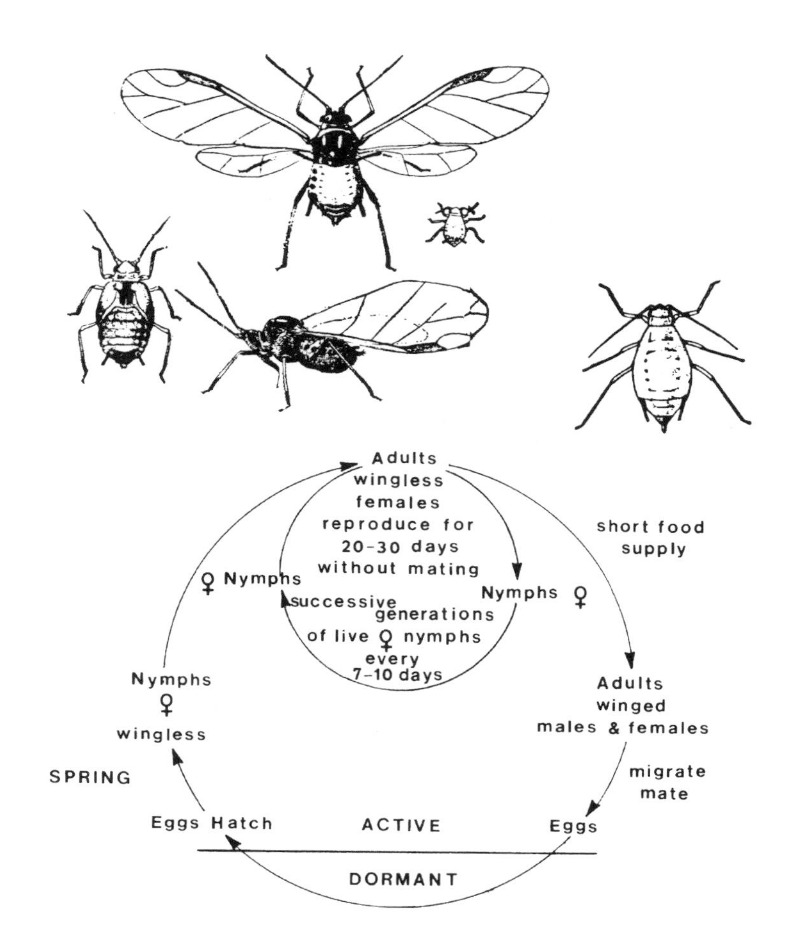

Figure 12.29. Life cycle of aphid. (Insect drawings courtesy of J. R. Baker, North Carolina Agricultural Extension Service, Raleigh, NC.)

on plants often indicates aphid infestation. Aphids feed by sucking out plant sap with their tubelike, piercing mouthpart. This causes distorted leaves when they feed on young leaf buds. Under short food supply, winged females appear and migrate.

A number of types of aphids including pink, black and dark green feed on most greenhouse vegetables. They suck the juices out of the plant and cause the leaves to become distorted and sticky from a honeydew deposit. They can be controlled with a weekly spray program using chemicals such as pyrethrin, Malathion and thiodan.

Some biological control can be achieved using various lady beetles and the green lacewing, *Chrysopa carnea.* A predaceous midge larva, *Aphidoletes aphidimyza,* is being marketed in Finland and Canada. The adult midge, a small black, delicate fly lives only a few days. The female lays 100–200 eggs on the underside of leaves close to aphid colonies, which hatch in three to four days. A large orange or red larva, up to three millimeters long matures in three to five days and drops to the ground where it forms a cocoon. Pupation takes about 10–14 days completing a total life cycle of three weeks.

The threshold temperature for larval development is approximately 6°C (43°F), with optimum temperatures between 23° and 25°C (73°–77°F) at 80–90 percent relative humidity. Adult midges feed on aphid honeydew excretions. The larva feeds on aphids, killing from 4–65 aphids each.

To use *Aphidoletes aphidimyza:*

a. Control ants which may protect the aphids.
b. Avoid using residual pesticides one month prior to introduction.
c. Reduce heavy aphid populations with Enstar (insect-growth regulator) or insecticidal soap. Areas of at least 10 aphids/plant or one aphid/square centimeter should be left to encourage egg laying by the midge.
d. Maintain temperatures between 20°–27° C (68°–81° F).
e. Introduce the predator at a rate of 1 pupa/3 aphids or 2–5 pupae/square meter (2–5 pupae/10 sq ft) of planted area.
f. Spread the pupae in shady areas near aphid-infested plants and repeat introductions in 7–14 days and thereafter as necessary.
g. Monitor plants for infestation weekly.

A parasitic fungus, *Verticillium lecanii,* trade name Vertalec, is available commercially as a biological-control agent of aphids.

Spot spraying in localized infestations may be done with chemical pesticides such as insecticidal soap, Enstar and Pirimor, but avoid spraying the *Aphidoletes* larvae. Reintroduction of the predatory larvae will be necessary in areas sprayed with chemical pesticides.

4. *Tomato leaf miner (Liriomyza bryoniae)* and the American serpentine leaf miner *(Liriomyza trifolii). Liriomyza bryoniae* is common on tomatoes, especially in the tropics. Adult flies are yellow-black in color and about two millimeters in length. The female deposits its eggs in the leaf, causing a small white puncture-protuberance. When the larvae hatch they eat "tunnels" through the leaf between the upper and lower leaf epidermis creating "mines" (Fig. 12.30). These tunnels may coalesce resulting in large areas of damage until the entire leaf dries up. The maturing larva drops from the leaves to the ground where it pupates. Within 10 days the adult emerges, the whole cycle from egg to adult fly takes three to five weeks, as shown in Figure 12.31.

The leaf miner has become resistant to many chemical pesticides. The leaf miner of lettuce in Venezuela *(Liriomyza huidobrensis)* is very resistant to most pesticides. While a number of biological-control agents have

Figure 12.30. Leaf-miner damage to tomato plant with distinct "mines" or "tunnels" appearing between the leaf epidermal tissues.

been identified and are presently available commercially for leaf miner of tomato and flower crops, there is no effective control agent for the lettuce leaf miner. The two insects, *Dacnusa sibirica* and *Diglyphus isaea* parasitize both leaf-miner species *Liriomyza bryoniae* and *L. trifolii*. A third, *Opius pallipes,* only parasitizes the tomato leaf miner. *Opius* and *Dacnusa* lay eggs in the leaf-miner larva. As the leaf-miner larva pupates, the parasite emerges instead of the leaf miner. *Diglyphus isaea* kills the leaf miner in its tunnel and lays an egg beside it. The wasp (parasite) develops in the tunnel, feeding on the dead larva.

Leaf samples must be taken from the crop and examined in a laboratory to identify the species of leaf miner, parasites present, and level of parasitism. If there are insufficient natural parasites present, *Dacnusa* or *Diglyphus* are introduced. The introduction depends on the season, the species of leaf miner, and the level of infestation.

5. Thrips (Heliothrips haemorrhoidalis) and *flower thrips (Frankliniella tritici* and F. *occidentalis).* Thrips *(Heliothrips haemorrhoidalis)* is becoming a problem on cucumbers. Adults are attracted to the yellow cucumber flower. Adult thrips are 0.75 millimeter long with feathery wings. They develop outside the greenhouse on weeds and invade the greenhouse. Feeding on leaf undersides, growing tips and flowers, they cause small, bleached dead spots on leaves and damage growing tips and flowers. Nymphs with rasping mouth parts scrape the leaf surface and suck the plant sap, causing a white, silvery, discoloration resulting in streaks. Thrips feed like spider mites by puncturing and sucking the leaf contents. Damage appears as narrow crevices and a silvery appearance on leaves. They feed in narrow crevices between the calyx and the newly forming fruit of the cucumbers, causing curled and distorted fruits. Thrips similarly damage peppers.

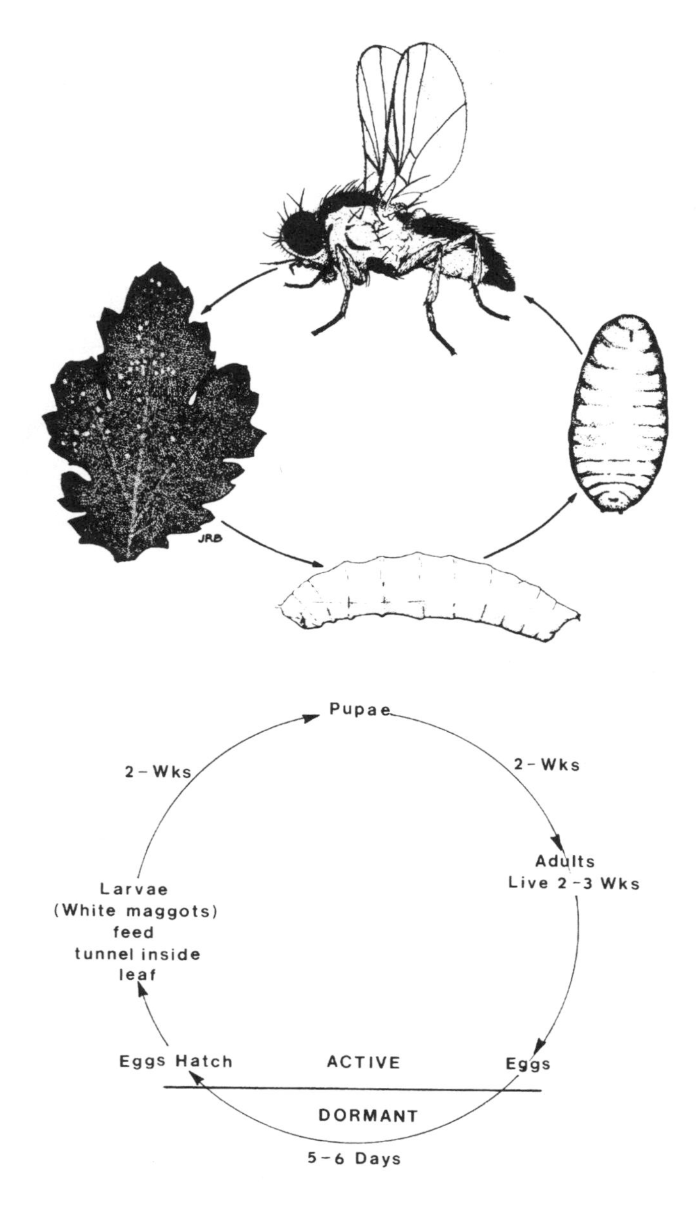

Figure 12.31. Life cycle of leaf miner. (Insect drawings courtesy of J. R. Baker, North Carolina Agricultural Extension Service, Raleigh, NC.)

Their two- to three-week life cycle begins with the adult female depositing eggs under the leaf surface (Fig. 12.32). After four days they hatch into nymphs which feed on the leaves for three days before molting into larger more active nymphs that feed for another three days before dropping to the ground to pupate, and emerge as adults within two days. They feed for about six days before beginning to lay from 50 to 100 eggs over 40 days.

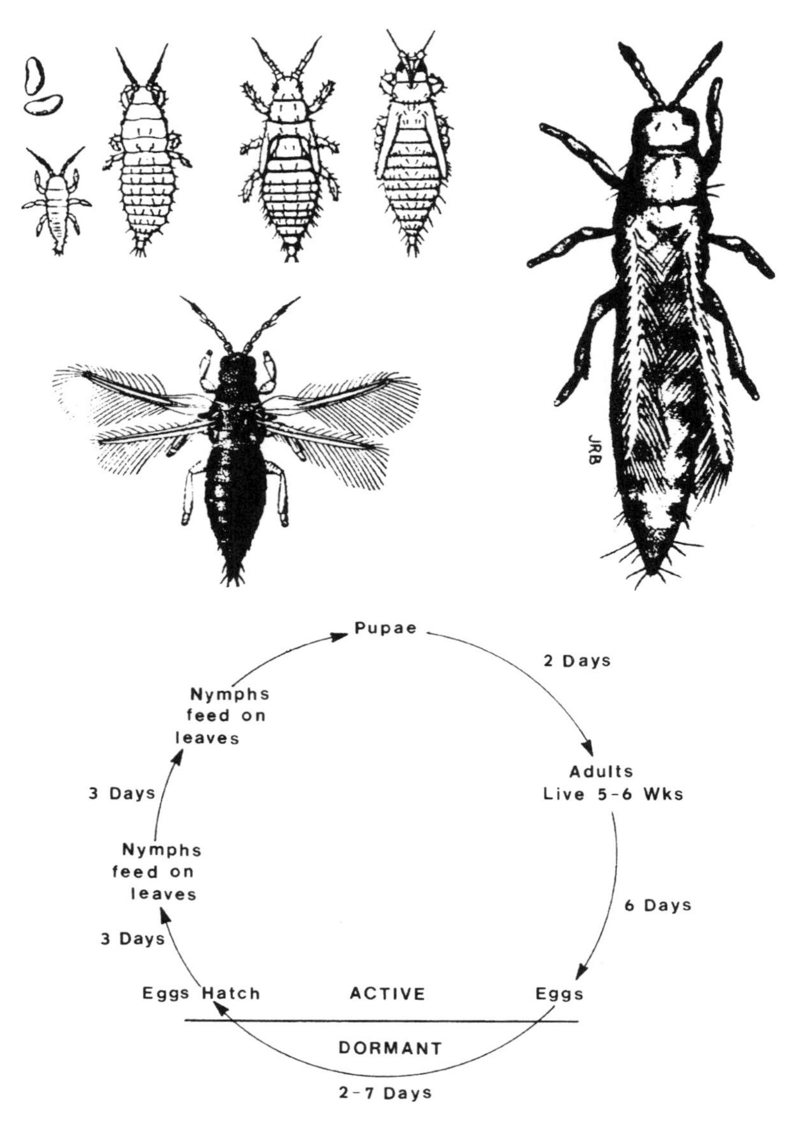

Figure 12.32. Life cycle of thrips. (Insect drawings courtesy of J. R. Baker, North Carolina Agricultural Extension Service, Raleigh, NC.)

Yellow sticky traps should be used for early detection and monitoring of the pest population. *Amblyseius cucumeris,* a predatory mite, is available commercially to control thrips. The predatory mite is similar in appearance to *Phytoseiulus,* differing only in its lighter pale-pink color and shorter legs. Its life cycle is similar to *Phytoseiulus.* The predator mite must be introduced in the crop at an early stage for preventive action so that a large population can be built up in order to control the thrips as soon as they appear. Their introduction and management is similar to that of *Phytoseiulus.* They should be introduced on a weekly basis until population levels reach 100 per plant. *Orius tristicolour,* a pirate bug, controls thrips in peppers and cucumbers. This predator will also feed on pollen, aphids, whitefly and spider mites when thrip populations are low. *Hypoaspis miles* feeds on thrips' pupae. It is suitable for cucumbers, tomatoes and peppers.

6. *Caterpillars and cutworms.* Caterpillars are larvae of butterflies, and cutworms are larvae of moths. These are common in most greenhouse crops. The larvae feed on aerial plant parts. Their presence is indicated by notches in leaves, and cut stems and petioles.

Cutworms climb up the plant and feed on the foliage during the night and are found in the soil or medium in the day. Caterpillars are not nocturnal like cutworms but feed on above-ground plant parts day and night.

Adult moths and butterflies fly into the greenhouse from outdoors and quickly lay eggs on the plants, where they hatch into feeding larvae within several days during high temperatures. Entry of adults should be prevented by using screens over shutters, vents, etc. The time of their life cycle varies with season, temperature, and the species (Fig. 12.33).

Control may be achieved by use of a number of effective chemical pesticides such as Lannate, Diazinon, Malathion, etc., as well as biological control with a parasitic bacterium, *Bacillus thuringiensis,* marketed as Dipel or Thuricide. This bacterium must be sprayed on a regular weekly basis, as new growth occurs to protect all surfaces. It is active only upon ingestion by the caterpillar or cutworm. The larvae are paralyzed so they stop eating a few hours after spraying. They die within one to five days. The bacterium is harmless to mammals, fishes, birds, and leaves no toxic residue in the environment. *Trichogramma evanescens,* a small wasp, effectively controls over 200 species of larvae by laying its eggs in the eggs of the butterflies and moths.

7. *Fungus gnats (Bradysia* species and *Sciara* species). The larvae of these small, dark grey or black flies feed normally on soil fungi and decaying organic matter, but as populations increase, they attack plant roots. They are legless white worms about six millimeters (¼ inch) in length with a black head. Adults have long legs and antennae, about three milli-

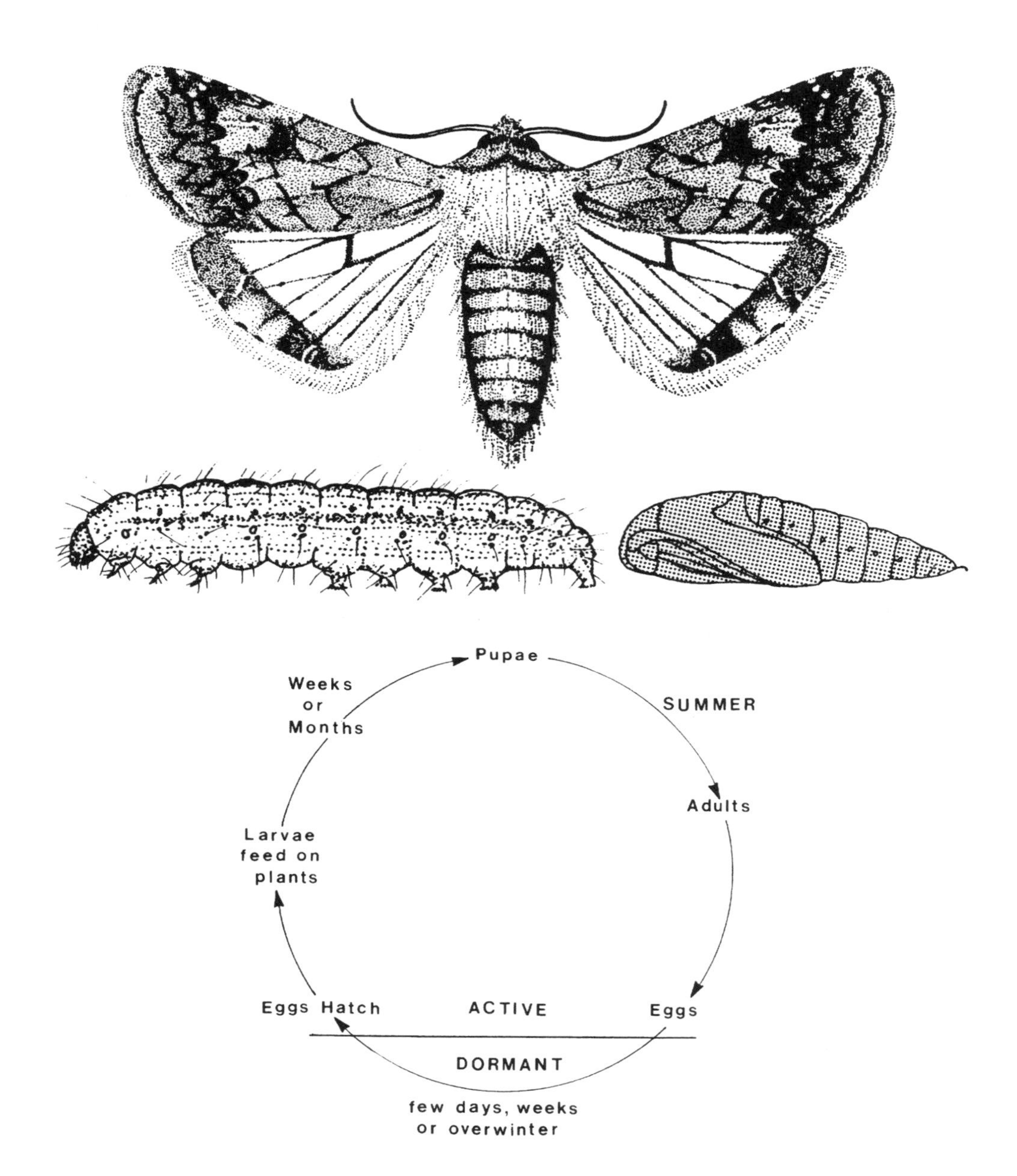

Figure 12.33. Life cycle of caterpillars and cutworms. (Insect drawings courtesy of J. R. Baker, North Carolina Agricultural Extension Service, Raleigh, NC.)

meters long, with one pair of clear wings. They attack all seedlings and are favored by the presence of moisture such as exists with capillary mats and the growth of algae. They also feed on the tap roots, stem cortex, and root hairs of mature cucumbers. They have a four-week life cycle as shown in Figure 12.34.

Control by use of yellow sticky traps and chemical pesticides such as Diazinon are effective. Also, avoid moist areas in the greenhouse; keep medium surfaces dry. Some success has been achieved using a predatory mite which feeds on the eggs and pupae in the soil. *Bacillus* bacterium may provide some control of the larvae. Vectobac, produced by Abbott Laboratories, is an effective *Bacillus* subspecies against fungus gnats. The predatory mite, *Hypoaspis miles,* feeds on fungus gnat eggs and small larvae. An insect-parasitizing nematode, *Steinernema carpocapsae*, controls fungus gnats by entering body openings in the larvae. The nematodes are mixed with water and applied as a drench, spray, or through the irrigation system.

In general, life cycles of insects can be shortened under optimal temperature and relative-humidity conditions. Control of insect pests must be carried out at the most susceptible stage of their life cycle. This is usually the adult or active nymph or larval stages that are feeding. For successful integrated pest control using biological agents, the predator and prey populations must be in balance. Use of selective chemical pesticides for control of localized population outbreaks of the pest is necessary. Before using any chemical pesticides, the grower must check with manufacturers, the supplier of biological agents and/or agricultural extension personnel to determine which may be used safely. Environmental conditions within the greenhouse must be maintained at levels favorable to the predator. Weekly monitoring of predator-prey populations is needed to maintain equilibrium and new introductions of predators made to achieve successful integrated pest management.

Natural insecticides are being introduced for use in all types of IPM programs. One such product is the active ingredient azadirachtin, an extract from the neem tree. It is marketed as Azatin or Neemix and is effective against whiteflies, aphids, thrips, leaf miners and fungus gnats.

Hydroponic culture minimizes pests and diseases in the growing medium by efficient sterilization between crops. The pests and diseases of the aerial part of the plant, however, are not influenced by hydroponic culture. Therefore, if proper preventive programs are not followed, severe infestations may occur similar to those in regular soil-culture field conditions.

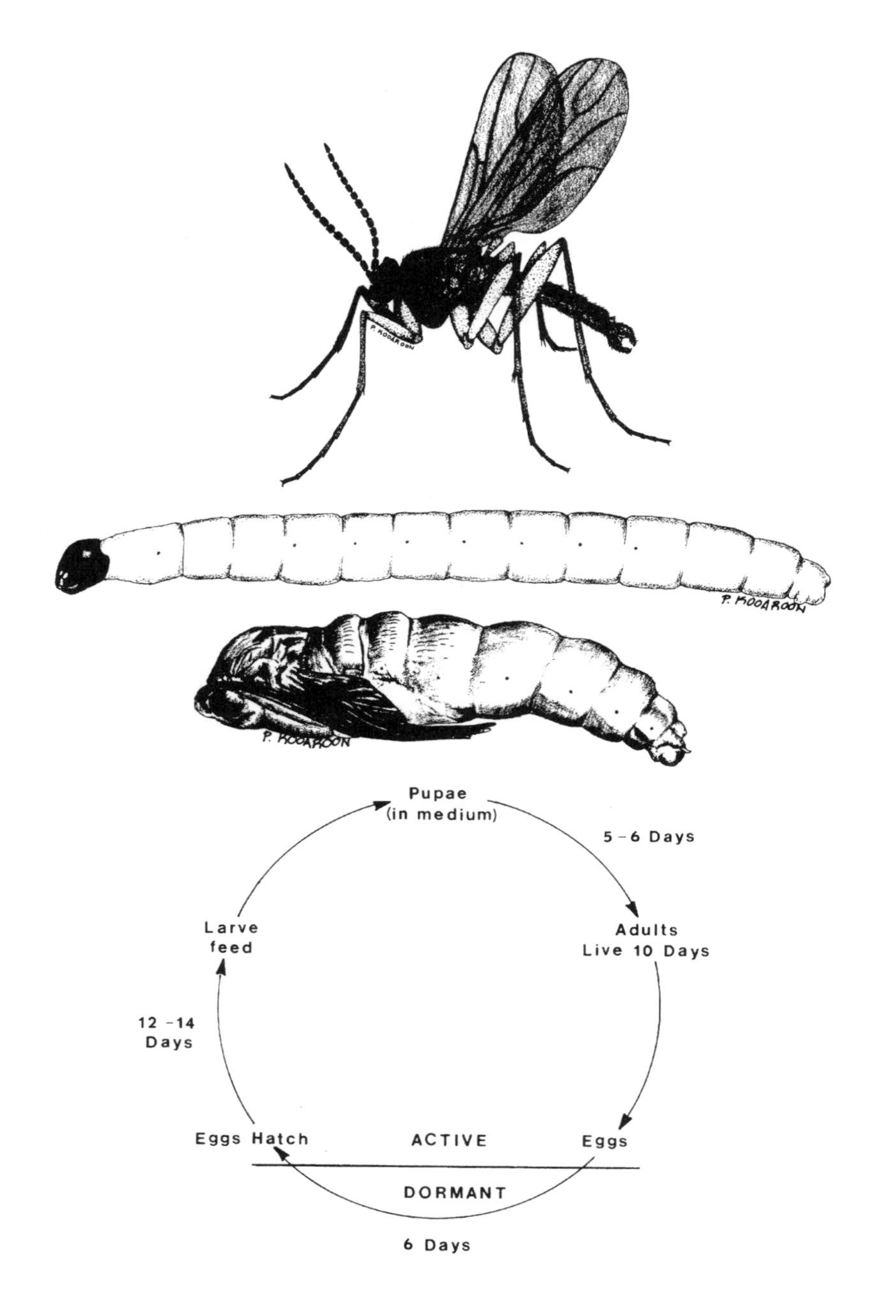

Figure 12.34. Life cycle of fungus gnat. (Insect drawings courtesy of J. R. Baker, North Carolina Agricultural Extension Service, Raleigh, NC.)

12.15 Vegetable Varieties

Many varieties of vegetables are available from seed houses. While both field and greenhouse varieties can be grown in a greenhouse, it is advantageous to use greenhouse varieties whenever possible since they are often bred to yield very heavily under controlled environment conditions. That is, in a greenhouse higher yields generally could be expected from greenhouse varieties than from field varieties. In most cases greenhouse varieties cannot be easily grown under field conditions since they are unable to withstand the temperature fluctuations encountered there.

A number of greenhouse varieties of vegetables perform particularly well in hydroponic culture. These along with other acceptable greenhouse varieties are given in Table 12.3.

Older tomato varieties such as Vendor, Vantage, Tropic and Manapel have been replaced by the Dutch varieties Dombito, Caruso, Larma, Perfecto, Trend, Belmondo and Trust which are superior in vigor, yields, and disease resistance. As discussed in Chapter 10, many growers are grafting the varieties such as Belmondo onto a rootstock resistant to root rots.

TABLE 12.3 Recommended Vegetable Varieties for Greenhouse and Hydroponic Culture

Vegetable	*Varieties*
Cucumbers: European	*Toska 70, Rocket, Brilliant, *Pandex, Farbio, La Reine, Sandra, *Uniflora D, *Corona, *Farona, *Marillo, *Fidelio, *Bronco, *Mustang, *Exacta, *Ventura,"1289," *Jessica, *Optima, *Flamingo
Lettuce: European	*Deci-Minor, *Ostinata, *Cortina
Novelty	Buttercrunch
Looseleaf	*Domineer, *Black Seeded Simpson, Grand Rapids, Waldmann's Dark Green
Tomatoes	Vendor, Vantage, Tropic, Manapel, *Dombito, *Caruso, *Larma, *Perfecto, *Belmondo, *Trend, *Trust, *Apollo
Peppers, Bell types:	
Green to red	*Delphin, *Plutona, *Tango, Cubico, *Mazurka, Val Valeta
Green to yellow	*Luteus, *Goldstar, *Samantha, *Gold Flame
Green to orange	*Wonder, *Eagle
Green to purple	*Violetta

*Varieties particularly suitable to hydroponic culture

The industry in British Columbia is now switching to Trust and Apollo as they are resistant to root rots and therefore would not have to be grafted. All of these Dutch varieties are resistant to tobacco mosaic virus (TMV), *Cladosporium* leaf mold, *Fusarium,* and *Verticillium* crown rots. Seed companies give symbols to each variety, indicating which diseases it is resistant to. For example, the tomato variety Caruso is given the symbol "TmC_5VF_2" indicating it is resistant to: TMV, *Cladosporium* races A,B,C,D and E, *Verticillium,* and *Fusarium* races 1 and 2.

The most popular cucumber varieties now used are Farona, Corona, Farbio, Marillo, Bronco, Mustang, Ventura and "1289." There are many other varieties available from the various seed producers. Some varieties are more suitable for spring/early-summer season while others are better for summer/fall crops. They are all female, seedless varieties that set fruit without pollination.

Peppers are becoming an important greenhouse crop. These are the sweet bell peppers which are green, maturing to red, orange, yellow or purple. The more popular green-to-red varieties are Delphin, Plutona and Tango, while the green-to-yellow ones are Luteus, Goldstar and Samantha.

12.15.1 Lettuce

Lettuce grows very well in hydroponic culture. There are four basic types of lettuce: European or bibb lettuce, looseleaf lettuce, head lettuce or iceberg, and cos or romaine lettuce. While many varieties of each type are available from seed houses, there are a few of each type which are particularly well suited to hydroponic culture and greenhouse environments. The most suitable varieties of European or bibb lettuce are Deciminor, Capitan and Ostinata. They should be planted at 20 cm by 20 cm (8 in. by 8 in.) spacing. Bibb lettuce requires a night temperature of 18° C (65° F), a day temperature of 17° to 19° C (63° to 66° F) on dull days and 21° to 24° C (70° to 75° F) on sunny days. They take about 60 days from seeding to maturity.

Looseleaf varieties are generally the easiest to grow. Varieties such as Slo Bolt, Black Seeded Simpson, Grand Rapids, and Waldmann's Dark Green are very vigorous in hydroponic culture. They require between 45 and 50 days to mature. Their spacing and temperature requirements are similar to European lettuce. Looseleaf varieties require night temperatures of 10° to 13° C (50° to 55° F), and 13° to 21° C (55° to 70° F) during the day, depending upon sunlight. However, they will tolerate higher temperatures to 27° C (75° to 80° F) without wilting, bolting or slowing growth. But if higher temperatures occur, "burning" of the leaf tip or margin may result. Some varieties are more tolerant to higher temperatures and resist "tipburn."

Of the many head lettuce varieties available, Great Lakes 659 grows best in hydroponics and greenhouse environments. It takes between 80 to 85 days to mature and will tolerate higher day temperatures up to 27° to 28° C (77° to 78° F) without bolting, provided that it receives full sunlight. Head lettuce should be spaced slightly wider apart at 25 centimeters by 25 centimeters (10 inches by 10 inches).

Valmaine Cos and Parris Island Cos have similar temperature requirements to looseleaf lettuce. They take about 75 days to mature and require similar spacing as looseleaf lettuce.

Lettuce should be seeded in peat pellets, rockwool cubes, Oasis blocks or flats of artificial medium as discussed earlier. Transplanting into growing beds should take place once the seedlings are 5–6 centimeters (2–2½ inches) in height (about 20 to 23 days after seeding).

All lettuce is susceptible to rot on the lower leaves when inadequate ventilation causes the lower leaves to remain moist. Since the fibrous root system of lettuce does not penetrate the medium very deeply, good drainage with frequent, uniform, moderate applications of the nutrient solution is better than longer periods of less frequent cycles.

12.16 Planting Schedules

A number of planting schedules are possible depending upon what crop or combination of crops is to be grown over the entire year. If only tomatoes are to be grown, a spring and fall crop system is used, as shown in Table 12.4. Particularly when growing on a backyard scale, intercropping of lettuce is recommended for the spring crop of tomatoes. This intercropping is not practical for commercial growers. Lettuce should be seeded and transplanted into the beds at the same time. One to two lettuce plants can be placed between each pair of tomato plants. The tomatoes will soon grow above the lettuce so they will not be shaded by it. While the tomatoes are still small, 12–18 inches (30–45 cm) in height, the lettuce will receive sufficient light to produce well. In this way a crop of lettuce can be harvested at least one month before the tomatoes will be ready. After this initial intercropping of lettuce, further intercrops can be placed under the tomatoes once the tomatoes are fully mature and several trusses of fruit and all the leaves up to the maturing truss have been removed. This will be about May, as shown in Table 12.4. One intercrop of lettuce could be grown with the fall crop through July and August (Table 12.4).

A combination crop of late spring tomatoes and two fall lettuce crops is practical on a commercial scale, as shown in Table 12.5.

A third schedule is one using a spring cucumber crop with a fall tomato crop, as shown in Table 12.6. Lettuce once again could be intercropped with the initial planting of both the cucumbers and the tomatoes.

TABLE 12.4 Planting Schedule for a Spring and Fall Crop of Tomatoes (Two Crops Annually) (Backyard Greenhouses Only)

Date	Activity
Dec. 20–31	Sow lettuce and tomato seeds in peat pellets
Feb. 1–15	Transplant seedlings into hydroponic beds
March 15	Harvest lettuce
April 15	Begin harvesting tomatoes
May 15	Terminate tomato plants, seed lettuce
June 1	Sow tomato seeds in peat pellets for fall crop; transplant lettuce intercrop under existing tomatoes
July 1	Harvest lettuce, pull tomato plants of spring crop, clean greenhouse, sterilize, etc.
July 15	Transplant fall tomato crop and lettuce intercrop into greenhouse beds
Aug. 15–30	Harvest lettuce
Sept. 15	Begin harvesting ripe tomatoes
Nov. 1	Terminate tomato plants
Dec. 20–31	Pull plants of fall crop, clean up, sterilize, etc., sow lettuce and tomato seeds of spring crop

TABLE 12.5 Combination Crop of Late Spring Tomatoes and Two Fall Lettuce Crops

Date	Activity
Late Spring Crop	
Tomatoes:	
Dec. 20–31	Sow tomato seeds
Feb. 1–15	Transplant tomatoes into beds
May–July	Harvest tomatoes
July 20–31	Remove plants, clean up, etc.
Fall Lettuce Crop	
First Crop:	
Aug. 10–15	Sow lettuce seeds
Sept. 5–10	Transplant lettuce into greenhouse beds
Oct. 10–20	Harvest crop
Second Crop:	
Sept. 15–20	Sow second crop lettuce seeds
Oct. 15–25	Transplant lettuce into beds
Dec. 12–15	Harvest crop

TABLE 12.6 Combination Crop of Spring Cucumbers and Fall Tomatoes

Date	*Activity*
Spring Cucumbers:	
Dec 25–31	Sow cucumber seeds (and lettuce seeds)
Feb. 1–15	Transplant cucumbers (and lettuce) to greenhouse beds
March 1–15	Harvest lettuce and begin harvesting cucumbers
June 25–15	Remove cucumbers, clean up, sterilize, etc.
Fall Tomatoes:	
June 15–30	Sow tomato (and lettuce) seeds
July 20–31	Transplant tomatoes (and lettuce)
Sept. 15	Harvest lettuce
Oct. 1	Begin harvesting tomatoes
Dec. 15–25	Remove tomato plants, clean up, sterilize, etc.

Cucumbers, if grown as the only crop year-round, may be scheduled in numerous ways. Some growers prefer to grow three to five crops a year, while others may grow only one long crop from February through October using the renewal umbrella system. Those using the single long crop often close their greenhouses down from November through January, saving on heating bills, and carrying out only a maintenance and repair program to their greenhouses during these winter months. Table 12.7 gives a schedule for a three-crop system of cucumbers. Backyard growers may intercrop lettuce during the first and second crops, but not during the third crop when lighting conditions are poor.

TABLE 12.7 A Three-Crop Schedule for Annual Cucumber Production

Date	*Activity*
Jan. 1	Sow cucumber (and lettuce) seeds (first crop)
Feb. 1–15	Transplant into beds (first crop)
March 1	Harvest lettuce
March 31	Begin harvesting cucumbers
May 1	Sow cucumber (and lettuce) seeds (second crop)
May 31	Pull cucumbers of first crop, clean up, etc.
June 1–10	Transplant cucumbers (and lettuce) into beds (second crop)
July 1	Harvest lettuce
July 15	Begin harvesting cucumbers (second crop)
Aug. 1	Sow cucumber seeds (third crop)
Aug. 31	Pull cucumber plants of second crop, clean up, etc.
Sept. 1–10	Transplant cucumbers into beds (third crop)
Oct. 15	Begin harvesting cucumbers
Dec. 20–31	Pull plants (third crop), clean up, sterilize, etc.

Over the past few years it has become common practice to grow a single crop of tomatoes, cucumbers or peppers per year. Tomatoes are sown in early November and grown through November of the following year, while cucumbers are sown a little later, in early December, and peppers are sown in early October as shown in Table 12.8.

TABLE 12.8 Single Crop of Tomatoes, Cucumbers, or Peppers

Date	*Activity*
Tomatoes:	
Early Nov.	Sow tomato seeds
Late Nov.	Transplant to larger rockwool blocks, keep under HID supplementary lighting with minimum 5,500 lux (510 foot candles) intensity in seedling area of greenhouse
Dec. 1–7	Set plants in rockwool blocks on top of slabs in greenhouse
Jan. 1–7	Transplant tomatoes into slabs (beds) when flower buds appear
Feb. 15–21	First harvest
Nov. 15	Last harvest
Nov. 15–30	Remove plants, clean up, etc.
Cucumbers:	
Dec. 1	Sow cucumber seeds in seedling house
Late Dec.	Transplant to slabs or beds in greenhouse
Feb, 1–7	Begin harvesting cucumbers
Nov. 15	Last harvest
Nov. 15–Dec. 15	Remove plants, clean up, etc.
Peppers:	
Oct. 1	Sow pepper seeds in seedling house
Late Oct.	Transplant peppers to rockwool blocks in seedling greenhouse
Early Dec.	Transplant to slabs or beds in greenhouse
Early March	Begin harvesting peppers
Nov. 15	Last harvest
Nov. 15–Dec. 7	Remove plants, clean up, etc.

Note: See Chapter 10 for more details on cropping procedures.

12.17 Crop Termination

In the growing of tomatoes, the growing point of each plant should be cut off 60 days prior to the expected date for pulling of the plants (Table 12.4). During this 60-day period, remove the many suckers which will develop at the tops of the plants.

Several days before starting the clean up, spray the plants to kill any insect infestations. Be careful not to use any pesticides having long-term residual effect which may harm the beneficial predators of the integrated pest-management program of subsequent crops. Stop the flow of water and nutrients to the slabs or beds a few days before physically removing the plants from the greenhouse. Pull out the roots, but leave the rest of the plant still supported by the string and clamps. This will result in the plants losing a lot of water and thus reduce the overall plant weight to be removed from the greenhouse. Wedge the plastic clamps apart with a spoon or knife edge when the plants are ready to be transported out of the greenhouse. The plastic support-clamps can be reused after soaking and washing them in a bleach solution. Dispose of the plants in the garbage, on a compost pile or bury them some distance from the greenhouse to avoid any disease or insect reinfestation of the new crop. After all the plants have been removed from the greenhouse, sweep all the floors, etc. clean, so that no plant debris remains. The beds, nutrient tank and growing medium must then be thoroughly sterilized according to measures given in earlier chapters. Once everything has been completely sterilized, the system is ready for its next crop.

A hydroponic system, if properly sterilized and cleaned between each crop, will continuously yield heavy crops over the years, giving its operator higher returns than could be achieved over the long run with soil.

12.18 Special Considerations

Whether plants are grown hydroponically or in soil, their cultural requirements are the same. Specific information on the growing of various plants is available from gardening books and various extension bulletins issued by universities and agricultural departments. A list of some university extension offices is given in Appendix 2.

Commercial greenhouses are now closely scrutinized by various environmental groups. As a result, use of water, runoff, pesticides, and fertilizer wastes are important aspects to be controlled. In Holland, greenhouses are being forced to re-use water, minimize the use of pesticides and fertilizers and recycle wastes. Recycling of plant materials could be achieved through their use in animal foods. Water and fertilizer usage can be minimized with recirculating hydroponic systems of rockwool and NFT cultures. Pesticide use is largely reduced by employing biological agents in an integrated pest-management program. The use of fungicides is being reduced through the breeding of disease-resistant varieties. Research in more efficient hydroponic methods, nutrient analyses and automatic solution adjustment with sterilization of the solution during each passage through the nutrient reservoir, all monitored and controlled by a central computer, are some of the immediate future needs of hydroponic culture.

It is this effective control which provides hydroponics with the potential of becoming the solution to intensive crop production throughout the world, and in man's future travels to other planets. Hydroponic experiments are now being conducted by several companies sponsored by NASA in its space program. It is to be the method of providing fresh vegetables to astronauts on the space station and future space travel. Experiments on the growing of plants in space are now scheduled for space flights in the near future. Hydroponic systems have been designed and tested to operate under micro-gravity environmental conditions of spaceflight. A biomass production chamber has been designed for NASA's "Controlled Ecological Life Support System" (CELSS). Many studies are currently underway to develop equipment and cultural procedures to grow crops hydroponically in space.

References

Johnston, H., Jr. 1973. *Cultural practices for tomatoes.* Univ. of Calif. Agric. Extension Bulletin, Nov. 1973.

———. 1975. *Greenhouse tomato production.* Univ. of Calif. Coop. Ext. Leaflet 2806.

———. 1975. *Greenhouse cucumber production.* Univ. of Calif. Ext. Leaflet 2775.

Larson, J. E. 1970. *Growing tomatoes in plastic greenhouses.* Texas A&M Univ. Agric. Ext. Bulletin, June 1970.

Loughton, A. 1975. The "How-To" of European cucumbers. *Am Veg. Grower,* Nov.1975, pp. 16, 18, 58 and 60.

Schales, F. D. and P. H. Massey, Jr. 1968. *Tomato production in plastic greenhouses.* Virginia Polytechnic Inst. Ext. Div. Publ. 154, July 1968.

———. 1969. *Starting early plants.* Virginia Polytechnic Inst. Ext. Div. Publ. 226, Jan.1969.

Smith, Dennis L. 1988. Peppers & Aubergines. *Grower Guide No. 3.* 92 pp. London: Grower Books.

Stoner, A. K. 1971. *Commercial production of greenhouse tomatoes.* U.S.D.A. Handbook No. 382, Washington, D. C.: Gov. Printing Office.

Tiessen, H., J. Wiebe and C. Fisher. 1976. *Greenhouse vegetable production in Ontario.* Ontario Min. of Agric. and Food Publ. 526.

Wittwer, S.H. and S. Honma. 1969. *Greenhouse tomatoes: guidelines for successful production.* East Lansing: Michigan State Univ. Press.

Wittwer, S. H., S. Honma and W. M. Robb. 1964. *Practices for increasing yields of greenhouse lettuce.* East Lansing: Michigan State Univ. Research Rept. 22.

Appendix 1

Hydroponic and Soilless-Culture Societies

The International Society for Soilless Culture (ISOSC) is an organization of people (growers, hobbyists, scientists) active in some way in various forms of soilless culture. The purpose of ISOSC is worldwide promotion of research and application of soilless culture. It operates as an information center on such culture. Every three to four years ISOSC organizes an international congress on soilless culture and publishes the proceedings of these congresses.

Any person who is actively engaged in research, advisory work or commercial application of soilless culture is eligible to become a member of ISOSC. Information may be obtained from: Secretariat of ISOSC, P.O. Box 52, Wageningen, Netherlands.

A similar society exists in North America. It is the Hydroponic Society of America. Information and membership may be obtained from: Hydroponic Society of America, 2819 Crow Canyon Road, Suite 218, San Ramon, CA 94583.

Appendix 2

Greenhouse Production Resources

Research Extension Services for Publications

B.C. Ministry of Agriculture
32916 Marshall Road
Abbotsford, B.C.
Canada V2S 1K2

Agricultural Engineering
Cooperative Extension Service
Pennsylvania State University
University Park, PA 16802

Cooperative Extension Service
Dept. of Agricultural Engineering
University of Kentucky
Lexington, KY 40506

Department of Vegetable Crops
New York State College of Agriculture
Cornell University
Ithaca, NY 14850

Superintendent of Documents
U.S. Government Printing Office
Washington, DC 20402

Agricultural Extension Service
University of California
Riverside, CA 92502

Agricultural Experiment Station
Mississippi State University
State College, MS 39762

Cooperative Extension Service
College of Agriculture
University of Illinois
Urbana, IL 61801

Agricultural Extension Service
University of California
Davis, CA 95616

Cooperative Extension Service
Rutgers University
New Brunswick, NJ 08903

Cooperative Extension Service
College of Agriculture and
Natural Resources
University of Connecticut
Storrs, CT 06268

Ontario Department of
Agriculture and Food
Parliament Buildings
Toronto, Ontario

Agricultural Extension Service
The University of Arizona
4201 East Broadway
Phoenix, AZ 85040

Ontario Dept. of Agriculture and Food
Horticultural Research Institute
Vineland Station, Ontario

Agricultural Experiment Station
Michigan State University
East Lansing, MI 48823

Agricultural Extension Service
Purdue University
Lafayette, IN 47907

Agricultural Extension Service
Texas A&M University
College Station, TX 77843

Department of Horticulture
Oregon State University
Corvallis, OR 97331

Cooperative Extension Service
Ohio State University
Wooster, OH 44691

Department of Horticultural Science
North Carolina State University
Raleigh, NC 27607

Department of Horticulture
Oklahoma State University
Stillwater, OK 74074

Vegetable Crops Department
University of Florida
Gainesville, FL 32611

Some Soil and Plant-Tissue Testing Laboratories

Soil and Plant Laboratory, Inc.
P.O. Box 11744, Santa Ana, CA 92711
P.O. Box 153, Santa Clara, CA 95052
P.O. Box 1648, Bellevue, WA 98009

Peninsu-LAB (Disease and Pest Lab)
23976 N.E. Newellhurst Court
Kingston, WA 98346

Norwest-Priva Plant Laboratories Inc.
203-20771 Langley Bypass
Langley, B.C., Canada

Soil Testing Laboratory
Purdue University
Agronomy Dept.
Lafayette, IN 47907

Griffin Laboratories
1875 Spall Rd.
Kelowna, B.C. Canada
V1Y 4R2

Department of Land Resource Science
University of Guelph
Guelph, Ontario, Canada

Soil Testing Laboratory
Texas A&M University
College Station, TX 77843

Ohio State University
Ohio Ag. Research and Dev. Center
Research-Extension Analytical Lab.
Wooster, OH 44691

Biological-Control Agents

Note: This is not a complete list of companies selling biological-control agents.

Applied Bio-Nomics Ltd.
P.O. Box 2637
Sidney, B.C.
Canada V8L 4C1

Beneficial Insectary
14751 Oak Run Rd.
Oak Run, CA 96069

Bio Insect Control
710 South Columbia
Plainview, TX 79072

Biosis
1057 East Meadow Circle
Palo Alto, CA 95303

G.B. Systems
P.O. Box 39063
North Ridgeville, OH 44039

Koppert (UK) Ltd.
Biological Control
P.O. Box 43
Tunbridge Wells, Kent TN2 5BX
England

Organic Pest Management
P.O. Box 55267
Seattle, WA 98155

Richters
P.O. Box 26
Goodwood, Ontario
Canada L0C 1A0

Safer Ltd.
6761 Kirkpatrick Cres.
Victoria, B.C.
Canada V8X 3X1

Beneficial Bugs
P.O. Box 1627
Apopka, FL 32703

Better Yield Insects
13310 Riverside Dr. East
Tecumseh, Ontario
Canada N8N 1B2

Bio-Resources
P.O. Box 92
Santa Paula, CA 93060

Biotactics Inc.
7765 Lakeside Dr.
Riverside, CA 92509

Hydro-Gardens Inc.
P.O. Box 9707
Colorado Springs, CO 80932

M & R Durango, Inc.
P.O. Box 886
Bayfield, CO 81122

Nature's Control
P.O. Box 35
Medford, OR 97501

Praxis
P.O. Box 4164
Auburn Heights, MI 48051

Rincon-Vitova Insectories Inc.
P.O. Box 95
Oak View, CA 93022

Unique Insect Control
P.O. Box 15376
Sacramento, CA 95851

Special Hydroponic Equipment

1. NFT Troughs:

Rehau Plastics, Inc.
P.O. Box 1706
Leesburg, VA 22075

Hydro-Gardens, Inc.
P.O. Box 9707
Colorado Springs, CO 80932

CropKing, Inc.
P.O. Box 310
Medina, OH 44258

Clover Greenhouses
200 Weakley Lane
Smyrna, TN 37167

Troy Hygro-Systems, Inc.
4096 Hwy, ES
East Troy, WI 53120

Westbrook Greenhouse Systems Ltd.
270 Hunter Rd.
Grimsby, Ontario
Canada L3M 5G1

2. Reko Double-Row NFT Troughs:

ADJ Horti-Projects Inc.
P.O. Box 3004
Langley, B.C.
Canada V3A 4R3

Reko bv
P.O. Box 191
6190 AD Beek (L)
Holland

Other Related Equipment

1. UV Sterilizers:

Trojan Technologies
845 Consortium Court
London, Ontario
Canada N6B 2S8

Aquafine Corporation
25230 West Ave. Stanford
Valencia, CA 91355

2. Water Chiller Units:

Frigid Units, Inc.
3214 Sylvania Ave.
Toledo, OH 43613

3. Hobby Units:

For a list of companies marketing hydroponic hobby units refer to other books by Resh published by Woodbridge Press: *Hydroponic Home Food Gardens* and *Hydroponic Tomatoes for the Home Gardener.*

Appendix 3

Units of Measurement – Conversion Factors

Units of	To Convert →	Into →	Multiply by
Length:			
25.401	Millimeters	Inches	0.0394
2.5401	Centimeters	Inches	0.3937
0.3048	Meters	Feet	3.2808
0.9144	Meters	Yards	1.0936
1.6093	Kilometers	Miles (statute)	0.6214
Area:			
645.160	Sq. Millimeters	Sq. Inches	0.001550
6.4516	Sq. Centimeters	Sq. Inches	0.1550
0.0929	Sq. Meters	Sq. Feet	10.7639
0.8361	Sq. Meters	Sq. Yards	1.1960
0.004046	Sq. Kilometers	Acres	247.105
2.5900	Sq. Kilometers	Sq. Miles	0.3861
0.4046	Hectares	Acres	2.4710
Volume:			
16.3872	Cubic Centimeters	Cubic Inches	0.0610
0.0283	Cubic Meters	Cubic Feet	35.3145
0.7646	Cubic Meters	Cubic Yards	1.3079
0.003785	Cubic Meters	Gallons (U.S.)	264.178
0.004545	Cubic Meters	Gallons (U.K.)	219.976
0.01639	Liters	Cubic Inches	61.0238
28.3205	Liters	Cubic Feet	0.03531
3.7850	Liters	Gallons (U.S.)	0.2642
4.5454	Liters	Gallons (U.K.)	0.2200
Weight:			
28.3495	Grams	Ounces (Av.)	0.0353
31.1035	Grams	Ounces (Troy)	0.0321
0.4536	Kilograms	Pounds (Av.)	2.2046
0.0004535	Metric Tons	Pounds (Av.)	2204.62
0.907185	Metric Tons	Tons (U.S.)	1.1023
1.016047	Metric Tons	Tons (U.K.)	0.9842
Multiply by	**← Into**	**← To Convert**	

Appendix 4

Physical Constants of Inorganic Compounds

Name	Formula	Density or Sp. Gravity	Solubility (gm/100 ml) Cold Water	Hot Water
Ammonium Nitrate	NH_4NO_3	1.725	118.3	871
Ammonium dihydrogen phosphate	$NH_4H_2PO_4$	1.803	22.7	173.2
Ammonium molybdate	$(NH_4)_6Mo_7O_{24}\cdot 4H_2O$	43		
Ammonium monohydrogen phosphate	$(NH_4)_2HPO_4$	1.619	57.5	106.0
Ammonium sulfate	$(NH_4)_2SO_4$	1.769	70.6	103.8
Boric acid	H_3BO_3	1.435	6.35	27.6
Calcium carbonate	$CaCO_3$	2.710	0.0014	0.0018
Calcium chloride	$CaCl_2$	2.15	74.5	159
Calcium chloride, hexahydrate	$CaCl_2\cdot 6H_2O$	1.71	279	536
Calcium hydroxide	$Ca(OH)_2$	2.24	0.185	0.077
Calcium nitrate	$Ca(NO_3)_2$	2.504	121.2	376
Calcium nitrate tetrahydrate	$Ca(NO_3)_2\cdot 4H_2O$	1.82	266	660
Calcium oxide	CaO	3.25–3.38	0.131	0.07
Calcium monophosphate	$Ca(H_2PO_4)_2\cdot H_2O$	2.220	1.8	decomposes
Calcium sulfate	$CaSO_4$	2.960	0.209	0.1619
Calcium sulfate dihydrate	$CaSO_4\cdot 2H_2O$	2.32	0.241	0.222
Copper sulfate pentahydrate	$CuSO_4\cdot 5H_2O$	2.284	31.6	203.3
Iron (II) hydroxide	$Fe(OH)_2$	3.4	0.00015	—
Iron (II) nitrate	$Fe(NO_3)_2\cdot 6H_2O$	1.6	83.5	166.7
Iron (II) sulfate heptahydrate	$FeSO_4\cdot 7H_2O$	1.898	15.65	48.6
Magnesium oxide	MgO_2	3.58	0.00062	0.0086
Magnesium orthophosphate	$Mg_3(PO_4)_2$	—	insoluble	insoluble
Magnesium monohydrogen phosphate heptahydrate	$MgHPO_4\cdot 7H_2O$	1.728	0.3	0.2
Magnesium orthophosphate tetrahydrate	$Mg_3(PO_4)_2\cdot 4H_2O$	1.64	0.0205	—
Magnesium sulfate heptahydrate	$MgSO_4\cdot 7H_2O$	1.68	71	91
Manganese dichloride tetrahydrate	$MnCl_2\cdot 4H_2O$	2.01	151	656

Appendix 4

Physical Constants of Inorganic Compounds

Name	Formula	Density or Sp. Gravity	Solubility (gm/100 ml) Cold Water	Hot Water
Manganous (II) hydroxide	$Mn(OH)_2$	3.258	0.0002	—
Manganous nitrate	$Mn(NO_3)_2 \cdot 4H_2O$	1.82	426.4	∞
Manganous dyhydrogen phosphate	$Mn(H_2PO_4)_2 \cdot 2H_2O$	—	soluble	—
Manganous monohydrogen phosphate	$MnHPO_4 \cdot 3H_2O$	—	slightly soluble	decomposes
Manganous sulfate	$MnSO_4$	3.25	52	70
Manganous sulfate tetrahydrate	$MnSO_4 \cdot 4H_2O$	2.107	105.3	111.2
Nitric acid	HNO_3	1.5027	∞	∞
Phosphoric acid (ortho)	H_3PO_4	1.834	548	very sol.
Phosphoric anhydride	P_2O_5	2.39	decomposes	decomposes to H_3PO_4
Potassium carbonate	K_2CO_3	2.428	112	0.156
Potassium carbonate dihydrate	$K_2CO_3 \cdot 2H_2O$	2.043	146.9	331
Potassium hydrogen carbonate	$KHCO_3$	2.17	22.4	60
Potassium carbonate trihydrate	$2K_2CO_3 \cdot 3H_2O$	2.043	129.4	268.3
Potassium chloride	KCl	1.984	34.7	56.7
Potassium hydroxide	KOH	2.044	107	178
Potassium nitrate	KNO_3	2.109	13.3	47
Potassium ortho-phosphate	K_3PO_4	2.564	90	soluble
Potassium dihydrogen phosphate	KH_2PO_4	2.338	33	83.5
Potassium monohydrogen phosphate	K_2HPO_4	—	167	very soluble
Potassium sulfate	K_2SO_4	2.662	12	24.1
Zinc carbonate	$ZnCO_3$	4.398	0.001	—
Zinc chloride	$ZnCl_2$	2.91	432	615
Zinc orthophosphate	$Zn_3(PO_4)_2$	3.998	insoluble	insoluble
Zinc dihydrogen phosphate	$Zn(H_2PO_4)_2 \cdot 2H_2O$	—	decomposes	—
Zinc orthophosphate tetrahydrate	$Zn_3(PO_4)_2 \cdot 4H_2O$	3.04	insoluble	insoluble
Zinc sulfate heptahydrate	$ZnSO_4 \cdot 7H_2O$	1.957	96.5	663.6

Appendix 5

Greenhouse and Hydroponic Suppliers

Biocontrol Agents

Microbials:

AgBio Development Inc.
9915 Raleigh Street
Westminster, CO 80030

Carolina Seeds Inc.
P.O. Box 2625, Hwy. 105 By-Pass
Boone, NC 28607

Grace-Sierra Horticultural Products Co.
1001 Yosemite Dr.
Milpitas, CA 95035

Harris Seeds/Garden Trends Inc.
60 Saginaw Dr., P.O. Box 22960
Rochester, NY 14692-2960

Rigo Co.
P.O. Box 189
Buckner, KY 40010

Sandoz Agro, Inc. (Steward)
1300 E. Touhy Ave.
Des Plaines, IL 60018

Sharp & Son
900 Lind Ave. S.W.
Renton, WA 98055

United Horticultural Supply
4564 Ridge Dr. N.E.
Salem, OR 97303

Predators/Parasites:

Beneficial Resources
P.O. Box 327
Danville, PA 17821

CropKing Inc.
P.O. Box 310
Medina, OH 44258

G.B. Systems
P.O. Box 39063
N. Ridgeville, OH 44039

IPM Laboratories, Inc.
P.O. Box 300
Locke, NY 13092-0300

Plant Sciences, Inc.
342 Green Valley Rd.
Watsonville, CA 95076

Plantco Inc.
314 Orenda Rd.
Brampton, ON L6T 1G1

Sharp & Son Ltd.
900 Lind Ave. S.W.
Renton, WA 98055

Pollinators (Bombus sp.):

Bees West, Inc.
P.O. Box 1378
Freedom, CA 95019

Beneficial Resources
P.O. Box 327
Danville, PA 17821

Biobest Trading bvba
Ilse Velden 18
B-2260 Westerlo, Belgium

G.B. Systems
P.O. Box 39063
N. Ridgeville, OH 44039

Chemicals & Pesticides

AgriDyne Technologies Inc.
2401 So. Foothill Dr.
Salt Lake City, UT 84109

BFG Supply Co.
P.O. Box 479, 14500 Kinsman Rd.
Burton, OH 44021

Grace-Sierra Horticultural Products Co.
1001 Yosemite Dr.
Milpitas, CA 95035

Hummert International
2746 Chouteau Ave.
St. Louis, MO 63103

OFE International, Inc.
P.O. Box 16130
Miami, FL 33116

Plant Products Corp.
P.O. Box 1149
Vero Beach, FL 32961-1149

Reddick Fumigants
P.O. Box 391
Williamston, NC 27892

Rigo Co.
P.O. Box 189
Buckner, KY 40010

Sharp & Son Ltd.
900 Lind Ave. S.W.
Renton, WA 98055

Uniroyal Chemical Co., Inc.
World Headquarters
Middlebury, CT 06749

United Horticultural Supply
4564 Ridge Dr. N.E.
Salem, OR 97303

Waldo & Associates, Inc.
28214 Glenwood Rd.
Perrysburg, OH 43551

Computer Hardware & Software

Acme Engineering & Mfg.
P.O. 978
Muskogee, OK 74402

Automata, Inc.
16216 Brooks Rd.
Grass Valley, CA 95945-8816

Canadian Hydrogardens Ltd.
1386 Sandhill Dr.
Ancaster, ON L9G 4V5

Conley's Greenhouse Mfg.
4344 Mission Blvd.
Pomona, CA 91766

CropKing Inc.
P.O. Box 310
Medina, OH 44258

Growth Zone Systems
1719 Hwy 99 S.
Mount Vernon, WA 98273

Frank Jonkman & Sons Ltd.
R.R. 4
Bradford, ON L3Z 2A6

Micro Grow Greenhouse Systems, Inc.
26111 Ynez Rd., Suite B-26
Temecula, CA 92591

Micro Vane, Inc.
8135 Cox's Dr.
Kalamazoo, MI 49002

National Greenhouses, Inc.
400 E. Main, P.O. Box 500
Pana, IL 62557

Neogen Corp.
620 Lesher Pl.
Lansing, MI 48912-1509

Nexus Greenhouse Corp.
10983 Leroy Dr.
Northglenn, CO 80233

Oglevee Computer Systems
150 Oglevee Lane
Connellsville, PA 15425

Priva Computers, Inc.
3468 S. Service Rd.
Vineland Station, ON L0R 2E0

Q-Com Corp.
2050 S. Grand Ave.
Santa Ana, CA 92705

Techmark, Inc.
P.O. Box 80835
Lansing, MI 48908-0835

United Greenhouse Systems
708 Washington St.
Edgerton, WI 53534

Jack Van Klaveren Ltd. (JVK)
P.O. Box 910
St. Catharines, ON L2R 6Z4

V & V Noordland Inc.
16 Commercial Blvd.
Medford, NY 11763

Van Wingerden Greenhouse Co.
4078 Haywood Rd.
Horse Shoe, NC 28742

Wadsworth Control Systems
5541 Marshall St.
Arvada, CO 80002

Westbrook Greenhouse Systems Ltd.
270 Hunter Rd.
Grimsby, ON L3M 4G1

Fertilizers

Beneficial Resources
P.O. Box 327
Danville, PA 17821

BFG Supply Co.
P.O. Box 479, 14500 Kinsman Rd.
Burton, OH 44021

AG RX
3250 Somis Rd.
Somis, CA 93066

CropKing, Inc.
P.O. Box 310
Medina, OH 44258

Fruit Growers Supply
111 N. Palm Ave.
Santa Paula, CA 93060

Grace-Sierra Horticutural Products Co.
1001 Yosemite Dr.
Milpitas, CA 95035

Hummert International
2746 Chouteau Ave.
St. Louis, MO 63103

Hydro Agri
303 Twin Dolphin Dr., Suite 500
Redwood City, CA 94065

McConkey Co.
P.O. Box 1690
Sumner, WA 98390

OFE International, Inc.
P.O. Box 161302
Miami, FL 33116

Plantco Inc.
314 Orenda Rd.
Brampton, ON L6T 1G1

Plant Products Corp.
P.O. Box 1149
Vero Beach, FL 32961-1149

Rigo Co.
P.O. Box 189
Buckner, KY 40010

Sharp & Son Ltd.
900 Lind Ave. S.W.
Renton, WA 98055

Sharp & Son Ltd.
9643-186th St.
Surrey, B.C. V3T 4W2

Southern Importers, Inc.
P.O. Box 8579
Greensboro, NC 27419

United Horticultural Supply
4564 Ridge Dr. N.E.
Salem, OR 97303

Van Waters & Rogers Inc.
1910 Lockwood
Oxnard, CA 93030-2603

Jack Van Klavern Ltd.
P.O. Box 910
St. Catharines, ON L2R 6Z4

Waldo & Associates, Inc.
28214 Glenwood Rd.
Perrysburg, OH 43551

Westcan Horticultural Ltd.
Bay 5, 6112-30 St. S.E.
Calgary, AB T2C 2A6

Greenhouse Structures and Coverings

Agra Tech Inc.
2131 Piedmont Way
Pittsburg, CA 94565

Atlas Greenhouse Systems, Inc.
Route 1, Box 339
Alapaha, GA 31622

ATTKO Greenhouse Inc.
5243 Stilesboro Rd.
Kennesaw, GA 30144

BFG Supply Co.
P.O. Box 479
Burton, OH 44021

Paul Boers Greenhouse Construction Co.
P.O. Box 134
St. Davids, ON L0S 1P0

Clearfield Greenhouse Co., Inc.
320 Wetmore St.
Manteca, CA 95336

Conley's Mfg. & Sales
4344 E. Mission Blvd.
Pomona, CA 91766

CropKing Inc.
P.O. Box 310
Medina, OH 44258

DACE
1937 High St.
Longwood, FL 32750

Dalsem Greenhouses USA Inc.
P.O. Box 54039
Jacksonville, FL 32245

De Cloet Greenhouse Mfg. Ltd.
R.R. 1
Simcoe, ON N3Y 4J9

Greenhouse System USA Inc.
P.O. Box 777
Watsonville, CA 95076

Hamilton Steel Fabrications
3290 M-40
Hamilton, MI 49419

Hove International, Inc.
1900 The Exchange, Suite 230
Atlanta, GA 30339

Hummert International
2746 Chouteau Ave.
St. Louis, MO 63103

Hydro-Gardens, Inc.
P.O. Box 25845
Colorado Springs, CO 80936

Industries Harnois Inc.
1044 Principale
St Thomas, PQ J0K 3L0

Jacobs Greenhouse Mfg. Ltd.
371 Talbot Rd.
Delhi, ON N4B 2A1

JVK Ltd.
P.O. Box 910, 1894 Seventh St.
St. Catharines, ON L2R 6Z4

Frank Jonkman & Sons Ltd.
R.R. 4
Bradford, ON L3Z 2A6

Keeler-Glasgow Co., Inc.
P.O. Box 158, 80444 C.R. 687
Hartford, MI 49057

Lincoln Greenhouse Construction Ltd.
R.R. 3, Mud St.
Smithville, ON L0R 2A0

Ludy Greenhouse Mfg. Corp.
143 W. Washington St.
New Madison, OH 45346

McCalif Grower Supplies, Inc.
P.O. Box 310, 2905 Railroad Ave.
Ceres, CA 95307

McConkey Co.
P.O. Box 1690
Sumner, WA 98390

National Greenhouses, Inc.
P.O. Box 500, 400 E. Main
Pana, IL 62557

Nexus Greenhouse Corp.
10983 Leroy Dr.
Northglenn, CO 80233

Ochmsen Plastic Greenhouse Mfg.
50 Carlough Rd.
Bohemia, NY 11716

Omni Growing Systems
145 Cushman Rd.
St Catharines, ON L2M 6T2

Poly Grower Greenhouse Co.
P.O. Box 359
Muncy, PA 17756

Poly-Tex, Inc.
P.O. Box 458
Castle Rock, MN 55010

Rough Brothers
P.O. Box 16010
Cincinnati, OH 45216

Sharp & Son Ltd.
900 Lind Ave. S.W.
Renton, WA 98055

Sharp & Son Ltd.
9643-186th St.
Surrey, B.C. V3T 4W2

X.S. Smith Inc.
P.O. Drawer X
Red Bank, NJ 07701

Structures Unlimited
2122 Whitfield Park Ave.
Sarasota, FL 34243

Stuppy Greenhouse Mfg. Inc.
P.O. Box 12456
North Kansas City, MO 64116

United Greenhouse Systems
708 Washington St.
Edgerton, WI 53534

Vary Industries
P.O. Box 248
Lewiston, NY 14092

V & V Noordland Inc.
16 Commercial Blvd.
Medford, NY 11763

Van Wingerden Greenhouse Co.
4078 Haywood Rd.
Horse Shoe, NC 28742

Waldo & Associates, Inc.
28214 Glenwood Rd.
Perrysburgh, OH 43551

Westbrook Greenhouse Systems Ltd.
270 Hunter Rd.
Grimsby, ON L3M 4G1

Westcan Horticultural Ltd.
Bay 5, 6112-30 St. S.E.
Calgary, AB T2C 2A6

Winandy Greenhouse Co., Inc.
2211 Peacock Rd.
Richmond, IN 47374

Greenhouse Equipment

Acme Engineering & Mfg. Corp.
P.O. Box 978
Muskogee, OK 74402

Aerotech, Inc.
929 Terminal Rd.
Lansing, MI 48906

Agrotec Inc.
P.O. Box 49
Pendleton, NC 27862-0049

American Coolair Corp.
P.O. Box 2300
Jacksonville, FL 32203

American Horticultrual Supply Inc.
4045 Via Pescador
Camarillo, CA 93012

A-roo Co.
P.O. Box 360050
Strongsville, OH 44136

ATTKO Greenhouse Inc.
5243 Stilesboro Rd.
Kennesaw, GA 30144

Automata, Inc.
16216 Brooks Rd.
Grass Valley, CA 95945-8816

Argus Control Systems Ltd.
No. 10, 1480 Foster St.
White Rock, B.C. V4B 3X7

BFG Supply Co.
P.O. Box 479, 14500 Kinsman Rd.
Burton, OH 44021

Biotherm Hydronic, Inc.
P.O. Box 750967
Petaluma, CA 94975

Bio-Energy Systems Inc.
221 Canal St.
Ellenville, NY 12428

Paul Boers Greenhouse
Construction Ltd.
P.O. Box 134
St. Davids, ON L0S 1P0

Bouldin & Lawson Inc.
Route 10, Box 208
McMinnville, TN 37110

Brighton By-Products Co. Inc.
P.O. Box 23
New Brighton, PA 15066

Campbell O'Brian Limited
#1 Thompson Crescent
Erin, ON N0B 1T0

Canadian Hydrogardens Ltd.
1386 Sandhill Dr.
Ancaster, ON L9G 4V5

Climate Control Systems Inc.
R.R.#5, #509 Hwy. 77
Leamington, ON N8H 3V8

Conley's Mfg. & Sales
4344 E. Mission Blvd.
Pomona, CA 91766

Cravo Equipment Ltd.
White Swan Rd., R.R.#1
Brantford, ON N3T 5L4

Crofton Grower Services Ltd.
7311 Vantage Way, Unit 110
Delta, B.C. V4G 1C9

CropKing Inc.
P.O. Box 310
Medina, OH 44258

Engineered Systems & Designs
119A Sandy Dr.
Newark, DE 19713

En Tech Control Systems, Inc.
P.O. Box 205, 155 S. 3rd St.
Montrose, MN 55363

Environmental Engineering
Concepts Inc.
1229 S. Gene Autry Trail
Palm Springs, CA 92264

Galaxy Agri-Products Inter. Inc.
44775 Yale Rd. West
Sardis, B.C. V2R 1A9

GEC Alsthom Inter. Canada Inc.
2 Paxman Rd.
Etobicoke, ON M9C 1B6

Gleason Equipment
28055 S.W. Boberg Rd.
Wilsonville, OR 97070

Growing Systems, Inc.
2950 N. Weil St.
Milwaukee, WI 53212

Growers Greenhouse Supplies Inc.
3559 North Service Rd.
Vineland Station, ON L0R 2E0

Growth Zone Systems
1719 Hwy. 99 South
Mount Vernon, WA 98273

Halifax Seed Co. Ltd.
5860 Kane St.
Halifax, NS B3K 5L8

Hamilton Boiler Works Limited
105 Cascade St.
Hamilton, ON L8E 3B7

Hamilton Engineering, Inc.
32615 Park Lane
Garden City, MI 48195

Hove International, Inc.
1900 The Exchange, Suite 220
Atlanta, GA 30339

Hydro-Gardens, Inc.
P.O. Box 25845
Colorado Springs, CO 80936

ICE Mfg. Ltd.
670 King Edward St.
Winnipeg, MB R3H 0P2

Industries Harnois Inc.
1044 Principale
St. Thomas, PQ J0K 3L0

Javo USA, Inc.
1900 Albritton Dr., Suites G & H
Kennesaw, GA 30144

JVK Ltd.
P.O. Box 910, 1894 Seventh St.
St. Catharines, ON L2R 6Z4

Jacobs Greenhouse Mfg. Ltd.
371 Talbot Rd.
Delhi, ON N4B 2A1

Jaybird Mfg., Inc.
RD 1, Box 489A
Centre Hall, PA 16828

Frank Jonkman & Sons Ltd.
R.R. 4
Bradford, ON L3Z 2A6

Keeler-Glasgow Co., Inc.
P.O. Box 158, 80444 C.R. 687
Hartford, MI 49057

Ken-Bar Inc.
25 Walkers Brook Dr.
Reading, MA 01867-0704

Lander Control Systems Inc.
129A Watson Rd. S.
Guelph, ON N1H 6H8

Leader Fan Industries Limited
130 Clairville Dr.
Etobicoke, ON M9W 5Y3

Ludy Greenhouse Mfg. Corp.
143 W. Washington St.
New Madison, OH 45346

McConkey Co.
P.O. Box 1690
Sumner, WA 98390

Mee Industries, Inc.
4443 N. Rowland Ave.
El Monte, CA 91731

Munters Corp.
P.O. Box 6428
Ft. Myers, FL 33911

Modine Mfg. Co.
1500 De Koven Ave.
Racine, WI 53403

National Greenhouses, Inc.
P.O. Box 500, 400 E. Main
Pana, IL 62557

Nexus Greenhouse Corp.
10983 Leroy Dr.
Northglenn, CO 80233

Omni Growing Systems
145 Cushman Rd.
St Catharines, ON L2M 6T2

Plantech Control Systems
3466 South Service Rd.
Vineland Station, ON L0R 2E0

Poly-Tex, Inc.
P.O. Box 458
Castle Rock, MN 55010

Schaefer Fan Co., Inc.
P.O. Box 647
St. Cloud, MN 56387

Sharp & Son Ltd.
900 Lind Ave. S.W.
Renton, WA 98055

Sharp & Son Ltd.
9643 - 186th St.
Surrey, B.C. V3T 4W2

X.S. Smith Inc.
Drawer X
Red Bank, NJ 07701

Structures Unlimited
2122 Whitfield Park
Sarasota, FL 34243

Stuppy Greenhouse Mfg. Inc.
P.O. Box 12456
North Kansas City, MO 64116

United Greenhouse Systems
708 Washington St.
Edgerton, WI 53534

V & V Noordland Inc.
16 Commercial Blvd.
Medford, NY 11763

Waldo & Associates, Inc.
28214 Glenwood Rd.
Perrysburgh, OH 43551

Westbrook Greenhouse Systems Ltd.
270 Hunter Rd.
Grimsby, ON L3M 4G1

Westcan Horticultural Ltd.
Bay 5, 6112-30 St. S.E.
Calgary, AB T2C 2A6

Winandy Greenhouse Co., Inc.
2211 Peacock Rd.
Richmond, IN 47374

Weaver Distributing
RD 2, Box 470
Fredericksburg, PA 17026-9545

Growing Media and Supplies

Agro Dynamics Inc.
12 Elkins Rd.
East Brunswick, NJ 08816

Agro Dynamics Inc.
415 Industrial Dr.
Milton, ON L9T 5A6

Agro Dynamics Inc.
3891 N. Ventura Ave., B1B
Ventura, CA 93001

Agromin Horticultural Soils
1501 Las Posas Rd.
Camarillo, CA 93010

ASB Greenworld Inc.
R.R.#1, Site 178, Point Sapin
Kent County, NB E0A 2A0

Berger Peat Moss Inc.
121 R.R.#1
St. Modeste, PQ G0L 3W0

BFG Supply Co.
P.O. Box 479, 14500 Kinsman Rd.
Burton, OH 44021

Brighton By-Products Co. Inc.
P.O. Box 23
New Brighton, PA 15066

Canadian Hydrogardens Ltd.
1386 Sandhill Dr.
Ancaster, ON L9G 4V5

Crofton Grower Services Ltd.
7311 Vantage Way, Unit 110
Delta, B.C. V4G 1C9

CropKing Inc.
P.O. Box 310
Medina, OH 44258

Robert G. Eckel Greenhouses
P.O. Box 12
Brooklin, ON L0B 1C0

Fisons Horticulture Inc.
110-110th Ave. N.E., Suite 490
Bellevue, WA 98004

Fines and Company
R.R.#1
Scotland, ON N0E 1R0

Global Horticultural Inc.
11 Commerce Court
Stoney Creek, ON L8E 4G3

Grace-Sierra Horticultural Products Co.
1001 Yosemite Dr.
Milpitas, CA 95035

HJS Wholesale Ltd.
1505 Molson St.
Winnipeg, MA R2G 3S6

Halifax Seed Co. Ltd.
5860 Kane St.
Halifax, NS B3K 5L8

Hummert Intenational
2746 Chouteau Ave.
St. Louis, MO 63103

J-M Trading Corp.
241 Frontage Rd., Ste. 47
Burr Ridge, IL 60521

JVK Ltd.
P.O. Box 910, 1894 Seventh St.
St. Catharines, ON L2R 6Z4

Jiffy Products of America Inc.
1119 Lyon Rd.
Batavia, IL 60510

Kim International, Inc.
1862 Enterprise Dr.
Norcross, GA 30093

Kube-Pak Corp. & Irrigation
RD 3, Box 2553, Route 526
Allentown, NJ 08501

Labon Inc.
1350 Newton
Boucherville, QC J4B 5H2

McCalif Grower Supplies, Inc.
P.O. Box 310, 2905 Railroad Ave.
Ceres, CA 95307

McConkey Co.
P.O. Box 1690
Sumner, WA 98390

Pargro Ltd.
401 Westpark Dr., Suite 202
Peachtree City, GA 30269

Pargro Ltd.
1 Geendale Dr.
Caledonea, ON N0A 1A0

Paygro, Inc.
11000 Huntington Rd., P.O. Box W
South Charleston, OH 45368

Pefferlaw Peat Products Inc.
R.R.#1
Cannington, ON L0E 1E0

Plant Products Co. Ltd.
314 Orenda Rd.
Brampton, ON L6T 1G1

Premier Peat Moss Ltd.
1785, 55e Ave.
Dorval, QC H9P 2W3

Premier Brands, Inc.
733 Yonkers Ave.
Yonkers, NY 10704

Sharp & Son Ltd.
9643 - 186th St.
Surrey, B.C. V3T 4W2

Sharp & Son Ltd.
900 Lind Ave. S.W.
Renton, WA 98055

Shemin Nurseries
R.R.#4, Fifth Line S.
Milton, ON L9T 2X8

Smithers-Oasis Co.
P.O. Box 118
Kent, OH 44240

Specialties Robert Legault Ltd.
204, boul. Labelle, Ste. 234
Ste-Therese, QC J7E 2X7

Southern Importers, Inc.
P.O. Box 8579
Greensboro, NC 27419

Strong-Lite Products Corp.
P.O. Box 8029
Pine Bluff, AR 71611

The Professional Gardener Co. Ltd.
915 - 23rd Avenue S.E.
Calgary, AB T2G 1P1

Otis S. Twilley Seed Co., Inc.
P.O. Box 65
Trevose, PA 19053

Vanhof and Blokker Ltd.
6745 Pacific Circle
Mississauga, ON L5T 1N8

Waldo & Associates, Inc.
28214 Glenwood Rd.
Perrysburg, OH 43551

Westcan Horticultural Ltd.
Bay 5, 6112 - 30th St. S.E.
Calgary, AB T2C 2A6

Whittemore Perlite Co., Inc.
30 Glenn St.
Lawrence, MA 01843

Hydroponic Equipment & Supplies

American Horticultrual Supply Inc.
4045 Via Pescador
Camarillo, CA 93012

Agro Dynamics Inc.
12 Elkins Rd.
East Brunswick, NJ 08816

Agro Dynamics Inc.
415 Industrial Dr.
Milton, ON L9T 5A6

Blue Water Hydroponics
191 Sackville Dr.
LR Sackville, NS B4C 2R5

Canadian Hydrogardens Ltd.
1386 Sandhill Dr.
Ancaster, ON L9G 4V5

CropKing Inc.
P.O. Box 310
Medina, OH 44258

Davis Engineering
8217 Corbin Ave.
Canoga Park, CA 91306

Engineered Systems & Designs
119A Sandy Dr.
Newark, DE 19713

Myron L. Company
6115 Corte Del Cedro
Carlsbad, CA 92009

Plant Products Co. Ltd.
314 Orenda Rd.
Brampton, ON L6T 1G1

Rambridge Structure & Design Ltd.
1316 Centre St. N.E.
Calgary, AB T2E 2A7

Sharp & Son Ltd.
900 Lind Ave. S.W.
Renton, WA 98055

Sharp & Son Ltd.
9643 - 186th St.
Surrey, B.C. V3T 4W2

SGI - SEL Group International Inc.
P.O. Box 220
Plainview, NY 11803

The Professional Gardener Co. Ltd.
915 - 23rd Ave. S.E.
Calgary, AB T2G 1P1

Westcan Horticultural Ltd.
Bay 5, 6112 - 30 St. S.E.
Calgary, AB T2C 2A6

Hydro-Gardens, Inc.
P.O. Box 25845
Colorado Springs, CO 80936

Westbrook Greenhouse Systems Ltd.
P.O. Box 99, 270 Hunter Rd.
Grimsby, ON L3M 4G1

Irrigation Equipment

American Horticultural Supply Inc.
4045 Via Pescador
Camarillo, CA 93012

API
P.O. Box 3974
Haines City, FL 33845

Advanced Irrigation Systems
P.O. Box 86881
North Vancouver, B.C. V7L 4P6

AGRITECH
2885 Lee Ave.
Sanford, NC 27330

Agrotec Inc.
P.O. Box 49
Pendleton, NC 27862-0049

Agroponic Industries Ltd.
204 4712-16 Ave. N.W.
Calgary, AB T3B 0N1

Amiad Filtration Systems
P.O. Box A
Reseda, CA 91337

H.E. Anderson
P.O. Box 1006, 2100 Anderson Dr.
Muskogee, OK 74402-1006

Andpro Ltd.
P.O. Box 399
Waterford, ON N0E 1Y0

BFG Supply Co.
P.O. Box 479, 14500 Kinsman Rd.
Burton, OH 44021

Brighton By-Products Co. Inc.
P.O. Box 23
New Brighton, PA 15066

Brushking
4173 Domestic Ave.
Naples, FL 33942

Canadian Hydrogardens Limited
1386 Sandhill Dr.
Ancaster, ON L9G 4V5

Carolina Seeds, Inc.
P.O. Box 2625
Boone, NC 28604

Climate Control Systems Inc.
R.R.#5, #509 Hwy. 77
Leamington, ON N8H 3V8

CropKing Inc.
P.O. Box 310
Medina, OH 44258

Dosatron International Inc.
1610 N. Fort Harrison Ave.
Clearwater, FL 34615

Dosmatic USA Inc.
896 N. Mill St., Ste. 201
Lewisville, TX 75057

Dramm Corporation
P.O. Box 1960
Manitowoc, WI 54221-1960

Robert G. Eckel Greenhouses & Irrig.
P.O. Box 12
Brooklin, ON L0B 1C0

Foley Flo-Marketing
237 Stadley Rough Rd.
Danbury, CT 06811-3235

Global Horticultural Inc.
11 Commerce Court
Stoney Creek, ON L8E 4G3

Greenhouse Systems USA Inc.
P.O. Box 777
Watsonville, CA 95076

Growing Systems, Inc.
2950 N Weil St.
Milwaukee, WI 53212

Hakmet Ltd.
P.O. Box 248, 881 Harwood Blvd.
Dorion, QC J7V 7J5

Halifax Seed Co. Ltd.
5860 Kane St.
Halifax, NS B3K 5L8

Hummert International
2746 Chouteau Ave.
St. Louis, MO 63103

Hydro-Gardens, Inc.
P.O. Box 25845
Colorado Springs, CO 80936

Industries Harnois Inc.
1044 Principale
St. Thomas, QC J0K 3L0

Jaybird Mfg., Inc.
RD 1, Box 489A
Centre Hall, PA 16828

J-M Trading Corp.
241 Frontage Rd., Ste. 47
Burr Ridge, IL 60521

JVK Ltd.
P.O. Box 910, 1894 7th St.
St. Catharines, ON L2R 6Z4

Frank Jonkman & Sons Ltd.
R.R.#4
Bradford, ON L3Z 2A6

Keeler-Glasgow
P.O. Box 158, 80444 C.R. 687
Hartford, MI 49057

Labon Inc.
1350 Newton
Boucherville, QC J4B 5H2

McConkey Co.
P.O. Box 1690
Sumner, WA 98390

Netafim Irrigation Inc.
10 E. Merrick Rd., Ste. 205
Valley Stream, NY 11580

Niagrow Systems Inc.
4300 Stanley Ave.
Niagara Falls, ON L2E 6T7

OFE International, Inc.
P.O. Box 161302
Miami, FL 33116

Plant Products Co. Ltd.
314 Orenda Rd.
Brampton, ON L6T 1G1

Rain For Rent
333 S. 12th
Santa Paula, CA 93060

Sharp & Son Ltd.
9643 - 186th St.
Surrey, B.C. V3T 4W2

Sharp & Son Ltd.
900 Lind Ave. S.W.
Renton, WA 98055

Shemin Nurseries
R.R.#4, Fifth Line S.
Milton, ON L9T 2X8

Submatic Irrigation Systems
25 W. Beaver Creek Rd., Unit # 10
Richmond Hill, ON L4B 1K4

The Professional Gardener Co. Ltd.
915 - 23rd Ave. S.E.
Calgary, AB T2G 1P1

Vanden Bussche Irrig. & Equip. Ltd
970 James St., Hwy. 3, Box 304
Delhi, ON N4B 2X1

Westcan Horticultural Ltd.
Bay 5, 6112 - 30 St. S.E.
Calgary, AB T2C 2A6

Waldo & Associates, Inc.
28214 Glenwood Rd.
Perrysburg, OH 43551

Windsor Factory Supply Ltd.
213 Talbot St. W.
Leamington, ON N8H 1N8

Young Products, Inc.
499 A Edison Ct.
Suisun, CA 94585

Zwart Systems
623 S. Service Rd., Unit # 1
P.O. Box 235
Grimsby, ON L3M 4G3

Lighting Equipment and Supplies

Agra Tech Inc.
2131 Piedmont Way
Pittsburg, CA 94565

BFG Supply Co.
P.O. Box 479, 14500 Kinsman Rd.
Burton, OH 44021

Brighton By-Products Co. Inc.
P.O. Box 23
New Brighton, PA 15066

Canadian Hydrogardens Ltd.
1386 Sandhill Dr.
Ancaster, ON L9G 4V5

Climate Control Systems Inc.
RR#5, #509 Hwy. 77
Leamington, ON N8H 3V8

CropKing Inc.
P.O. Box 310
Medina, OH 44258

Distributions Corbeil Bigras Ltee
3398 Boul. Industriel
Laval, QC H7L 4R9

Duro-Lite Lamps
419 Attwell Dr.
Rexdale, ON M9W 5W5

Galaxy Agri-Products International Inc.
44775 Yale Rd. West
Sardis, B.C. V2R 1A9

Global Horticultural Inc.
11 Commerce Court
Soney Creek, ON L8E 4G3

GTE Sylvania Canada Limited
54 Atomic Ave.
Etobicoke, ON M8Z 5L4

Halifax Seed Co. Ltd.
5860 Kane St.
Halifax, NS B3K 5L8

Hummert International
2746 Chouteau Ave.
St. Louis, MO 63103

Hydro-Gardens, Inc.
P.O. Box 25845
Colorado Springs, CO 80936

Frank Jonkman and Sons Ltd.
RR#4
Bradford, ON L3Z 2A6

JVK Ltd.
P.O. Box 910, 1894 Seventh St.
St. Catharines, ON L2R 6Z4

Labon Inc.
1350 Newton
Boucherville, QC J4B 5H2

Light-Tech Systems
429 Dewitt Rd., Unit 13
Stoney Creek, ON L8E 4C3

Phillips Lighting
601 Milner Ave.
Scarborough, ON N1B 1M8

P.L. Light Systems Canada Inc.
P.O. Box 206
Grimsby, ON L3M 4G3

Sharp & Son Ltd.
900 Lind Ave. S.W.
Renton, WA 98055

The Professional Gardener Co. Ltd.
915 - 23rd Ave. S.E.
Calgary, AB T2G 1P1

Waldo & Associates, Inc.
28214 Glenwood Rd.
Perrysburg, OH 43551

Westbrook Greenhouse Systems Ltd.
270 Hunter Rd.
Grimsby, ON L3M 4G1

Westcan Horticultural Ltd.
Bay 5, 6112 - 30 St. S.E.
Calgary, AB T2C 2A6

Zwart Systems
623 S. Service Rd., Unit #1
P.O. Box 235
Grimsby, ON L3M 4G3

Seeds

Alf Christianson Seed Co.
P.O. Box 98
Mt. Vernon, WA 98273

American Takii, Inc.
301 Natividad Rd.
Salinas, CA 93906

Asgrow Seed Co./Bruinsma Seed Co.
RR#3, Site 39C
Summerland, B.C. V0H 1Z0

Asgrow Seed Co.
7000 Portage Rd.
Kalamazoo, MI 49001

Ball Seed Co.
622 Town Rd.
West Chicago, IL 60185

Ball Superior Ltd.
1155 Birchview Dr.
Mississauga, ON L5H 3E1

Crofton Grower Services Ltd.
7311 Vantage Way, Unit 110
Delta, B.C. V4G 1C9

CropKing Inc.
P.O. Box 310
Medina, OH 44258

De Ruiter Seeds, Inc.
P.O. Box 20228
Columbus, OH 43220

Ferry-Morse Seed Co.
P.O. Box 4938
Modesto, CA 95352

H.G. German Seeds
201 West Main St.
Smethport, PA 16749

Halifax Seed Co. Ltd.
5860 Kane St.
Halifax, NS B3K 5L8

Harris Seeds/Garden Trends Inc.
P.O. Box 22960
Rochester, NY 14692-2960

Harris Moran Seed Co.
4511 Willow Rd., Suite 3
Pleasanton, CA 94588

A.H. Hummert Seed Co.
2746 Chouteau Ave.
St. Louis, MO 63103

Hydro-Gardens, Inc.
P.O. Box 25845
Colorado Springs, CO 80936

Johnny's Selected Seeds
Foss Hill Rd., Dept. 532
Albion, ME 04910

Labon Inc.
1350 Newton
Boucherville, QC J4B 5H2

Henry F. Michell Co.
P.O. Box 60160
King of Prussia, PA 19406-0160

Northrup King Co.
P.O. Box 959
Minneapolis, MN 55440

Nunhems Seed Corp.
P.O. Box 18, 221 E. Main St.
Lewisville, ID 83431

Pan American Seed Co.
P.O. Box 438
West Chicago, IL 60186-0438

Park Seed Co., Wholesale Div.
FS237 Cokesbury Rd.
Greenwood, SC 29647-0001

Penn State Seed Co.
Rte. 309, Box 390
Dallas, PA 18612

W.H. Perron & Co. Ltd.
2914 Labelle Blvd.
Laval, QC H7P 5R9

Rogers Seed Co.
P.O. Box 4188
Boise, ID 83711

Royal Sluis Inc.
1293 Harkins Rd.
Salinas, CA 93901

Sakata Seed America, Inc.
P.O. Box 880
Morgan Hill, CA 95038

Sharp & Son Ltd.
900 Lind Ave. S.W.
Renton, WA 98055

Sharp & Son Ltd.
9643 - 186th St.
Surrey, B.C. V3T 4W2

Shepherd's Garden Seeds
30 Irene St.
Torrington, CT 06790

Sluis & Groot America, Inc.
5405 Illinois Rd.
Fort Wayne, IN 46804

Stokes Seeds Inc.
P.O. Box 548
Buffalo, NY 14240

Stokes Seeds Ltd.
P.O. Box 10, 39 James St.
St. Catharines, ON L2R 6R6

Sunseeds
18640 Sutter Blvd.
Morgan Hill, CA 95037

United Genetics Seeds Co.
P.O. Box 2323
Gilroy, CA 95021

Vanhof & Blokker Ltd.
6745 Pacific Circle
Mississauga, ON L5T 1N8

Vaughan's Seed Co.
5300 Katrine Ave.
Downers Grove, IL 60515-4095

Vilmorin Inc.
P.O. Box 707
Empire, CA 95319

Rijk Zwaan Zaadteelt en
Zaadhandel B.V.
Burgemeester Crezeelaan 40,
P.O. Box 40
DeLier, Netherlands 2678 ZG

Nickerson-Zwaan B.V.
P.O. Box 19
2990 AA Barendrecht
Netherlands

Bibliography

Hydroponics

General

Publications:

Bentley, M. 1974. *Hydroponics plus.* Sioux Falls, South Dakota: O'Connor Printers.

Blanc, Denise. 1985. *Growing without soil.* Institut National de la Recherche Agronomique, Paris

Bridwell, R. 1990. *Hydroponic gardening.* rev. ed. Santa Barbara, CA: Woodbridge Press.

Broodley, J.W. and R. Sheldrake Jr. 1973. *Cornell peat-lite mixes for commercial plant growing.* Inform. Bull. 43. Ithaca, NY: Cornell Univ.

Butler, J.D. and N.F. Oebker. 1962. *Hydroponics as a hobby growing plants without soil.* Ext. Circ. #844. Urbana, IL: Univ. of Illinois, Ext. Service in Agric. and Home Economics.

Dalton, L. and R. Smith. 1984. *Hydroponic gardening.* Auckland: Cobb Horwood Publ.

Day, D. 1991. *Growing in perlite.* Grower Digest 12, London: Grower Books.

Douglas, J.S. 1975. *Hydroponics: The Bengal system with notes on other methods of soilless cultivation.* 5th ed. London: Oxford Univ. Press.

Douglas, J.S. 1984. *Beginner's guide to hydroponics.* new ed. London: Pelham Books.

Douglas, J.S. 1985. *Advanced guide to hydroponics.* new ed. London: Pelham Books.

Ellis, C. and M.W. Swaney. 1947. *Soilless growth of plants.* New York: Prentice-Hall.

Ellis, N.K., M. Jensen, J. Larsen and N. Oebker. 1974. *Nutriculture systems—growing plants without soil.* Bull. No. 44. West Lafayette, ID: Purdue Univ.

Gerber, John M. 1985. *"Hydroponics," Horticulture Facts.* VC–19–82. rev. Urbana, IL: Univ. of Illinois at Urbana-Champaign.

Gericke, W.F. 1940. *The complete guide to soilless gardening.* New York: Prentice-Hall.

Gooze, J. 1986. *The hydroponic workbook: A guide to soilless gardening.* Stamford, NY: Rocky Top Publ.

Harris, D. 1988. *Hydroponics: The complete guide to gardening without soil: A practical handbook for beginners, hobbyists and commercial growers.* London: New Holland Publ.

Hewitt, E.J. 1966. *Sand and water culture methods in plant nutrition.* Bucks, England: Commonwealth Agric. Bur.

Hoagland, D.R. and D.I. Arnon. 1950. *The water-culture method for growing plants without soil.* Circ. 347. Berkeley, CA: Agric. Exp. Stn., Univ. of Calif.

Hollis, H.F. 1964. *Profitable growing without soil.* London: The English Univ. Press.

Hudson, J. 1975. *Hydroponic greenhouse gardening.* Garden Grove, CA: National Graphics, Inc.

Hydroponic Society of America. 1979–1995. *Proceedings of the annual conferences on hydroponics.* Hydroponic Society of America, 2819 Crow Canyon Rd., Suite 218, San Ramon, CA 94583.

International Society for Soilless Culture (ISOSC). 1973–92. *Proceedings of the international congresses on soilless culture.* Secretariat of ISOSC, P.O. Box 52, Wageningen, Netherlands.

Jensen, M.H. 1971. *The use of polyethylene barriers between soil and growing medium in greenhouse vegetable production.* Tucson, AZ: Environ. Research Laboratory, Univ. of Arizona.

Johnson, H. Jr., G.J. Hochmuth and D.N. Maynard. 1985. *Soilless culture of greenhouse vegetables.* Cooperative Ext. Service, Univ. of Florida, IFAS Bulletin 218. Gainesville, FL: C.M. Hinton, Publ. Distrib. Center.

Jones, J. Benton. 1983. *A guide for the hydroponic & soilless culture grower.* Portland, OR: Timber Press.

Jones, J. Benton, B. Wolf and H. Mills. 1991. *Plant analysis handbook.* Athens, GA: Micro-Macro Publ., Inc.

Jones, L., P. Beardsley and C. Beardsley. 1990. *Home hydroponics ... and how to do it!* rev. ed. New York, NY: Crown Publishers.

Jutras, M.W. 1979. *Nutrient solutions for plants (hydroponic solutions). Their preparation and use.* Circ. 182. Clemson, SC: South Carolina Agric. Expt. Sta.

Kenyon, S. 1992. *Hydroponics for the home gardener.* rev. ed. Key Porter Books, Ltd.

Kramer, J. 1976. *Gardens without soil: house plants, vegetables, and flowers.* New York, NY: Scribner.

Larsen, J.E. 1971. *Formulas for growing tomatoes by nutriculture methods (hydroponics).* (Mimeo) College Station, TX: Texas A & M University.

Larsen, J.E. 1971. *A peat-vermiculite mix for growing transplants and vegetables in trough culture.* College Station, TX: Texas A & M Univ.

Maas, E.F. and R.M. Adamson. 1971. *Soilless culture of commercial greenhouse tomatoes.* Publ. 1460. Information Div., Canada Dept. of Agric., Ottawa, ON, Canada.

Mason, J. 1990. *Commercial hydroponics.* Kenthurst, NSW, Australia: Kangaroo Press.

Maynard, D.N. and A.V. Baker. 1970. *Nutriculture—A guide to the soilless culture of plants.* Publ. No. 41. Amherst: Univ. of Massachusetts.

Mittleider, J.R. 1982. *More food from your garden.* Santa Barbara, CA: Woodbridge Press.

Muckle, M.E. 1982. *Basic hydroponics.* Princeton, B.C.: Growers Press.

Muckle, M.E. 1990. *Hydroponic nutrients—easy ways to make your own.* rev. ed. Princeton, B.C.: Growers Press.

Nicholls, R.E. 1990. *Beginning hydroponics: Soilless gardening: A beginner's guide to growing vegetables, house plants, flowers, and herbs without soil.* Philadelphia, PA: Running Press.

Philipsen, D.J., J.L. Taylor and I.E. Widders. 1985. *Hydroponics at home.* Ext. Bull. E-1853. East Lansing, MI: Michigan State Univ., Cooperative Ext.Service.
Resh, H.M. 1990. *Hydroponic home food gardens.* Santa Barbara, CA: Woodbridge Press.
Resh, H.M. 1993. *Hydroponic tomatoes for the home gardener.* Santa Barbara, CA: Woodbridge Press.
Saunby, T. 1974. *Soilless culture.* 3rd printing. Levittown, NY: Transatlantic Arts.
Savage, Adam J., editor. 1985. *Hydroponics worldwide: State of the art in soilless crop production.* Honolulu, HI: International Center for Special Studies.
Savage, Adam J. 1989. *Master guide to planning profitable hydroponic greenhouse operations.* rev. ed. Honolulu, HI: International Center for Special Studies.
Schales, J.E. *Soilless culture of greenhouse tomatoes.* (Mimeo) Blacksburg, VA: Virginia Polytechnic Institute.
Schwarz, M. 1968. *Guide to commercial hydroponics.* Jerusalem, Israel: Israel Univ. Press.
Shive, J.W. and W.R. Robbins. 1938. *Methods of growing plants in solution and sand cultures.* New Jersey Agr. Expt. Sta. Bull. 636.
Smith, D.L. 1987. *Rockwool in horticulture.* London: Grower Books.
Stoughton, R.H. 1969. *Soilless cultivation and its application to commercial horticultural crop production.* Doc. No. MI/95768. Admin. Unit, Distribution and Sales Section, FAO of the United Nations, Via delle Terme de Caracalla, Rome 00100, Italy.
Stout, J.G. and M.E. Marvel. 1966. *Hydroponic culture of vegetable crops.* Gainsville, FL: Circ. 192-A. Florida Agric. Ext. Service.
Sundstrom, A.C. 1989. *Simple hydroponics—for Australian and New Zealand gardeners.* 3rd ed. South Yarra, Victoria, Australia: Viking O'Neil.
Sutherland, S.K. 1986. *Hydroponics for everyone.* South Yarra, Victoria, Australia: Hyland House.
Taylor, J.D. 1983. *Grow more nutritious vegetables without soil: New organic method of hydroponics.* Santa Ana, CA: Parkside Press.
Taylor, J.D. and R.L. Flannery. 1970. *Growing greenhouse tomatoes in a peat-vermiculite media.* Veg. Crops Offsets Series #33. New Brunswick, NJ: College of Agric. and Environ. Sci., Rutgers Univ.
Universiy of Arizona. 1973. Annual report, Environ. Res. Lab., Univ. of Arizona, Arid Lands Res. Center, Abu Dhabi. Tucson, AZ: Environ. Research Laboratory.
Van Patten, G.F. 1990. *Gardening: The rockwool book.* Portland, OR: Van Patten Publ.
Wallace, T. 1961. *The diagnosis of mineral deficiencies in plants.* 3rd ed. New York: Chemical Publ.
Whiting, A. 1985. *Lettuce from Eden: Hydroponic growing systems for small or large greenhouses.* Cargill, ON: A Whiting.

Withrow, R.B. and A.P. Withrow. 1948. *Nutriculture.* Circ. 328. Purdue Univ. Agric. Expt. Sta.

Articles:

Alafifi, M.A. 1977. The use of the sea coastal areas in Abu Dhabi as a growing medium for vegetables. *Proc. of 4th International Congr. on Soilless Culture*, Las Palmas, Oct. 25–Nov. 1, 1976, pp. 377–84.
Apel, A. and S. Levi. 1966. Growing vegetables on straw bales. *New Zealand Gardener* 22:577.
Barker, A.V. and R. Bradfield. 1963. An outdoor gravel culture set-up for plant growth studies. *Agron. Journal* 55(5):420–25.
Boodley, J.W. 1986. Phenolic foam, a unique plastic, its characteristics and use in hydroponics. *Proc. of 19th National Agric. Plastics Congr.*, Peoria, IL, pp. 203–209.
Cain, J.C. 1963. Automatic sub-irrigation equipment for sand culture. *Amer. Soc. Hort. Sci. Proc.* 82:631–36.
Carpentier, D.R. 1991. Hydroponics on a budget. *The Agric. Education Magazine* 63(7):15–16, 23.
Carr, D. 1979. From hose to hydroponics (current practices in the United Kingdom). *Gardeners Chronicle/HTJ* 185(17):34–35.
Collett, R. 1989. Hydroponic gardening. *Flower and Garden* 33(2):70–72, 74, 84.
Collett, R.K. 1982. Lazy man's garden—hydroponically. *Flower and Garden* 26(4):36–39.
Creaser, G. 1978. Hydroponic growing (Soilless gardening). *The Planter* 3(5):13.
Edson, S.N. 1958. Florida sawdust for home hydroponics. *Florida State Hort. Soc. Proc.* 71:63–67.
Ehrlich, K.F. and M.C. Cantin. 1983. Aeroponic production of lettuce in Quebec during winter in a solar assisted greenhouse. *Proc. of 17th National Agric. Plastics Congr.*, Peoria, IL., pp. 88–97.
Fontes, M.R. 1973. Controlled-environment horticulture in the Arabian Desert at Abu Dhabi. *Hort. Science* 8(1):13–16.
Fox, J.P. 1987. Hydroponics at work. *Agrologist* 16(1):12–14.
Gilbert, H. 1983. Hydroponics/nutrient film techniques, 1979–1983 (Bibliography). USDA, Beltsville, MD: *Quick bibliography series—National Agric. Library* (83–31):13 p.
Gilbert, H. 1984. Hydroponics/nutrient film technique. USDA, Beltsville, MD: *Quick bibliography series—National Agric. Library* (84–56):13 p.
Gilbert, H. 1985. Hydroponics/nutrient film technique: 1979–85. USDA, Beltsville, MD: *Quick bibliography series—National Agric. Library* (86–22):16 p.
Gilbert, H. 1987. Hydroponics/nutrient film technique, 1981–1986. USDA, Beltsville, MD: *Quick bibliography series—National Agric. Library* (87–36):19 p.
Gilbert, H. 1992. Hydroponics/nutrient film technique: 1983–1991. USDA, Beltsville, MD: *Quick bibliography series—National Agric. Library* (92–43):49 p.

Gilbert, Henry. 1979. Hydroponics and soilless cultures, 1969–May 1978. USDA, Beltsville, MD: *Quick bibliography series—National Agric. Library* (79–02):18 p.

Grasgreen, I., H. Janes and G. Giacomelli. 1986. The growth of hydroponic lettuce under tomatoes with supplementary lighting. *Proc. of 19th National Agric. Plastics Congr.*, Peoria, IL., pp. 193–202.

Greene, R.E., J.S. Bullock and R.H. Maier. 1962. Plastic beds as a supporting medium in nutriculture systems. *Agron. Journal* 54(4):363.

Hall, D.A. and G.C.S. Wilson. 1986. The development of hydroponic culture in Scotland. *Research and Development in Agric.* 3(2):61–69.

Handley, L.L., L.S. Casey, J.L. Lopez, J.M. Sutija, H.I. Abdel-Shafy and S.B. Colley. 1986. Gravel bed hydroponics for wastewater renovation and biomass production. *Biomass energy development* – ed. by Wayne H. Smith. Plenum Press, N.Y. pp. 287–302.

Handwerker, T.S. and M. Neufville. 1989. Hydroponics—spaceage agriculture. *The Agric. Education Magazine* 61(9):12–13.

Hanger, B.C. 1983. Preliminary results with Australian rockwool used for plant propagation and in hydroponic systems. *Inter. Plant Propagators' Society* 32:151.

Harris, D.A. 1977. A modified drop-culture method for the commercial production of tomatoes in vermiculite. *Proc. of 4th Int. Congr. on Soilless Culture*, Las Palmas, Oct. 25–Nov.1, 1976, pp. 85–90.

Hashimoto, Y., T. Morimoto, T. Fukuyama, H. Watake, S. Yamaguchi and H. Kikuchi. 1989. Identification and control of hydroponic system ion sensors. *Acta Horticulturae* 245:490–497.

Hershey, D.R. 1989. Easily constructed, inexpensive, hydroponic propagation system. *HortScience* 24(4):706.

Hershey, D.R. 1990. Pardon me, but your roots are showing. *Soilless Grower* 11(5):4–5.

Hicklenton, P.R. and M.S. Wolynetz. 1987. Influence of light- and dark-period air temperatures and root temperatures on growth of lettuce in nutrient flow systems. *Jour. of the Amer. Soc. for Hort. Sci.* 112(6):932–935.

Hodges, C.N. and C.O. Hodge. 1971. An integrated system for providing power, water and food for desert coasts. *HortScience* 6(1):10–16.

Hydro harvest revisited. 1985. *Blair & Ketchum's Country Journal* 12(3):10–11.

Jensen, M.H. and M.A. Teran. 1971. Use of controlled environment for vegetable production in desert regions of the world. *HortScience* 6:33–36.

Jensen, M.H. and N.G. Hicks. 1973. Exciting future for sand culture. *Am. Veg. Grower*, Nov. 1973, pp.33, 34, 72, 74.

Jensen, M.H. and W.L. Collins. 1985. Hydroponic vegetable production. *Horticultue Reviews* 7:483–558.

Kautz, G.W. 1979. Basement garden (hydroponic gardening under lights). *Popular Science* 215: (Dec. 1979).

Knight, S.L. and C.A. Mitchell. 1983. Enhancement of lettuce yield by manipulation of light and nitrogen nutrition. *Jour. of Amer. Soc. for Hort. Science* 108(5):750–754.

Knight, S.L. and C.A. Mitchell. 1987. Stimulating productivity of hydroponic lettuce in controlled environments with triacontanol. *HortScience* 22(6):1307–1309.

Kobayashi, K., Y. Monma, S. Keino and M. Yamada. 1988. Financial results of hydroponic farmings of vegetables in the central Japan. *Acta Horticulturae* 230:337–341.

Kratky, B.A., H. Imai and J.S. Tsay. 1989. Non-circulating hydroponic systems for vegetable production. *Proc. of 21st National Agric. Plastics Congr.*, Peoria, IL., pp. 22–27.

LaFavore, M. 1982. Gardens without ground. *Organic Gardening* 29(7):72–74.

MacFadyen, J.T. 1984. The allure of hydroponics. *National Audubon Soc.* 86(4):12–15.

Marfa, O., L.Serrano and R. Save. 1987. Lettuce in vertical and sloped hydroponic bags with textile waste. *Soilless Culture* 3(2):57–70.

Marsh, L.S., L.D. Albright, R.W. Langhans and C.E. McCulloch. 1987. Economically optimum day temperatures for greenhouse hydroponic lettuce production. *Amer. Soc. of Agric. Engineers.* 37 pp.

Marshall, N. 1985. Commercial hydroponic vegetable growers in Massachusetts. *New Alchemy Quarterly* 19:12.

Massantini, F. 1977. Floating hydroponics; a new method of soilless culture. *Proc. 4th Int. Congr. on Soilless Culture*, Las Palmas, Oct.25–Nov.1, 1976, pp. 91–98.

Massantini, F. 1985. The light and dark sides of aeroponics. *Soilless Culture* 1(1):85–96.

Maxwell, K. 1986. Soilless (hydroponic) culture—the past—present—and future. An Australian viewpoint. *Soilless Culture* 2(1):27–34.

McCoy, D. 1982. Nature's "hydroponic" harvest. *The New Farm* 9(5):38–40.

Mohyuddin, M. 1987. Hydroponics. *Agrologist* 16(1):10–11.

Moneysmith, M. 1981. Hydroponics: growing wave of the future. *Country Gentleman* 132(3):20–21,80.

Morgan, J.V., AHORA, A.L. Tan. 1983. *Acta Horticulturae* 133:39–46.

Munsuz, N., G. Celebi, Y. Ataman, S. Usta and I. Unver. 1989. A recirculating hydroponic system with perlite and basaltic tuff. *Acta Horticulturae* 238:149–156.

Namioka, H. 1977. Kuntan as a substrate for soilless culture. *Proc. 4th Int. Congr. on Soilless Culture*, Las Palmas, Oct.25–Nov.1, 1976, pp. 289–302.

Okano, T., T. Hoshi and H. Terazoe. 1988. Development of hydroponic system and adaptation of microcomputers for a commercial size vegetable factory. *Acta Horticulturae* 230:343–348.

Olson, R.L., M.W. Oleson and T.J. Slavin. CELSS for advanced manned mission. *HortScience* 23(2):275–286.

Peck, N.H. and G.E. MacDonald. 1987. Outdoor sand-nutrient culture system. *HortScience* 22(5):949.

Penningsfeld, F. 1977. Soilless culture, using ion-exchange resins. *Proc. 4th Int.Congr. on Soilless Culture*, Las Palmas, Oct.25–Nov.1, 1976, pp. 247–259.

Peterson, L.A. and A.R. Krueger. 1988. An intermittent aeroponics system. *Crop Science* 28(4):712–713.

Pomeroy, D. 1990. Growing herbs the hydroponic way. *The Business of Herbs* 8(2):1–2,19.

Prokakis, G. and R. Gonzales. 1988. Effect of gibberellic acid (GA3) on nitrate and oxalate levels in greenhouse-grown spinach. *Proc. 15th Plt. Growth Regulating Soc. of Amer.*, Lake Alfred, FL, pp 78–84.

Rault, P. 1985. Hydroponics for vegetable production in a desert climate. Account of the Sarir-Libya project. *L'Agronomie tropicals* 40(2):115–123.

Ricketts, R.T. Jr. 1979. Hydroponics, what a way to grow. *Light Garden* 16(2):50–53.

Robbins, S.R. 1958. Commercial experiences in soilless culture. *Int. Hort. Congr. Proc.* 15(1):112–117.

Robbins, W.R. 1946. Growing plants in sand cultures for experimental work. *Soil Science* 62:3–22.

Rodale, R. 1979. Gardening without soil, is it practical? *Organic Gardening* 26(5):26–32.

Salad garden on a deck rail ... it's hydroponic. *Sunset (Central West edition)* 180(4):126–128.

Sardinsky, R. 1985. Water farms: integrated hydroponics in Maine. *New Alchemy Quarterly* 19:13–14.

Schippers, P.A. 1986. Hydroponic growing systems. *Proc. 19th National Agric. Plastics Congr.* Peoria, IL., pp. 121–133.

Schwarz, M. 1963. The use of brackish water in hydroponic systems. *Plant and Soil* 19(2):166–172.

Schwarz, M. 1985. The use of saline water in hydroponics. *Soilless Culture* 1(1):25–34.

Sheen, T.F. 1987. Nutrient absorption of leafy lettuce grown in root floating hydroponics. *Jour. of Agric. Res. of China* 36(4):372–380. The Taiwan Agric. Research Institute.

Smay, V.E. 1978. How to grow lush gardens without soil: hydroponics. *Popular Science* 212:104+ (March 1978).

Smay, V.E. 1978. Modular hydroponics—plant growing systems that fit any space. *Popular Science* 212:118+ (May 1978).

Soffer, H. and D.W. Burger. 1988. Effects of dissolved oxygen concentration in aero-hydroponics on the formation and growth of adventitious roots. *Jour. of Amer. Soc. of Hort. Science* 113(2):218–221.

Soffer, H. and D.W. Burger. 1989. Plant propagation using an aero-hydroponics system. *HortScience* 24(1):154.

Steiner, A.A. 1961. The future of soilless culture: its possibilities and restrictions under various conditions all over the world. *Int. Hort. Congr. Proc.* 15(1):112–117.

Steiner, A.A. 1961. A universal method for preparing nutrient solutions of a certain desired composition. *Plant and Soil* 15(2):134–154.

Steiner, A.A. 1968. Soilless culture. *Proc. of the 6th Colloquium of the Int. Potash Instit.*, Florence, pp. 324–341.

Stoner, R. and S. Schorr. 1983. Aeroponics versus bed and hydroponic propagation. *Florists' Review* 173(4477):49–51.
Ter Hoeven, T. and L.A.J. Lamers. 1977. Hydroponic gardens in offices. *Proc. 4th Int. Congr. on Soilless Culture*, Las Palmas, Oct.25–Nov.1, 1976, pp. 57–60.
Thorndike, J. 1980. Hydroponics at home. *Horticulture* 58(11):28–29.
Thorndike, J. 1980. The unearthly world of hydroponics. *Horticulture* 58(11):24–28,30–31.
Tropea, M. 1977. The controlled nutrition of plants II—A new system of "vertical hydroponics." *Proc. 4th Int. Congr. on Soilless Culture*, Las Palmas, Oct.25–Nov.1, 1976, pp. 75–83.
Vercellino, C. 1987. Producing herbs hydroponically. *Proc. 2nd National Herb Growing and Marketing Conference*, West Lafayette, IN., pp. 104–106.
Vincenzoni, A. 1977. "Colonna de coltura," a contribution to the develop ment of hydroponics. *Proc. 4th Int. Congr. on Soilless Culture*, Las Palmas, Oct.25–Nov.1, 1976, pp. 99–105.
Wallace, A., S.M. Soufi and A.M. El Gazzar. 1979. Wastewater as a source of nutrients and water for hydroponic crop production in arid regions (California). *Advances in Desert and Arid Land Technology and Development* 1:263–270.
Wallace, G.A. and A. Wallace. 1984. Maintenance of iron and other micronutrients in hydroponic nutrient solutions (Tomatoes, cucumbers). *Journal of Plant Nutrition* 7(1/5):575–585.
Wolkomir, R. 1987. Astrocrops. *Omni* 9(10):16,101.
Wright, B.D., W.C. Bausch and W.M. Knott. 1988. A hydroponic system for microgravity plant experiments. *Transactions of the ASAE* 31(2):440–446.
Yamaguchi, F.M., JBPHB, and A.P. Krueger. 1983. Electroculture of tomato plants in a commercial hydroponics greenhouse. *Journal of Biological Physics* 11(1):5–10.
Zeroni, M., J. Gale and J. Ben-Asher. 1983. Root aeration in a deep hydroponic system and its effects on growth and yield of tomato (*Solanum lycopersicum*).

Nutrient Film Technique (NFT)

Publications:

Cooper, A. 1979. *The ABC of NFT.* London: Grower Books.
Molyneux, C.J. 1994. *A practical guide to NFT.* Preston, Lancashire, England: T. Snape & Co. Ltd.
Schippers, P.A. 1977. *Construction and operation of the nutrient flow technique for growing plants.* Veg. Crops Mimeo 187. Cornell Univ. Ithaca, NY.
Schippers, P.A. 1977. *Annotated bibliography on nutrient film technique.* Veg. Crops Mimeo 186. Cornell Univ., Ithaca, NY.
Schippers, P.A. 1979. *The nutrient flow technique.* Veg. Crops Mimeo 212. Cornell Univ., Ithaca, NY.

Wilcox, G.E. 1981. *The nutrient film hydroponic system.* Bull. 166. West Lafayette, IN: Purdue Univ. Agric. Expt. Sta.
Wilcox, G.E. 1981. *Growing greenhouse tomatoes in the nutrient film hydroponic system.* Bull. 167. West Lafayette, IN: Purdue Univ. Agric. Expt. Sta.
Wilcox, G.E. 1987. *NFT and principles of hydroponics.* Bull. 530. West Lafayette, IN: Purdue Univ. Agric. Expt. Sta.

Articles:

Adams, P. and A.M. El-Gizawy. 1988. Effect of calcium stress on the calcium status of tomatoes grown in NFT. *Acta Horticulturae* 222:15–22.
Adams, P. 1989. Plant growth in NFT and other soilless substrates. *Aspects of Applied Biology* 22:341–348.
Bailey, B.J., B.G.D. Haggett, A. Hunter, W.J. Albery and L.R. Savenberg. 1988. Monitoring nutrient film solutions using ion-selective electrodes. *Jour. of Agric. Engineering Research.* 40(2):129–142.
Bedasie, S. and K. Stewart. 1987. Effect of intermittent flow on seasonal production of NFT lettuce. *Soilless Culture* 3(1):11–19.
Bedasie, S. and K. Stewart. 1987. Effect of watering regime on the growth and development of NFT lettuce. *Soilless Culture* 3(1):3–9.
Bedasie, S. and K. Stewart. 1987. Effect of watering regime on the growth and development of NFT lettuce. *Soilless Culture* 3(2):3–7.
Benoit, F., AHORA, and N. Ceustermans. 1981. Energy input in two NFT (Nutrient film technique). *Acta Horticulturae*115:227–233.
Benoit, F. and N. Ceustermans. 1986. Survey of a decade of research (1974–1984) with nutrient film technique (NFT) on glasshouse vegetables. *Soilless Culture* 2(1):5–17.
Benoit, F. and N. Ceustermans. 1986. Growth control of tomatoes and cucumbers in NFT by means of rockwool and polyurethane blocks. *Soilless Culture* 2(2):3–9.
Benoit, F. and N. Ceustermans. 1989. Growing lamb's lettuce (*Valerianella olitoria L.*) on recycled polyurethane (PUR) hydroponic mats. *Acta Horticulturae* 242:297–304.
Buley, N. 1984. Innovations in greenhouse growing challenge conventional methods (Hydroponics, aeroponics, nutrient film technique, ornamental plants, growing more and better plants in less space and at lower costs). *American Nurseryman* 159(1):95–99.
Charbonneau, J., A. Gosselin and M.J. Trudel. 1988. Influence of electrical conductivity and intermittent flow of the nutrient solution on growth and yield of greenhouse tomato in NFT. *Soilless Culture* 4(1):19–30.
Charlesworth, R. 1977. Soil and nutritional problems; NFT problems. *The Grower*, Jan. 13, 1977, p. 65.
Clayton, A. 1980. Nutrient film technique: practicalities and problems. *Horticulture Industry*, Feb. 1980, pp. 18–20.
Cooper, A.J. 1973. Rapid turn-round is possible with experimental nutrient film technique. *The Grower*, May 5, 1973.
Cooper, A.J. 1974. Improved film technique speeds growth. *The Grower*, March 2, 1974.

Cooper, A.J. 1974. Hardy nursery stock production in nutrient film. *The Grower*, May 4, 1974.
Cooper, A.J. 1974. Soil? Who needs it? Part I. *Am. Veg. Grower* 22(8):18,20.
Cooper, A.J. 1974. Soil? Who needs it? Part II. *Am. Veg. Grower* 22(9):13,64.
Cooper, A.J. 1975. Rapid progress through 1974 with nutrient film trials. *The Grower*, Jan. 25, 1975.
Cooper, A.J. 1975. Nutrient film technique—early fears about nutrition unfounded. *The Grower*, Aug. 23, 1975, pp. 326–327.
Cooper, A.J. 1975. Crop production in recirculating nutrient solution. *Scientia Horticulturae* 3:251–258.
Cooper, A.J. 1975. Comparing a nutrient film tomato crop with one grown in the soil. *The Grower*, Dec. 12, 1975.
Cooper, A.J. 1977. Crop production with nutrient film technique. *Proc. of 4th Int. Congr. on Soilless Culture*, Las Palmas, Oct.25–Nov.1, 1976, pp. 121–136.
Cooper, A.J. 1985. 22 new ABC's of NFT. *Hydroponics Worldwide: State of the Art in Soilless Crop Production*, Ed., Adam J. Savage. pp. 180–185. Int. Center for Special Studies, Honolulu, HI.
Cooper, A.J. 1986. NFT cropping from the beginning to the present day. *Proc. of 19th National Agric. Plastics Congr.*, Peoria, IL., pp. 105–120.
Cowan, I. 1979. NFT (nutrient film technique) simplicity (Designs and costs for amateur greenhouse growers, Great Britain). *Greenhouse* 3(9):24–25.
Cowan, I. 1979. D.I.Y.—N.F.T. (Nutrient film technique, homemade equipment for growing greenhouse plants successfully without soil). *Greenhouse* 3(4):24,26-27.
Deen, J. 1979. The simplest growing medium. *Gardeners Chronicle/HTJ*, 186(15):35, 36.
Devonald, V.G. and A. Tapp. 1987. The effect of warming the nutrient solution on the early growth of tomatoes in NFT in a heated and unheated environment. *Soilless Culture* 3(1):31–38.
Doolan, D.W., AHORA, M.J. Hennerty and J.V. Morgan. 1983. Culture of micropropagated strawberry plants in nutrient film technique. *Acta Horticulturae* 133:103–109.
Douglas, J. Shalto. 1976. Hydroponic layflats. *World Crops*, March/April 1976, pp. 82–87.
Giacomelli, G.A., AHORA, H.W. Janes, D.R. Mears and W.J. Roberts. 1983. A cable supported NFT tomato production system for the greenhouse. *Acta Horticulturae* 133:89–102.
Giacomelli, G.A. and H.W. Janes. 1981. The growth of greenhouse tomatoes by the nutrient film technique at various nutrient solution temperatures. *Proc. of 16th National Agric. Plastics Congr.* Peoria, IL., pp. 1–13.
Giacomelli, G.A. and H.W. Janes. 1984. NFT (nutrient film technique) greenhouse tomatoes grown with heated nutrient solution. *Acta Horticulturae* 148(2):827–834.
Gormley, T.R., M.J. Maher and P.E. Walshe. 1983. Quality and performance of eight tomato cultivars in a nutrient film technique system. *Crop Research* 23(2):83–93.

Graves, C.J. 1983. The nutrient film technique. *Horticultural Reviews* 5:1–44.
Graves, C.J., AHORA, and R.G. Hurd. 1983. Intermittent solution circulation in the nutrient film technique. *Acta Horticulturae* 133:47–52.
Guernsey cautions on NFT crops. When you fail, you lose the lot, warns adviser. *The Grower*, Feb. 24, 1977, p. 397.
How NFT compares. *The Grower*, Oct. 7, 1976.
Jackson, M.B., P.S. Blackwell, J.R. Chrimes and T.V. Sims. 1984. Poor aeration in NFT and a means for its improvement. *The Jour. of Horticultural Science* 59(3):439–448.
Khudheir, G.A., AHORA and P. Newton. 1983. Water and nutrient uptake by tomato plants grown with the nutrient film technique in relation to fruit production. *Acta Horticulturae* 133:67–88.
Lauder, K. 1976. GCRI hope to make low cost crops worth growing by NFT. *The Grower*, July 15, 1976.
Lauder, K. 1977. Lettuce on concrete. Supplement to *The Grower*, Feb. 24, 1977, pp. 40–46.
Lim, E.S. 1985. Development of an NFT system of soilless culture for the tropics. *Pertanika* 8(1):135–144.
Lim, E.S. 1986. Hydroponic production of vegetables in Malaysia using the nutrient film technique. *Soilless Culture* 2(2):29–39.
Lovelidge, B. 1976. Kent grower to go the whole hog with NFT tomato crops. *The Grower*, Aug. 19, 1976.
Lovelidge, B. 1976. Better working life is extra benefit, says Sussex grower. *The Grower*, Nov. 4, 1976.
Maher, M.J. 1977. The use of hydroponics for the production of greenhouse tomatoes in Ireland. *Proc. of 4th Int. Congr. on Soilless Culture*, Las Palmas, Oct. 25–Nov. 1, 1976, pp. 161–169.
Making NFT commercial. *World Crops*, May/June 1977, pp. 125–126.
Massantini, F. and G. Magnani. 1983. Mobile hydroponics for energy saving. *Acta Horticulturae* 148(1):81–88.
Miliev, K. 1977. Nutrient film experiments in Bulgaria. *Proc. of 4th Int. Congr. on Soilless Culture*, Las Palmas, Oct. 25–Nov. 1, 1976, pp. 149–159.
Nelson, R and S.H. Mantell. 1988. Growth performance of micropropagated plantlets of sweet potato (*Ipomoea batatas* (L.) Lam.) established in a nutrient film technique system. *Crop Research* 28(2):145–156.
New twist for hydroponics. *Am. Veg. Grower* 24(11):21–23.
NFT package deal should be on offer next year, say ICI. *The Grower*, Nov. 4, 1976.
Noguera, V., M. Abad, J.J. Pastor, A.C. Garcia-Codoner, J. Mora and F. Armengol. 1988. Growth and development, water absorption and mineral composition of tomato plants grown with the nutrient film technique in the East Mediterranean Coast region of Spain. *Acta Horticulturae* 221:203–211.
Nutrient film technique: Cropping with the Hydrocanal commercial system. *World Crops*, September/October 1976, pp. 212–218.
Nutrient film techniques; PP's seven factors for success. *The Grower*, April 21, 1977, p. 907.

Pardossi, A., F. Tognoni, R. Tesi, M. Bertolacci and P. Grossi. 1984. Root zone warming in tomato plants in soil and NFT. *Acta Horticulturae* 148(2):865–870.

Research is lagging behind NFT's development—Cooper. *The Grower*, March 31, 1977, pp. 735–736.

Rheinberg, P and D.S. Shaw. 1977. The microbial ecology of a nutrient film hydroponic system. *Proc. of 4th Int. Congr. on Soilless Culture*, Las Palmas, Oct. 25–Nov. 1, 1976, pp. 137–148.

Richardson, F.I. 1981. Nutrient film technique. *Agriculture in Northern Ireland* 55(11):349–351.

Rober, R. 1983. Nutrient film technique and substrates. *Acta Horticulturae* 133.

Rozek, S., W. Sady, J. Myczkowski and T. Wojtaszek. 1984. Certain aspects of nourishment of tomatoes grown by the nutrient film technique: I. The effect of various nitrate levels in the nutrient solution on nitrate reductase activity and tomato yield. *Acta Physiologiae Plantarum* 6(4):203–214.

Rozek, S., W. Sady, J. Myczkowski and T. Wojtaszek. 1985. Certain aspects of nourishment of tomatoes grown by the nutrient film technique: II. Some indices of plant metabolism under selected conditions of nitrate fertilization. *Acta Physiologiae Plantarum* 7(2):71–84.

Rozek, S., W. Sady, J. Myczkowski and T. Wojtaszek. 1986. Some indices of nitrate metabolism in lettuce grown by the nutrient film technique on varying nutrient solutions. *Acta Physiologiae Plantarum* 8(1):43–52.

Schippers, P.A. 1978. Soilless growing in Europe. *Amer. Veg. Grower and Greenhouse Grower* 26(12):17–18.

Schippers, P.A. 1979. The nutrient flow technique—a versatile and efficient hydroponic growing method. *Proc. of 1st Ann. Conf. of the Hydroponic Society of America*, Brentwood, CA., Oct. 20, 1979.

Schippers, P.A. 1979. Hydroponics is the answer. *Amer. Veg. Grower and Greenhouse Grower* 27(2):16, 18, 34.

Schippers, P.A. 1980. Hydroponic lettuce: the latest. *Amer. Veg. Grower and Greenhouse Grower* 28(6):22–23, 50.

Schlagnhaufer, B.E., E.J. Holcomb and M.D. Orzolek. 1987. Effects of supplementary light, solution heating, and increased solution Ca levels on lettuce production in the nutrient film technique. *Applied Agricultural Research* 2(2):124–129.

Sparkes, B. 1977. Variation on NFT theme. *The Grower*, April 21, 1977, pp. 905–906.

Spensley, K. and G.W. Winsor. 1978. Nutrient film technique, crop culture in flowing nutrient solution. *Outlook on Agriculture* 9(6):299–305.

Toop, E.W., G.H. Silva and G. Botar. 1988. Comparison of 24 lettuce cultivars in a controlled environment with extra CO_2 in NFT and stagnant solution. *Soilless Culture* 4(1):51–64.

Verwer, F.L.J.A.W. 1977. Growing horticultural crops in rockwool and nutrient film. *Proc. of 4th Int. Congr. on Soilless Culture*, Las Palmas, Oct. 25–Nov. 1, 1976, pp. 107–119.

Wees, D. and K. Stewart. 1986. The potential of NFT for the production of six herb species. *Soilless Culture* 2(2):61–70.
Wees, D. and K. Stewart. 1987. The influence of bicarbonate enrichment and aeration on dissolved carbon dioxide and oxygen in NFT nutrient solutions used for lettuce production. *Soilless Culture* 3(1):51–62.
Wilcox, G.E. 1980. High hopes for hydroponics. *Amer. Veg. Grower and Greenhouse Grower* 28(11):11–12, 14.
Wilcox, G.E. 1984. Nutrient uptake by tomatoes in nutrient film technique hydroponics. *Acta Horticulturae* 145:173–180.

Insect and Disease Control

Daughtrey, M. 1983. Hydroponic systems are vulnerable to disease. *Ag Impact*, March 1983, pp. 11–12.
Funck-Jensen, D., AHORA and J. Hockenhull. 1983. The influence of some factors on the severity of *Pythium* root rot of lettuce in soilless growing systems. *Acta Horticulturae* 133:129–136.
Gold, S.E. and M.E. Stranghellini. 1985. Effects of temperature on *Pythium* root rot of spinach grown under hydroponic conditions. *Phytopathology* 75(3):333–337.
Hockenhull, J., AHORA and D. Funck-Jensen. 1983. Is damping-off, caused by *Pythium*, less of a problem in hydroponics than in traditional growing systems. *Acta Horticulturae* 133:137–145.
Jarvis, W.R. and C.D. McKenn. 1984. *Tomato diseases*. Canada Dept. of Agric. Publ. 1479/E. Agric. Canada, Ottawa, Ontario.
Malais, M. and W.J. Ravensberg. 1992. *Knowing and recognizing—the biology of glasshouse pests and their natural enemies.* Koppert B.V., Berkel en Rodenrijs, Netherlands.
Nelson, Paul V. 1985. *Greenhouse operation and management.* 3rd ed. Reston, Virginia: Reston Publ. Co.
Ontario Dept. of Agric. and Food. 1975. *Greenhouse vegetable production recommendations.* Publ. 365. Ontario Dept. of Agric. and Food, Parliament Buildings, Toronto, ON, Canada.
Partyka, R.E. and L.J. Alexander. 1973. *Greenhouse tomatoes—disease control.* Bull. SB-16. Columbus, OH: Ohio State Univ. Cooperative Ext. Service.
Roorda van Eysinga, J.P.N.L. and K.W. Smilde. 1981. *Nutritional disorders in glasshouse tomatoes, cucumbers, and lettuce.* Wageningen: Center for Agric. Publ. and Documentation.
Schwartzkopf, S.H., D. Dudzinski and R.S. Minners. 1987. The effects of nutrient solution sterilization on the growth and yield of hydroponically grown lettuce. *HortScience* 22(5):873–874.
Stanghellini, M.E., and R.L. Gilbertson. 1988. *Plasmopara lactucae-radicis*, a new species of rots of hydroponically grown lettuce. *Mycotaxon* 31(2):395–400.
Stanghellini, M.E., and W.C. Kronland. 1986. Yield loss in hydroponically grown lettuce attributed to subclinical infection of feeder rootlets by *Pythium dissotocum. Plant Disease* 70(11):1053–1056.

Stanghellini, M.E., L.J. Stowell and M.L. Bates. 1984. Control of root rot of spinach caused by *Pythium aphanidermatum* in a recirculating hydroponic system by ultraviolet irradiation. *Plant Disease* 68(12):1075–1076.

Steiner, M.Y. and D.P. Elliott. 1983. *Biological pest management for interior plantscapes.* 30 pp. The Publications Office, Min. of Agric. and Food, Victoria, B.C., Canada.

Zinnen, T.M. 1988. Assessment of plant diseases in hydroponic culture. *Plant Disease* 72(2):96–99.

Professional Publications and Research Journals

Horticultural Research. Scottish Academy Pess, 33 Montgomery St., Edinburgh.

HortScience. Publ. bimonthly by the American Society for Horticultural Science, 914 Main St., St.Joseph, MI 49085.

Journal of the American Society for Horticultural Science. Publ. bimonthly for ASHS. American Society for Horticultural Science, 914 Main St., St. Joseph, MI 49085.

The Journal of Horticultural Science. Headley Bros. Ltd., The Invicta Press, Ashford, Kent, England.

Proceedings of the Florida State Horticultural Society. Dr. H. J. Reitz, Secretary, Florida State Hort. Soc., P. O. Box 552, Lake Alfred, FL 33850.

Trade Magazines and Periodicals

American Horticulturist. Publ. by the American Horticulural Society, 7931 East Boulevard Dr., Alexandria, VA 22308.

American Nurseryman. 310 South Michigan Ave., Chicago, IL 60604.

American Vegetable Grower. Publ. by Meister Publ. Co., Willoughby, OH 44094.

Canadian Florist, Greenhouse and Nursery. 287 Queen St. South, Streetsville, Ont., Canada.

Florist, An FTD Publication. FTDA, 900 West Lafayette, Detroit, MI 48226.

Flower News, the Florists' National Weekly Newspaper. 549 West Randolph St., Chicago, IL 60606.

Grower Talks. Geo. J. Ball, Inc., W. Chicago, IL 60185.

Greenhouse Canada. Publ. by NCC Publ. Ltd., 222 Argyle Ave., Delhi, ON, Canada N4B 2Y2.

Greenhouse Grower. Publ. by Meister Publ. Co., Willoughby, OH 44094.

Greenhouse Management & Production. Box 1868, Fort Worth, TX 76101.

Horticulture. Horticulture Subscription Service, 125 Garden St., Marion OH 43302.

Seed World. 434 S. Wabash Ave., Chicago, IL 60605.

The Florist's Review. Florist's Publ. Co., 343 S. Dearborn St., Chicago, Ill.

The Grower. 49 Doughty St., London, WC IN 2BR, England.

The Growing Edge. P.O. Box 1027, Corvallis, OR 97339.

The Packer. Circulation Manager, One Gateway Center, Kansas City, KS 66101. Publ. by Vance Publ. Co., 300 W. Adams St., Chicago, IL 60606.

Yoder Grower Circle News. Yoder Bros. Inc., Barberton, Ohio.

General Subject Index